PLASTIC PRODUCT DESIGN

Second Edition

Ronald D. Beck
Senior Development Engineer, Fisher Body Division
General Motors Corp., Warren, Michigan

VNR VAN NOSTRAND REINHOLD COMPANY
New York

Library of Congress Catalog Card Number **79-15134**
ISBN **0-442-20632-1**

Printed in the United States of America

Van Nostrand Reinhold Company Inc.
115 Fifth Avenue
New York, New York 10003

Van Nostrand Reinhold Company Limited
Molly Millars Lane
Wokingham, Berkshire RG11 2PY, England

Van Nostrand Reinhold
480 La Trobe Street
Melbourne, Victoria 3000, Australia

Macmillan of Canada
Division of Canada Publishing Corporation
164 Commander Boulevard
Agincourt, Ontario M1S 3C7, Canada

15 14 13 12 11 10 9 8 7 6 5 4 3

Library of Congress Cataloging in Publication Data

Beck, Ronald D
Plastic product design.

Bibliography: p.
Includes index.
1. Plastics. I. Title.
TP1120.B42 1980 668′.4 79-15134
ISBN 0-442-20632-1

Preface

This book is intended to serve as a guide to the study and use of plastic product design. It is intended primarily for students in all fields of engineering. It should also be useful to qualified engineers who require a simple introduction to the subject. The book includes information on plastic materials and the processes by which the raw materials are converted into finished products.

The main topics discussed on product design are: mold design for part requirements; molded holes and undercuts; threads; inserts; fastening and joining plastics; decorating plastics; extrusion design and processing; reinforced plastics; and tests and identifications of plastics.

The sources from which the material in this book has been derived are so many and so diverse that it is impractical to attempt to acknowledge them individually. All dimensions are given in English and some in metric units. Conversion tables are available in Appendix Tables.

I trust this design book will go a long way toward giving engineers and students a positive confidence in planning to utilize all the advantages inherent in plastics today.

Questions and answers are available from the author for use by those persons teaching a product design course.

Ronald D. Beck

Contents

1

Plastic Materials

WHAT ARE PLASTICS?

Plastics are organic materials made from large molecules that are constructed by a chain-like attachment of certain building-block molecules. The properties of the plastic depend heavily on the size of the molecule and on the arrangement of the atoms within the molecule. For example, polyethylene is made from the ethylene building block which is initially a gas. Through a process called polymerization, a chain of ethylene molecules is formed by valence bonding of the carbon atoms within the ethylene molecule. The high molecular weight product which results is called a polymer. Hence, the designation polyethylene is used to distinguish the high-molecular-weight plastic from its gaseous counterpart, ethylene, which is the monomer that becomes polymerized. The "poly" refers to the "many" ethylene building block molecules or monomers, that join to form the polyethylene plastic molecule. Frequently, the term "resin" is used interchangeably with "polymer" to describe the backbone molecule of a plastic material. However, "resin" is sometimes used to describe a syrupy liquid of both natural and synthetic origin.

Plastics, in the finished product form, are seldom comprised exclusively of polymer but also include other ingredients such as fillers, pigments, stabilizers, and processing aids. However, designation of the plastic material or molding compound is always taken from the polymer designation.

TYPES OF PLASTICS

Broadly speaking plastics may be divided into two categories: thermoplastics and thermoset plastics. The classes of materials are so named because of the effect of temperature on their properties.

Thermosets

Thermoset plastics are polymers which are relatively useless in their raw states. Upon heating to a certain temperature a chemical reaction takes place which causes the molecules to bond together or cross-link. After vulcanization and polymerization, or curing, the thermoset material remains stable and cannot return to its original state. Thus, "thermo-set" identifies those materials that become set in their useable state resulting from the addition of heat. Normally, a thermoset polymer is mixed with fillers and reinforcing agents to obtain the properties of a molding compound.

Thermosets are the hardest and stiffest of all plastics, are chemically insoluble after curing, and their properties are less affected by changes in temperature than are the heat-sensitive thermoplastics. The closest non-plastic counterparts to thermosets in properties are ceramics. Common examples of thermoset plastics are: phenolics, melamine, urea, alkyds, and epoxies. Molding compounds made from these polymeric resins always contain additional fillers and reinforcing agents to obtain optimum properties.

Thermoplastics

Thermoplastic polymers are heat-sensitive materials which are solids at room temperature, like most metals. Upon heating, the thermoplastics begin to soften and eventually reach a melting point and become liquid. Allowing a thermoplastic to cool below its melting point causes resolidification or freezing of the plastic. Successive heating and cooling cycles cause repetition of the melting-freezing cycle just as it does for metals.

The fact that thermoplastics melt is the basis for their processing into finished parts. Thermoplastics may be processed by any method which causes softening or melting of the material. Examples of thermoplastic fabrication techniques using melting are: injection molding, extrusion, rotational casting, and calendering. Fabrication methods which take advantage of softening below the melting point are: thermoforming (vacuum or pressure), blow molding, and forging. Of course, normal metal-cutting techniques can also be applied to thermoplastics in the solid state. Common examples of thermoplastics are: polyethylene, polystyrene, polyvinyl chloride (PVC), and nylon (polyamide).

THERMOSETS

Alkyds

Alkyds are a family of thermoset molding compounds made from un saturated polyester resins. Like most thermosets, alkyds are hard and stiff and retain their mechanical and electrical properties at elevated temperatures.

Paramount with alkyds is their arc resistance and track resistance in the presence of an arc. Low moisture absorption and retention of electrical properties when wet permits alkyds to offer excellent performance in automotive ignition equipment. The non-tracking characteristic of alkyds permits their use in arc chutes and in heavy-duty switchgear. Low change in dielectric loss factor enable alkyds to be used in TV tuner segments because tuner capacitance remains unchanged throughout the operating temperature range of the TV set.

The raw material comes in four basic forms: granular, rope, putty, and bulk. Granular material is used primarily in compression molding, although some grades are processed by transfer molding. Normally, granulars are preformed into a pill which is then preheated before loading into the mold. Granular materials offer the lowest mechanical properties of all alkyds, but certain granulars possess exceptional electrical properties for selected applications.

Rope alkyd is a putty-like material containing reinforcement with glass or synthetic fibers. Primarily used in transfer molding, ropes are the workhorse of the alkyd family and offer a good balance of impact strength and other mechanical properties while retaining excellent electrical properties. Ropes cure rapidly and can be molded into intricate shapes.

Putty is a soft non-reinforced material that is primarily used to encapsulate certain types of capacitors. Low molding pressure permits putty to be used without chance of damage to the delicate capacitor foils. Bulk alkyd is highly reinforced and may be transfer or compression molded. Because of high reinforcement, bulk alkyds offer extremely high impact strength and superior arc resistance. However, moldability is more restricted to simpler shapes than for rope-type materials.

The automotive electric motor brush holder shown in Fig. 1–1 is a typical application for alkyds. Other applications are: electrical and electronic insulators, arc chutes, TV tuner segments, automotive coil tops, distributor caps and switch insulators that rely on the electrical and mechanical properties of alkyds, as well as their high dimensional stability in the presence of temperature and humidity changes. Alkyds are available in many colors, but are seldom used in decorative applications because of cost and generally a less-smooth surface finish when compared to aminos or phenolics. Many of the alkyd molding materials are self-extinguishing.

Allyl Resins and Compounds

Allyl resins are thermosetting resins that are most commonly used in molding compounds such as diallyl phthalate (DAP) and diallyl isophthalate (DIAP). Among the most costly thermosets, DAP and DIAP are used primarily in specialty applications which demand extreme dimensional accuracy and premium electrical and chemical resistance properties, particularly at elevated temperatures. Although both materials are close in most properties, DIAP can be used at higher temperatures (up to 500° F) than DAP which usually is restricted to continuous use at or below 300° F, although higher temperatures can be tolerated for short periods. DAP is usually selected

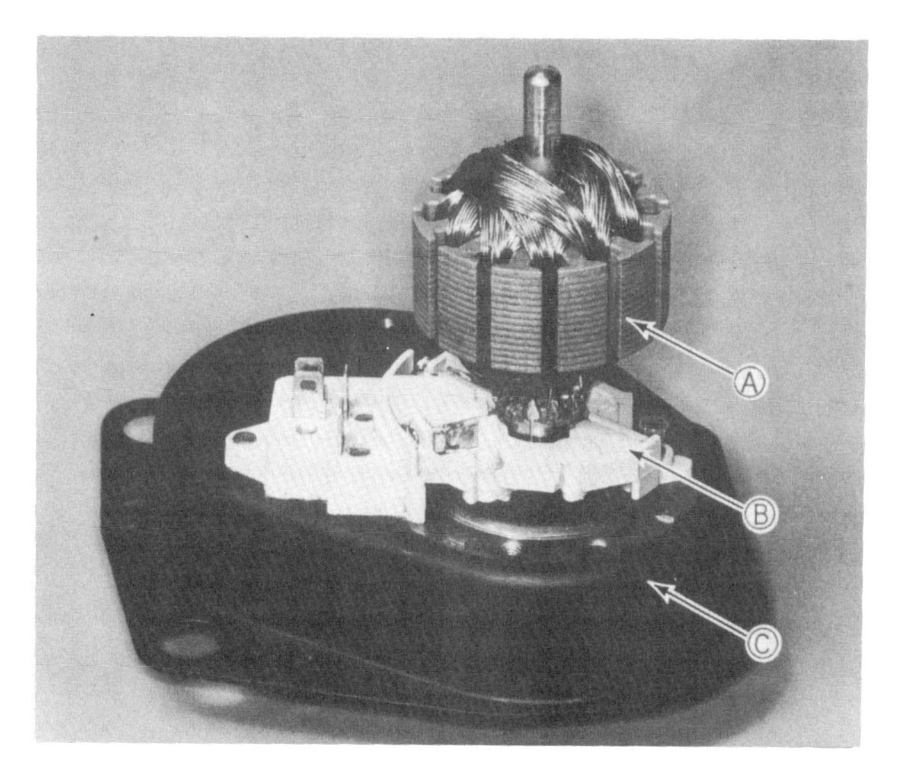

Figure 1–1. This picture shows some parts of an automobile windshield wiper motor. (A) Armature. (B) Alkyd brush holder assembly. (C) Metal motor housing mount. Alkyd plastic material was used in this brush holder because of its superior electrical and good mechanical properties.

when greater electrical properties, higher temperature, or greater dimensional stability are required than can be offered by an alkyd.

The excellent electrical properties of the allyls, particularly arc resistance, surface and volume resistivity, and negligible change in electrical properties with extreme variations in temperature and humidity, make the allyls well suited for arc chutes, insulators, terminals, and other electrical and electronic components such as those in Fig. 1–2. Resistance to fungi enables DAP to function well in special military communication gear.

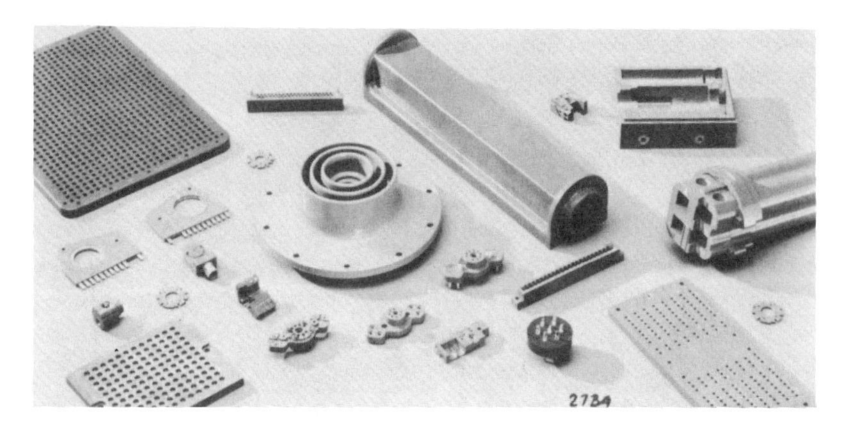

Figure 1–2. This picture illustrates many molded electrical and industrial parts made from diallyl phthalate (DAP) materials. (Courtesy Durez Plastics Div., Hooker Chemical Co.)

Like other thermosets, diallyl phthalate requires fillers and reinforcing agents to impart mechanical properties to the molding compound. Typical fillers and reinforcing agents used in diallyl phthalate compounds are: cellulose, asbestos, glass fibers, and synthetic fibers such as nylon, Orlon and Dacron. Filler selection is determined by the particular property combination desired.

Allylic molding compounds are available in many colors and may be transfer or compression molded. Soft flow grades are available for encapsulation of delicate inserts, and self-extinguishing or non-burning grades are available.

Amino Molding Compounds

In the thermoset family, the aminos are recognized by their unlimited range of colorability. Two materials, melamine formaldehyde and urea formaldehyde are classed as aminos. Both materials offer good electrical properties such as arc resistance, have high surface hardness, excellent gloss and appearance, resist solvents, and are non-burning. The chief drawbacks of aminos are their low impact strength and great dimensional change upon exposure to humidity.

Melamines

Melamine possesses greater temperature and humidity resistance than ureas, enabling it to be used successfully in dinnerware, circuit breakers and switch-gear, utensil handles, small appliance housings, and appliance knobs. Melamine-impregnated overlays are used to impart decorative details to items like dinnerware and utensil handles.

Ureas

With a maximum continuous-use temperature of 170° F, as opposed to 210° F for melamine, urea finds use in the less-demanding applications. Common uses of urea are in colored domestic electrical switches, outlets, and wall plates as shown in Fig. 1–3. Also shown is the

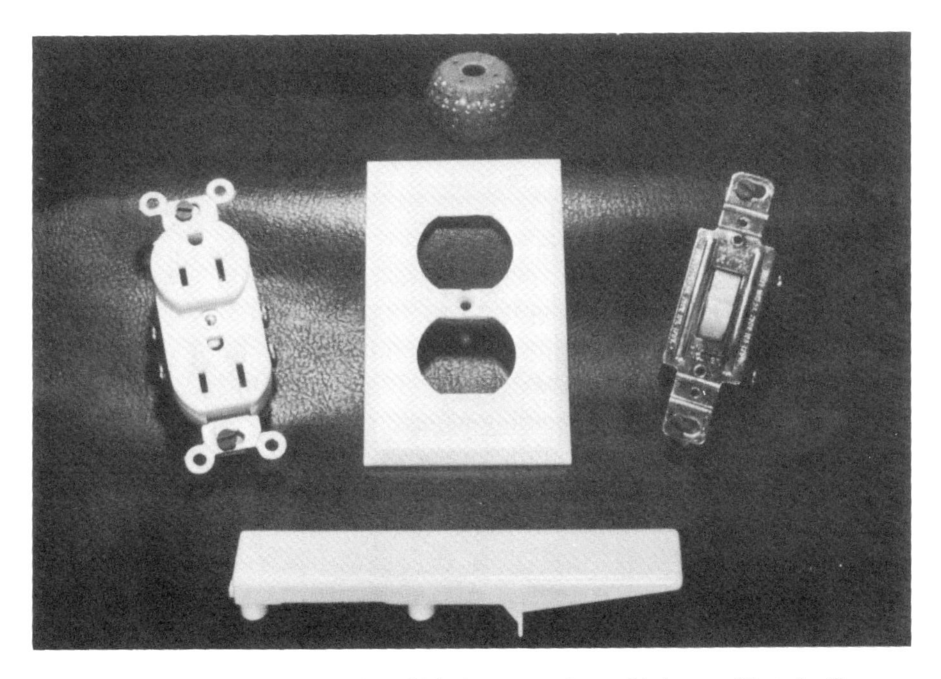

Figure 1–3. A group of urea formaldehyde compression molded parts. (Photo by George Freund)

unique IBM "Selectric" typewriter character sphere which uses the combination of lightness and surface hardness of urea in this hard-working, plated functional part.

Epoxies

Epoxy resins are the highest performance materials in the thermoset family, possessing greater chemical and heat resistance than the other thermosets, but usually at a cost penalty. Outdoor weatherability and virtually no change in electrical properties upon outdoor exposure have enabled epoxies to be used in high-voltage insulators in power distribution lines, a task previously performed by ceramics.

As with other thermosets, epoxies must be combined with fillers and reinforcing agents, such as glass, to provide the mechanical properties needed for molded parts. Epoxies also may be used in conjunction with glass rovings or filament windings to produce chemically resistant, high-temperature parts like nose cones and laboratory sinks. Glass reinforced epoxy parts offer higher strength, temperature resistance and greater durability than similar parts made from polyester resins.

Potting compounds that are poured at atmospheric pressure are used to encapsulate small electronic parts such as coils. Recently, epoxy molding compounds have been developed that can be compression or transfer molded at low pressures. This feature enables applications that were potted to be molded at rapid cycles. Applications such as relay coils can be encapsultated by transfer molding at low pressures while still fully impregnating the windings without fear of fracturing the delicate winding wires.

Principal areas of application for epoxy molding compounds are in the electronic and electrical industry. Epoxies are used whenever less-costly materials are not satisfactory.

Phenolics

Phenolics is among the oldest of the synthetic plastics, having been discovered in 1909. Despite age, phenolics are still high-volume engineering materials, and continue to find applications. Because of their maturity, phenolics are often used as the baseline for comparion in the thermoset family. Basically, phenolics possess the properties typical of all thermosets: stiffness, heat resistance, retention of properties at elevated temperatures, and solvent resistance.

Molding compounds made with phenolic resin can be formulated with a wide variety of fillers which impart an accordingly wide range of possible properties to the compound. Special formulations can be made for certain electrical properties, for impact, for heat resistance, or for chemical resistance. Naturally, special-purpose materials are more costly than general-purpose materials which are among the lest expensive of all plastics.

Filler and reinforcement materials include wood flour, asbestos, synthetic fiber, chopped cloth, glass fiber, and cotton flock. Each reinforcing medium is selected for a specific purpose.

One drawback of phenolics is their limited colorability. Black and brown are the most common colors for phenolics. However, some grades may be obtained in dark green or dark red-brown.

All thermoset processing methods may be used for phenolics, including recently developed screw-injection molding, and special thermoset extrusion. Machining and polishing are common post-molding operations performed on phenolics.

Applications for phenolic molding compounds are numerous, as shown in Fig. 1–4. They include a wide array of parts such as automobile distributor caps, automatic transmission valve bodies, switches, thrust washers, appliance pump impellers and housing, photographic processing tanks, encapsulated electronic components, small motor housings and brush holders, and electric tool housings.

Polyesters

Polyester resins are thermosetting resins that are derived from the polymerization of diabasic acids with poly-functional alcohols. These colorless liquid resins are normally used with fiber reinforcement and fillers. Parts made from polyester compounds are hard, rigid, and temperature resistant. By varying the type of resin and reinforce-

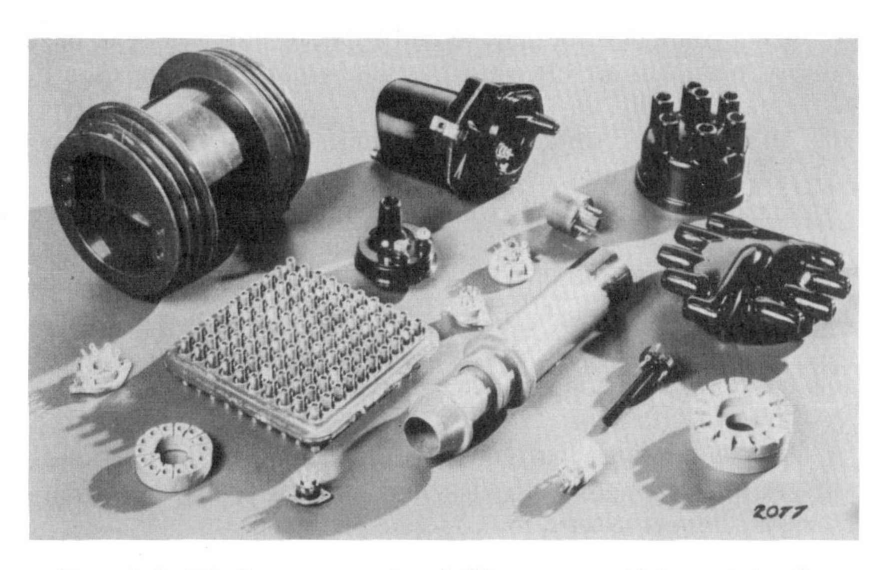

Figure 1–4. This illustrates a number of different parts molded out of phenolic compounds. The parts shown are used in electrical insulating applications. *(Courtesy Durez Plastics Div., Hooker Chemical Co.)*

ment, a wide range of properties and fabrication techniques is possible, including molded-in color.

Hand lay-up is the oldest fabrication technique. Here, a layer of glass cloth or mat is shaped to a buck (male mold), and catalyzed polyester resin is applied manually to the surface. After curing of the resin, additional layers of glass and resin can be applied. This is the technique that was initially used to fabricate large objects such as boats.

In order to speed up the fabrication process, spray-up techniques were developed. In this process, a special type of spray gun is used that dispenses catalyzed resin and chopped glass fibers which are then sprayed on a buck. After curing time, the finished part such as a boat or other large object is removed from the mold.

Mat molding is the technique in which a mat, or preform of glass fibers containing a small amount of resin is placed either on the male or female part of the mold. Resin is then poured atop the mat and the mold is closed to form the part and to cure the resin with heat. Articles such as molded chairs and containers are formed in this manner.

Molding compounds called "premixes" are comprised of resin, glass fibers or chopped glass mate, catalyst, and fillers. These compounds may be compression, transfer, or screw injection molded in the same fashion as other thermosetting molding compounds. In premix and allied compounds, a wide range of properties is possible, depending on formulation. Because of the long glass fiber reinforcement used here, much higher impact strengths are possible than are normally expected from a thermoset. Typical applications of these molding compounds are automotive heater housings (shown in Fig. 1–5), fender extensions, and hood scoops.

Polyester resins and their compounds find a wide variety of applications from automobile parts, building panels, chemical storage tanks, chemical piping, and even complete car bodies such as is used on the Corvette.

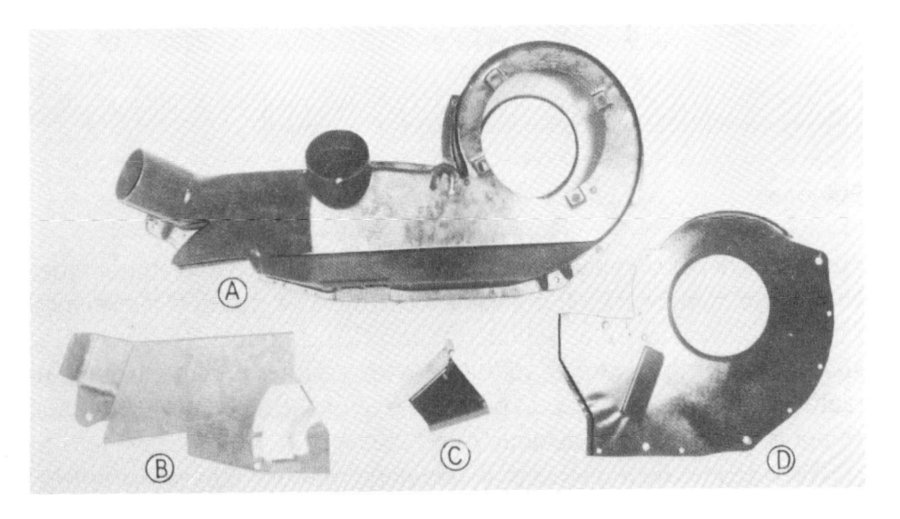

Figure 1–5. Automobile plastic blower assembly parts. (A) Polyester sisal premix assembly riveted together. (B) Formed fiberboard part with an injection molded polypropylene part riveted to formed fiber board. (C) Damper door. (D) Steel cover riveted to a polyester premix part.

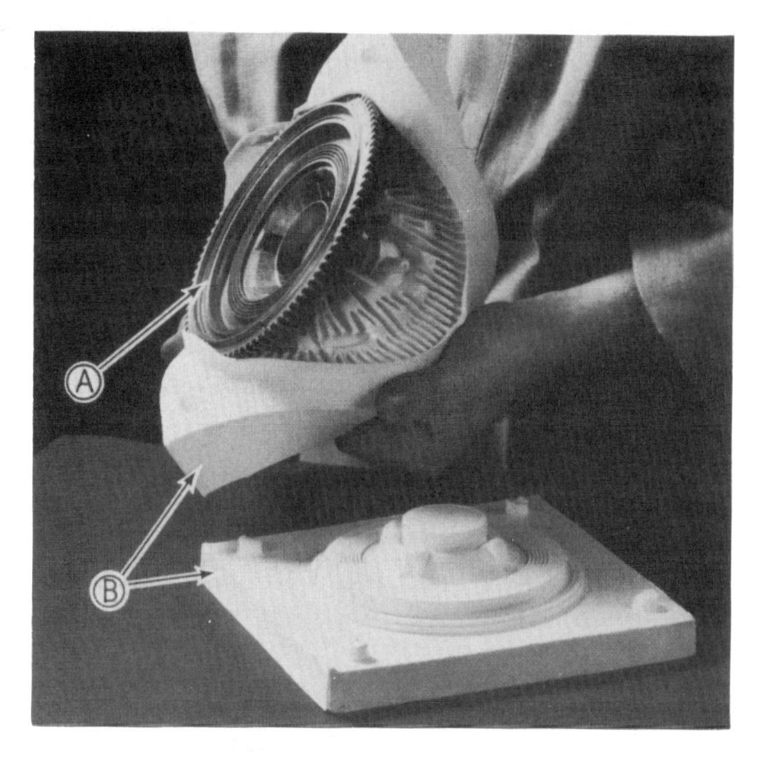

Figure 1–6. A plastic prototype part that has been cast into a flexible silicone mold. (A) Plastic cast part. (B) Flexible silicone mold. *(Courtesy Dow Corning)*

Silicones

Silicones are perhaps the most difficult of all to place and characterize within the framework of the usual definition of plastics. They are a family of semi-organic polymers comprising chains of alternating silicon and oxygen atoms, modified with various organic groups attached to the silicon atoms. The backbone of the molecule is common to many minerals such as quartz and mica.

The silicones include resins, elastomers, rubbers, adhesives, potting compounds, fluids, and oils. One of the most familiar applications of silicone is that of room-temperature-vulcanized (RTV) rubber. RTV is used widely in making flexible molds. The molds can be used for casting of models or prototypes (Fig. 1–6). Adhesives and caulking materials that cure when exposed to air are used for sealing around edges of bathtubs and windows.

Most silicone-rubber molded parts remain useful from $^{-}70$ to $+500°$ F. The silicone rubbers are especially useful for high-temperature service, because they do not employ plasticizers to maintain rubberiness. They are highly resistant to oxidation and deterioration caused by heat aging, in addition to their serviceability at very low and very high temperatures. Silicones are physiologically inert, odorless, tasteless, and non-toxic.

Typical applications are: bonding; waterproofing; oils and greases; insulating varnishes; potting and embedding compounds; rigid and rubber-molded parts; paint additives; release agents; textile finishes; and leather treatments.

THERMOPLASTICS

Acrylonitrile Butadiene Styrene (ABS)

ABS plastics are thermoplastics of the styrene family. They are comprised of monomers of styrene plastic and acrylonitrile and butadiene rubbers (Fig. 1–7). Because styrene is a non-crystalline or glassy material by itself, it lacks the toughness and impact strength required

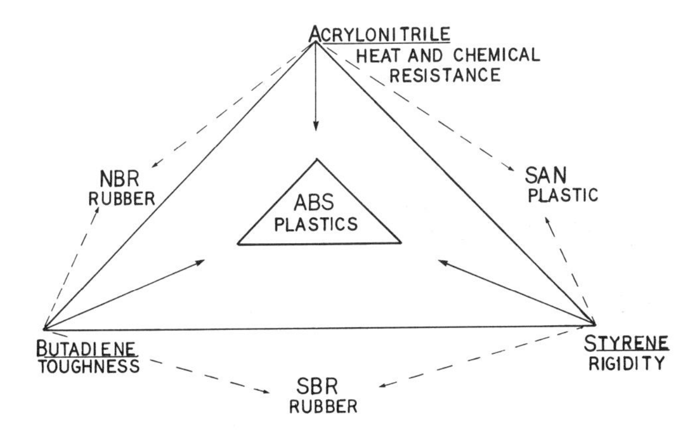

Figure 1–7. This illustrates the relationship of plastic to rubber compounds in making ABS plastics.

for many applications. By varying the content of both acrylonitrile and butadiene in the ABS, a wide range of specific materials with concurrent wide property ranges becomes available. ABS in all versions possesses greater impact strength and ductility than does polystyrene.

High gloss and unlimited colorability make ABS materials well suited in decorative applications, such as automotive interior trim parts, instrument clusters, and television and radio cabinets. The first plastic to be successfully chrome plated in large volume parts, ABS has found its way into demanding applications like automobile grills. High impact resistance and moderate cost enable ABS to be used in luggage, safety helmets, sports car underbodies, and some furniture parts. Some typical applications appear in Fig. 1–8.

Normally classified as slow burning, ABS can be blended with

Figure 1–8. ABS thermoplastic is one of the most versatile materials that the design engineer has to work with. This is illustrated by the many numerous items displayed in this picture. *(Courtesy Marbon Chemical Co.)*

PVC to become flame resistant for applications demanding this characteristic.

Chemical resistance to alkalies and some acids is good, although ABS is subject to stress cracking when exposed to certain acids and organic solvents.

Acetals

Acetals are crystalline polymers that are normally referred to as members of the engineering plastics family, because of their mechanical properties. High stiffness, tensile strength, and fatigue endurance have enabled acetal to replace metal in many applications. Injection molding and extrusion are the normal fabrication methods for acetal, although acetal parts can be machined from stock shapes. Fastening and joining of acetal parts can be done by spin welding or ultrasonic welding or by the use of conventional metal fasteners. Good snap-back characteristics enable acetal parts to be snap-fitted to metal parts or to other acetal components.

Strength combined with little change in impact resistance with changes in temperature, a low coefficient of friction, creep resistance, and dimensional stability enable acetals to replace metals in such applications as gears, cams, bearings, and switches in automobile and business-machine parts. Resistance to solvents and most alkalies permit acetals' use in pump impellers. However, contact with most acids should be avoided.

Appearance parts made from acetals can be pigmented for color, or they can be painted, provided a special primer and acid etch is used. Other decorating techniques that can be used are printing, metallizing, plating, silk screening, and dyeing.

The combination of all properties enable acetal to find use in a myriad of applications, such as the spray gun shown in Fig. 1–9.

Acrylics

Acrylics are non-crystalline polymers that are renowned for their clarity and light-transmission properties, combined with excellent re-

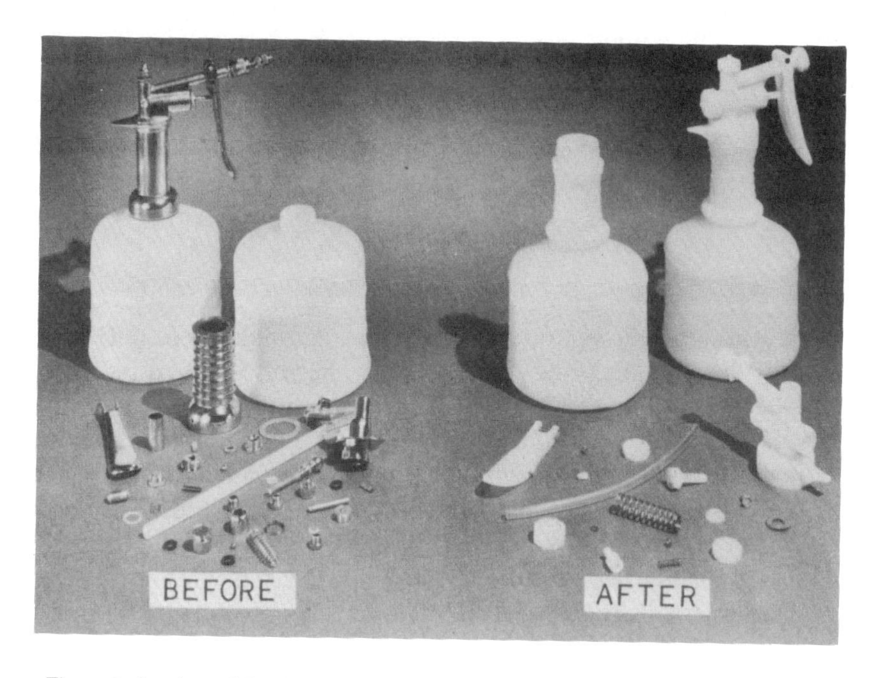

Figure 1–9. A small hand garden spray gun assembly. The gun has been redesigned and made from Celcon acetal copolymer. The container is blow molded from high density polyethylene. *(Courtesy Celanese Plastics Co.)*

tention of physical and optical properties after long exposure to the outdoors.

Although the strength and stiffness of acrylic are in the range of the engineering materials like nylon and acetal, its use temperature generally remains below 200° F, except for special grades. Electrical properties of acrylic are good at low frequencies, and acrylic will not track in the presence of an arc. While these materials do burn slowly, a low degree of smoke is emitted that makes acrylics acceptable in many building applications like decorative panels and light diffusers.

Acrylics may be modified and copolymerized to very certain properties such as impact strength or heat resistance. However, such modifications generally impair optical properties, which are the forte of acrylics, along with dimensional stability and color retention when the acrylics are exposed to ultraviolet light.

Primary fabrication methods for acrylics are: injection molding; extrusion; casting; and blow molding. Secondary methods are: thermoforming; machining; stamping; painting; and metallizing. Colors are possible in opaque, transparent, and translucent degrees.

The properties of acrylic enable them to be used in airplane canopies, window glazing, tail lamps, medallions for automobiles and appliances, reflectors, outdoor signs and lighting, building panels, watch crystals, and fiber optic-light pipes. Some typical lens applications are shown in Fig. 1–10.

Cellulosics

The oldest of the plastic family, cellulosics were discovered in 1833 and found commercialization in the celluloid collars that were popular before the turn of the century. Unlike other plastics that are pro-

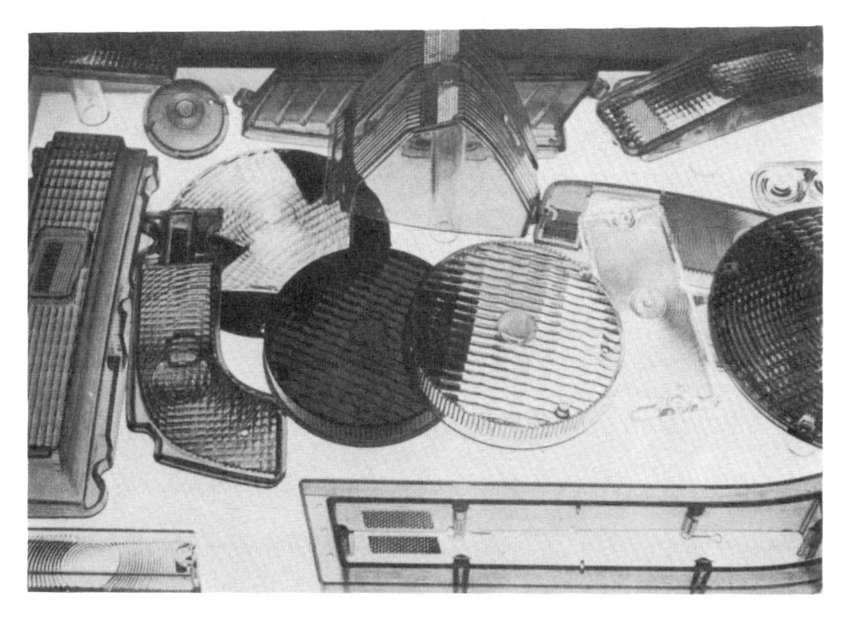

Figure 1–10. Automotive lenses are molded of methyl methacrylate. This plastic material has superior optical qualities and weather resistance. It has good decorative possibilities and is non-yellowing. *(Courtesy duPont)*

duced by polymerization, natural cellulose is found in many plants. However, the chief sources of cellulose for plastic are found in cotton linters and in wood pulp.

In natural form, cellulosics do not melt. Yet, with proper chemical treatment, cellulosics can be made into materials that are processed in standard thermoplastic equipment. These materials are competitive in properties with thermoplastics. The chief cellulosic materials used today are: cellulose acetate; cellulose propionate; cellulose acetate butrate; and ethyl cellulose. Figure 1–11, illustrates an excellent application for cellulose acetate butyrate. The snowmobile windshield has toughness, transparency, and good weatherability.

Figure 1–11. A snowmobile windshield made from cellulose acetate butyrate. This material has impact-resistant qualities and the ability to withstand icy temperatures.

In designing parts to be made from any of the cellulosics, it should be kept in mind that dimensions vary, due to changes in humidity and temperature. Cellulosics have a tendency to "creep" or "cold flow" if under any load.

Chlorinated Polyether

Chlorinated polyether is sometimes referred to as the "poor man's fluorocarbon." While chlorinated polyether does not contain fluorine, and thus is not a fluorocarbon, it contains chlorine that imparts many of the properties characteristic of fluorocarbons, but to a lesser degree.

Chlorinated polyether is close in properties to polyvinylidene fluoride, except PVF_2 has greater temperature and chemical resistance. Properties favoring chlorinated polyether are its low molding shrinkage and inherent dimensional accuracy in molded parts, along with stiffness. These properties permit chlorinated polyether to be used successfully in gears for delicate metering devices. Other applications include valve liners and pump heads.

Fabrication methods include injection and compression molding, extrusion, and fluidized bed coating. Because chlorinated polyether is more stable at processing temperatures than are melt-processable fluorocarbons, it is considered to be more easily processed than fluorocarbons.

Fluoroplastics

Fluoroplastics or fluorocarbon plastics are a group of paraffinic hydrocarbon polymers in which a fluorine atom replaces some or all of the hydrogen atoms in the molecule. Compared with other thermoplastics, the fluoroplastics offer superior chemical heat resistance and electrical properties. Materials in the fluorocarbon family are: polytetrafluoroethylene (TFE); fluorinated ethylene propylene (FEP); chlorotrifluoroethylene (CTFE); polyvinylidene fluoride (PVF_2); and polyvinylfluoride (PVF).

Polytetrafluoroethylene

TFE is the most distinguished member of the fluorocarbon family, since it contains the greatest fluorine content. Since fluorine content regulates properties, TFE possesses the greatest amount of the properties for which the fluoroplastics are noted. Unfortunately, this high fluorine content also makes TFE the most difficult thermoplastic to process.

While TFE is classed as a thermoplastic, it does not actually melt and cannot be processed on conventional thermoplastic equipment. Rather, TFE is processed more like ceramics or powdered metals. TFE may be fabricated directly into finished parts by a special molding process, or parts may be machined from molded billets or from an extruded rod. Thin films of TFE are normally skived from molded billets, using equipment similar to that employed in producing plywood laminations. Films as thin as 0.001 in. may be made by skiing.

The process for molding both parts and billets parallels that used for ceramics. First, powder is loaded into a mold, and the mold is closed gradually to eliminate air entrapment. After the powder has been pressed into a preform, it is removed from the mold and sintered in an oven at a carefully controlled temperature until the gel temperature of about 700° F is reached. The preform is gradually cooled to room temperature and is then ready to be machined into final form.

Small parts may be automatically preformed in a hydraulic press, similar to a metal stamping press. These parts must also be sintered in an oven before parts achieve their full properties.

Extrusion is accomplished by ram-extrusion, which could be described as continuous compression molding and sintering within a tube. Paste extrusion that uses a solvent and low-molecular-weight TFE is used for thin-wall tubes and wire coating.

TFE is noted for its low coefficient of friction (the lowest of all plastics), heat resistance, greatest chemical resistance of all plastics, and lowest change in critical electrical properties over a wide range of temperatures and environments. Because of its properties, TFE finds applications in bridge bearing pads, chemical piping equipment, electronic insulators, arch chutes, seal backup rings, seals, human heart valves, cyrogenic valves, and automobile power steering units, such as that shown in Fig. 1–12.

One shortcoming of TFE is its high creep under load. To overcome this problem, TFE can be filled with other materials such as glass, bronze, graphite, and molybdenum disulfide, either alone or in any combination. While fillers decrease tensile properties slightly, load-bearing capability is greatly enhanced by the addition of fillers.

Fluorinated Ethylene Propylene

FEP is the closest to TFE of the melt-processable fluoroplastics in properties. Conventional melt extrusion, injection, and compression molding may be employed using FEP. In addition, secondary opera-

Figure 1–12. Sealing rings (A) from glass filled TFE are used in automobile power steering units because of their low frictional coefficient, dimensional stability and resistance to automatic transmission fluids.

tions, such as vacuum forming, heat sealing, and spin welding can be used with FEP.

Compared with TFE, FEP offers somewhat lower temperature resistance (400° F), chemical resistance, and electrical properties. However, FEP enjoys the advantage of greater mechanical properties, particularly modulus and creep resistance, as well as ease of fabrication.

FEP finds application in lined valves and other equipment for the chemical industry, injection-molded electronic insulators, food-process equipment components, and containers subjected to strong chemicals.

Chlorotrifluoroethylene (CTFE)

CTFE is available in several formulations, containing both fluorine and chlorine in the molecular structure and can be copolymerized with VF_2. Similar to other fluoroplastics in general properties, CTFE is noted for its low permeability to water and gasses, making it the least permeable of all plastics.

Being melt-processable, CTFE can be fabricated into injection molded parts or extruded into tubing and thin films that are transparent. Transparency, combined with high resistance to radiation and zero moisture permeability, makes CTFE the prime choice for packaging of medical supplies that can be sterilized by radiation after packaging.

Chemical resistance to all reagents except certain halogen-containing substances and an upper temperature limit of 400° F allow CTFE to be used in chemical pipe fittings and equipment liners. Cryogenic piping and other equipment take advantage of the minus 400° F low temperature capability of CTFE.

The non-tracking and high dimensional stability of CTFE enable it to be used in super-critical electronic connectors, coil forms, terminals, and wire insulation. Low ultraviolet absorption assures retention of mechanical and electrical properties after long outdoor exposure.

Polyvinylidene Fluoride (PVF_2)

Polyvinylidene fluoride is melt-processable in conventional injection molding and extrusion equipment. While PVF_2 is generally considered to offer less chemical resistance than other fluoroplastics, it resists certain halogenated reagents, but is sensitive to polar solvents such as ketones and esters.

The high stiffness of PVF_2 enables it to be the only fluoroplastic to be used in rigid pipe. Common uses are in chemical process equipment in valves and pumps. Having a rather limited temperature range, compared with other fluorocarbons (-80 to + 300° F), PVF_2 is not used in cryogenic applications. However, its superior mechanical properties permit PVF_2 to be used in load-bearing applications that other fluoroplastics cannot sustain.

Ionomer Resins

The term "ionomer" is used because ionized carboxyl groups cross-link ionically in the intermolecular backbone. This cross-linking effect on the molecular chain makes ionomers perform nearly equal to true cross-linked materials, although processing is done at conventional temperatures. Properties of the material can be regulated by varying the degree of crystallinity, molecular weight, and degree of ionization.

Generally close to polyethylene in mechanical properties, ionomers excell in high impact strength (particularly at low temperatures) and high elongation. However, ionomers are limited in their upper temperature range to about 160° F.

Fabricated by all conventional thermoplastic processing methods, ionomers are used in wire coating, safety shields and similar high-impact molded parts, blow-molded containers, and in flexible packaging applications in film form.

Nylon

The term "nylon" is now the generic name given to the family of polymers known as polyamides. While there are numerous variations of polyamides, the most popular ones used in this country are:type 6; type 6/6; type 6/10; type 11; and type 12. The nylons are designated by the number of carbon atoms in the diamine, followed by the number of carbon atoms in the diacid.

All nylons share some property similarities. The common characteristics are: toughness; fatigue resistance; impact resistance; inertness to aromatic hydrocarbons; and low friction. While nylons 6/10, 11, and 12 exhibit lower total moisture absorption than do types 6 and 6/6, all nylons absorb some moisture. The absorbed moisture acts as a plasticizer, thus increasing impact strength and decreasing stiffness. Moisture absorption is a reversible phenomena and moisture will be rejected if parts are dried, with according property changes.

Nylon 6 and 6/6 are the most popular, because of their cost and property ranges. Molded parts of these materials generally take advantage of many of nylon's properties. Door latch hardware, automobile speedometer gears, windshield-wiper gears, seat sliders, and carburetor linkages utilize nylon's combination of low friction, abrasion resistance, strength, toughness, fatigue resistance, and temperature resistance.

All types of nylon are fabricated by conventional melt-process equipment, except calendering. Secondary operations including hot stamping, ultrasonic welding, spin welding, heat sealing and staking, as well as machining are possible with nylon. Pigmented formulations are available in a wide variety of colors.

Nylon 6/10 is more flexible than either 6 or 6/6, but exhibits lower moisture absorption than either of these. Thus, nylon 6/10 is used in vending-machine and home-laundry water-mixing valves, because it retains properties and dimensions when in contact with water better than do the other nylons. Other uses of 6/10 nylon include tubing and food packaging. Most common applications of 11 and 12 are in film or tubing, because of their flexibility.

All nylons may be copolymerized with other materials to arrive at specific property levels. For example some copolymers of type 6 offer flexibility similar to type 11 at lower cost. Other copolymers contribute greatly to impact strength with only slight loss in stiffness.

Fillers can be used to increase stiffness, particularly at elevated temperatures. Glass and asbestos have been used to improve stiffness, and molybdenum disulfide has been used to increase lubricity.

Fig. 1–13 shows a wide array of parts that use the multiplicity of nylon's properties, including coil forms, electrical connectors, gears, fuel filters, lamp lens, fasteners, and blow-molded bottles.

Polycarbonate

Polycarbonate is an engineering thermoplastic that is characterized by its high stiffness combined with high impact strength and high retention of mechanical properties at elevated temperatures.

Processable by injection molding, extrusion, and blow molding, polycarbonate remains stable at processing temperatures. Available in optically clear or colored formulations, polycarbonate is used in lenses, appliance housings, safety helmets, and goggles in which these properties are necessary.

Surprisingly high-impact resistance, even at low temperatures, for

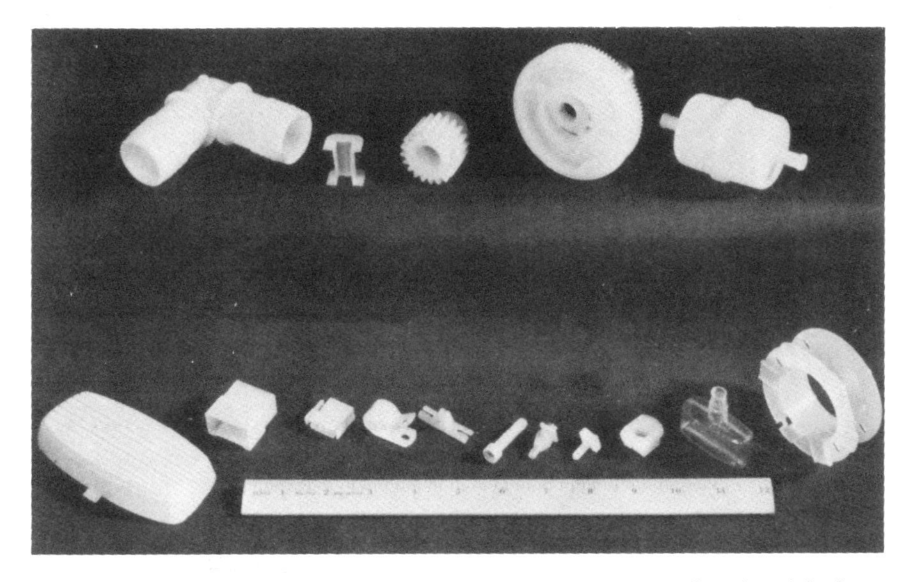

Figure 1–13. This picture illustrates many applications and parts that have been injection molded and blow molded from nylon. *(Photo by G. N. Freund)*

a material with a high moldulus, polycarbonate is used for gears, as shown in Fig. 1–14, and in outdoor air-conditioner housings and hand-tool housings to replace die castings and metal stampings.

Although dielectric strength is good at various temperatures and frequencies, polycarbonate is not normally recommended for use in the presence of an arc. Resistance to dilute acids and aliphatic hydrocarbons is good. However, polycarbonate is attacked by alkalies and aromatic hydrocarbons.

Thermoplastic Polyesters

Thermoplastic polyester molding materials are crystalline. They are listed chemically as PBT (polybutylene-terephthalate and PTMT

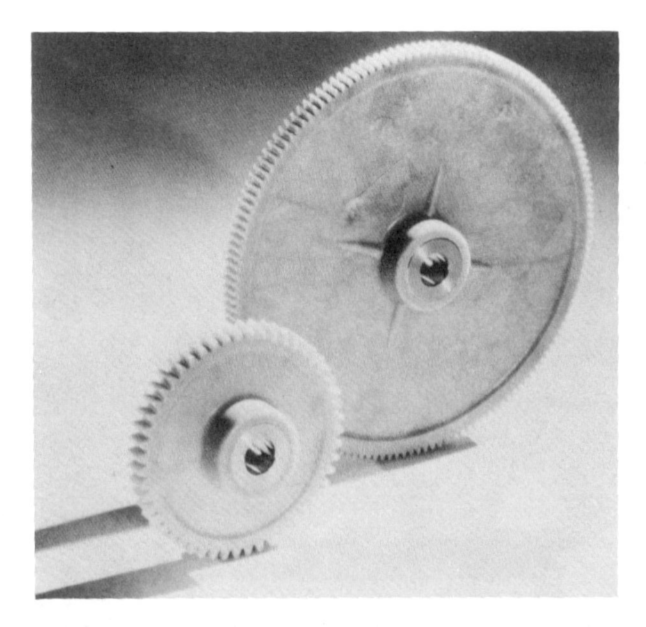

Figure 1–14. Polycarbonate is truly an engineering material as shown by the two gears that are used in a weather satellite. The material is a polycarbonate filled with 20% fiber glass and 22% TFE fluorocarbon. This application required a material with a very low coefficient of linear thermal expansion.

(polytetra-methylene-terephthalate). The chemical formulation and resulting properties vary only slightly between these materials.

Thermoplastic polyesters have many of the desirable qualities of both thermosets and thermoplastics. This material exhibits higher tensile, flexural and dielectric strengths than most plastic molding compounds. The overall molding cycle time and molding characteristics are better than most thermosets. Its rapid crystallization rate coupled with good melt flow results in fast molding cycles. Fiber-glass is added to the material 15–30%) in order to improve the heat-resistance capabilities and physical strength.

Due to the fact that thermoplastic polyesters have excellent chemical resistance they are difficult to paint. However, acrylic, epoxy, and enamel paints have been used to paint molded parts.

Thermoplastic polyesters can be fastened by standard thermoforming screws and ultrasonic bonding. Other methods of fastening include snap-fits, and molded-in inserts.

Epoxy resin can be used as an adhesive for bonding. End use applications of thermoplastic polyesters include switches, integrated circuit parts, ignition coil caps, distributor caps, rotors, gears, and pump impellers and housings. Figure 1–15 shows auto ignition parts made from thermoplastic polyester resins.

Polyethylene

Polyethylene is the oldest of the olefin thermoplastics and is one of the most widely used low-cost plastics available. Processable by all thermoplastic processes, polyethylene is noted for its flexibility, low-temperature impact resistance, low coefficient of friction, good dielectric strength, and generally good chemical resistance.

Not a single material, but a whole array of materials with varying properties, polyethylenes are broadly classed into low-density and high-density variants. Low-density materials exhibit branching of the chain, which minimize the degree of crystallinity possible and, hence, the spaces between the molecules cause a low density. Thus, low-crystalline, low-density polyethylene is flexible, transparent to translucent, and has a lower maximum temperature range than does high-

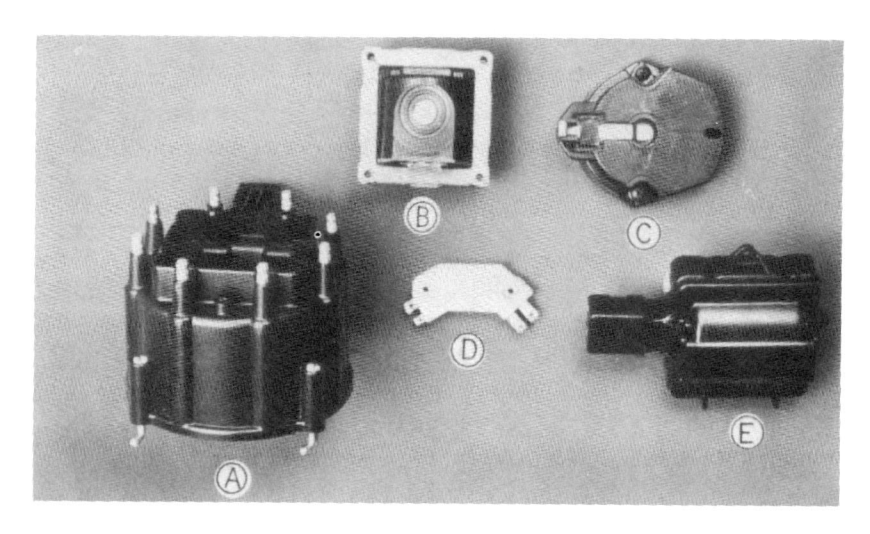

Figure 1–15. Injection molded thermoplastic polyester auto ignition system parts. (A) Distributor cap. (B) Coil case. (C) Rotor. (D) Electronic module. (E) Cover. *(Courtesy Celanese Plastic Co.)*

density polyethylene. Low-density polyethylene finds use in covers for refrigerator dishes, soft gaskets, flexible film for packaging, and squeeze bottles.

High-density polyethylene is comparatively free of branching and is, therefore called "linear polyethylene." Linear polyethylene is crystalline; therefore, compared with low-density materials, these polyethylenes are stiffer, stronger, more opaque, less resistant to impact, but more resistant to higher temperatures. Applications for high-density polyethylene include: dishpans; garbage pails; automobile cowl and trim panels; rigid detergent bottles; beverage cases; toys; refrigerator dishes; film for boil-in-bag packaging; and wire insulation.

Since density, or degree of branching, can be varied, there is an almost infinite possible combination of properties that can be obtained by varying the density of polyethylene. Table 1–1 provides a graphic representation of how properties of polyethylene vary with density. Each class of application demands a specific material with certain properties. Typical application of high and low-density polyethylene are shown in Fig. 1–16.

TABLE 1–1. THIS ILLUSTRATES THE RANGE OF SOME PROPERTIES OF POLYETHYLENE WITH DIFFERENT DENSITY

	Polyethylene			
Density	0.91	0.925	0.945	0.96
Crystallinity	65%	75%	85%	95%
Softening point	220°F	245°F	255°F	260°F
Elongation	500%	300%	100%	20%
Hardness Shore "D"	42	50	60	70

Within a given density range, another factor that controls properties is melt index. Melt index is an indication of molecular weight. A low-melt-index (high-melt-viscosity) material possesses high molecular weight. Low-melt-index materials exhibit high impact, better stress crack resistance, and better solvent resistance than do high-melt index materials. Generally, the highest-melt-index materials are used

Figure 1–16. This picture illustrates many articles made from polyethylene and also many methods of fabrication in making these articles. (A) Extruded film. (B) Injection blow molded lemon. (C) Injection molded toy. (D) Blow molded bottle.

for injection molding, and the lowest-melt-index materials are reserved for extrusion and blow-molding applications.

Ultrahigh-molecular-weight polyethylene is processed almost like TFE. Parts molded (by preforming and sintering) are used in textile and similar machinery in which extremely high impact and wear resistance is required. Extruded HMWPE is used in pressure piping, because it can sustain stress for an extended time without bursting.

Certain grades of high-density polyethylene can be processed by rotational molding. In this process, powdered material is introduced into a hollow mold which is then heated and rotated about two axes, causing the molten material to coat the mold interior. After cooling, the part is removed. Rotational molding is used to form parts that are too complicated to blow mold or in which low tooling cost is a predominant factor.

Polyethylene is vulnerable to environmental stress cracking. Thus, ordinary chemicals such as wetting agents, soaps, detergents, silicones, and certain hydrocarbons when contacting stressed polyethylene will, in time, cause cracking. Close tolerances on the dimensions of molded parts made from polyethylene are difficult to hold. Close tolerances should be avoided or kept to a minimum. As polyethylene expands about twice that of most common plastics, allowance should be made for the coefficient of thermal expansion when the polyethylene part is to be assembled to parts made from other materials.

Polyimide

Thermoset Polyimide. Polyimides exist both as thermosets and thermoplastics. Thermoset polyimide is one of the most heat-resistant polymers known. It is both molded and used in laminates. The laminates are made with woven glass fabrics, graphite, boron fibers, quartz fabrics and high modulus organic fibers. The molding polyimides are made with graphite powder, chopped glass fibers, molybdenum disulfide, PTFE, and asbestos fibers. Thermoset polyimides are also produced in films and wire enamels.

Molded polyimide parts exhibit high heat resistance, ranging to 500° F for several hours and intermittently to temperatures as high as 900° F. They are also inherently resistant to combuston. Glass fiber reinforced polyimide moldings show excellent physical properties at elevated temperatures (480° F). The coefficient of thermal expansion is closely matched to that of metals. Creep is almost nonexistent even at high temperatures. The thermoset polyimides have excellent electrical properties and are unaffected by most common chemicals. They are attacked by dilute alkali and concentrated inorganic acids.

Polyimide molded parts are fabricated by compression, transfer, injection, and powder metal sintering techniques. The molded parts include electronic aero-space, jet engine parts bushings, coil bobbins, bearings, etc. The current price of this material is approximately $4 per pound.

Thermoplastic Polyimide. This material is technically classed as a thermoplastic, although it does not melt and is difficult to injection mold. Polyimide compounds can be injection molded but require high melt temperatures (660° F or 350° C) and high injection pressures. The molding cycle time is long due to the time required for the molded part to cool to a temperature at which it can be ejected without distortion. Polyimides are available as resins and compounds. They are obtainable in dry powders, solvent-based solutions and solid forms that can be machined. The molding compounds contain fillers such as glass, fibers, graphite fibers, molybdenum disulfide, asbestos fibers and PTFE. The thermal expansion of a glass fiber polyimide is close to that of metals. The polyimides inherently have excellent resistance to burning. Graphite and PTFE filled polyimides exhibit low coefficient of friction. They have excellent stiffness, toughness and lubricity, low creep and abrasion, and high fatigue strength. The polyimides are used to make bearing sleeves, rings, seals, thrust washers, valve seats, etc. Phenolic and epoxy adhesive adhere very well to polyimide parts.

Polyphenylene Oxides (PPO)

Polyphenylene oxide is an amorphous type thermoplastic, It is known for its useful temperature range from −275° to 375° F. Processing PPO is performed on conventional injection molding and extrusion equipment. PPO has low mold shrinkage, low thermal coefficient of expansion, and negligible water absorption. With a modulus of elasticity and tensile strength above that of moisture-conditioned nylon and with nearly the same izod strength, PPO possesses the toughness needed in many engineering applications. PPO retains many of its physical properties at elevated temperatures. Chief applications for PPO take advantage of key properties. Food handling devices and containers utilize the resistance of PPO to food acids, detergents, and water. Medical instruments take advantage of steam and temperature resistance in autoclaving. Water treatment devices like filter plates, valves, and pumps require dimensional stability, creep resistance and hydrolytic stability. A typical pump cover molded from PPO is shown in Fig. 1–17.

Modified PPO falls short of its more costly counterpart in tensile strength and temperature resistance, while offering higher impact strength. Because of its available properties and reduced cost, modi-

Figure 1–17. A marine pump housing injection molded from Noryl, a modified polyphenylene oxide (PPO) thermoplastic material. This material has low water absorption and has resistance to water up to 165° F.

fied PPO is used in such applications as television yokes, tool handles, small appliance housings, business machine housings, sprinkler parts, and decorative grilles.

Polyphenylene Sulfide

Polyphenylene sulfide (PPS) is a crystalline aromatic polymer. It has a high melting point (550° F), good thermal stability, and good chemical resistance. Fiberglass (up to 40%) and mineral fillers are added to PPS to enhance the physical and electrical properties. The mechanical properties of PPS are unaffected by long-term exposure in air at 450° F. PPS is injected molded, compression molded, and used in coatings. A screw-ram type injection machine is recommended for effective temperature control and overall short cycle capability. PPS can be applied in film form or coatings by spray and fluidized-bed coating techniques.

Polypropylene

Polypropylene is a crystalline thermoplastic material that is stiffer than polyethylene and has one of the lowest densities of all plastics. It is translucent and has a milk-white natural color. The material has excellent colorability and can be pigmented to an unlimited array of colors.

The excellent physical and electrical insulating properties of polypropylene are due to its crystalline structure. It is immune to stress cracking. Low-temperature brittleness (its main disadvantage) has been overcome with the introduction of new copolymers. It can be blended with synthetic elastomers, such as polyisobutylene, and copolymerized with amounts of other monomers. Polypropylene can be modified to obtain improved properties by adding fillers such as asbestos or glass fibers.

Polypropylene is easy to fabricate and can be processed by injection molding, extrusion, and blow molding. A special feature of polypropylene is its ability to be molded with an integral hinge that has outstanding flex life. Polypropylene can be decorated in the conventional methods such as spray painting, paint-wipe, hot stamping, vacuum metalizing, etc.

Current uses for this material include housings with molded integral hinges joining two halves, luggage, appliance parts, housewares, toys, cable insulation, automobile interior panels, and party-picnicware, as seen in Fig. 1–18. It is also widely used as a packaging film and in fibers and monofilaments for carpets, rope, twine, textiles, and outdoor furniture. The film of polypropylene can be laminated to paper or cloth.

Figure 1–18. Party-picnicware molded of polypropylene. The containers are of a double walled construction to maintain the temperature of the warm or cold food placed in them. The containers are injection molded in two parts and spun-welded together.

Polystyrene and Copolymers

Polystyrene is undoubtedly the plastic with which most persons are familiar. Broadly speaking, polystyrene is a rigid, hard thermoplastic whose chief attribute seems to be its ability to be blended with other materials to arrive at a myriad of properties within the polystyrene family. General-purpose polystyrene is among the lowest-cost materials. It offers tensile strengths between that of polyethylene and polypropylene; although polystyrene exhibits about double the stiffness of polyethylenes, they are quite brittle. Appearance is surely an attribute of polystyrene, which is naturally clear and can be colored in opaque or translucent colors, all of which exhibits a brilliant surface gloss and appearance. However, being sensitive to ultra-violet light, polystyrene is not normally used in outdoor applications. Polystyrene resists mineral oils, alkalies, and water, but is soluble in hydrocarbon solvents. Hence, solvent welding is possible.

Polystyrene can be modified for improved impact strength by the addition of rubber polymer to the styrene. Rubbers such as polybutadiene and styrene-butadiene (SBR) are added to the polystyrene by blending or graft polymerization. Normally called "high-impact polystyrene," rubber-modified styrene exhibits lower strength, hardness, clarity, and chemical resistance when compared with general purpose polystyrene.

Impact polystyrene can be found in refrigerator door liners, appliance housings, camera housings, luggage, furniture components, closures for bottles, and shoe heels.

Styrene may also be copolymerized with other monomers to achieve significantly different properties from those found in conventional or rubber-modified styrene. Styrene copolymerized with acrylonitrile (SAN) exhibits higher impact and stiffness than polystyrene, as well as better solvent resistance, while retaining clarity.

By polymerizing styrene with acrylonitrile and butadiene, to result in the terpolymer acrylonitrile-butadiene-styrene (ABS), a completely new family of materials results. Since ABS has achieved such status as a singular family of materials, it is discussed separately earlier in this chapter.

Polyurethanes

Polyurethane resins are a unique family of materials that possess some of the properties of plastics and some of the properties of rubbers. Foamed flexible urethane is familiar in seat cushions for cars and furniture. Rigid urethane is principally used as an insulation material. Molding and casting versions of urethanes are sometimes called "elastoplastics," because their properties resemble both plastics and elastomers; although flexible urethanes are not cross-linked as are conventional elastomers.

The solid "elastoplastics" may be obtained in grades for liquid casting, calendering, injection molding, and extrusion. Known for ultrahigh abrasion resistance, toughness, and cut and tear resistance, these materials find use in solid tires, vibration cushions, aircraft fuel bladders, and automotive bumper parts.

Polyurethane elastomers are known for their extremely good mechanical strength and are perhaps unexcelled among elastomers in their abrasion resistance.

Vinyl Polymers

The most common of vinyl polymers is polyvinyl chloride or PVC. While many thermoplastics are touted for their versatility, no other plastic is found in such divergent applications as PVC, which is used in flexible raincoats, flooring, and rigid pipe. Such versatility is a product both of the base resin and of the availabilities of fillers, impact modifiers, plasticizers, and other additives that make the board property spectrum available to the vinyl resin.

Rigid PVC compounds are found in pipe and pipe fittings, as shown in Fig. 1–19, electrical connectors, domestic storm windows, rain gutters, down spouts, siding, and shutters. Flexible PVC has replaced rubber and other materials in areas like gasketing, automobile wire connectors and vacuum tubing, packaging, toys, floor mats, crash-pad covers, garden hose, shower curtains, and many more.

Vinyl dispersions. Polyvinyl chloride dispersions are suspensions of the vinyl resin in organic liquids. The organic liquid (plasticizers)

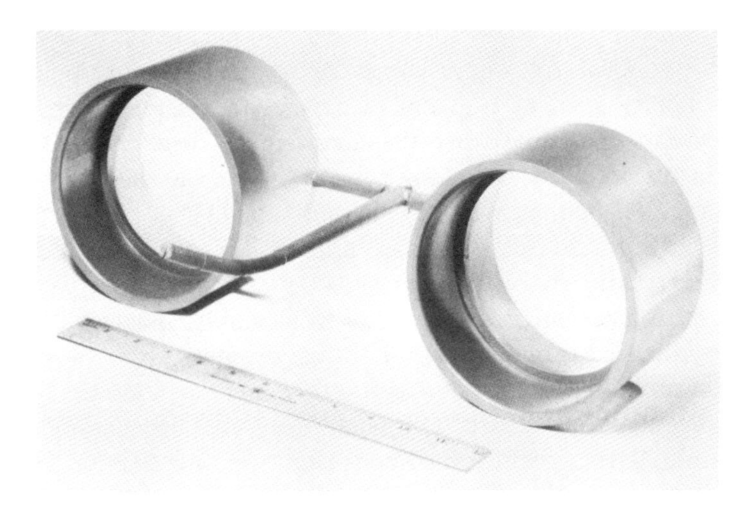

Figure 1–19. Rigid polyvinyl chloride (PVC) five inch diameter pipe sleeves. The sprue and runner system is still attached to the two injection molded sleeves. This is an excellent application for rigid PVC compounds.

have little or no tendency to dissolve the resin at normal temperatures, but becomes a solvent for the resin when heated. As additional heat is applied, the resin particles fuse into a solid mass. There are two types of vinyl dispersions–vinyl plastisols and vinyl organosols.

Vinyl plastisols. When a finely divided polyvinyl chloride powder (.02 to 2u) is dispersed in a plasticzer, it is called a plastisol. Plastisols will vary in consistency from a cream to a heavy paste. If the plastisol is thickened with gelling or thickening agents (so it will not run), it is called a "plastigel."

As heat is applied to a plastisol, the resin concentration increases until a final fusion temperature of approximately 350° F is reached (Fig. 1–20). Plastisols are used in slush and rotational molding, dipping, spraying, film casting, and coating.

Vinyl organosol. An organosol is a suspension of a finely divided resin (generally PVC) in a plasticizer with a volatile organic liquid. If a plastisol is modified with a volatile solvent or diluent, and the solvent or diluent evaporates upon heating, the plastisol is now considered an organosol. An organosol can be prepared from a plastisol, by merely adding a volatile diluent or solvent that serves to lower the viscosity and evaporates when the compound is heated (Fig. 1–20).

Organsols are used in coating substrates such as sheet metal. The substrate is cleaned, primed, coated, and baked. The solvents are first flashed off from the organosol, and final fusion (similar to the plastisols) takes place.

The wide array of products made from vinyls, suggests that they can be processed in many ways. Fabrication methods for vinyls are: injection molding; extrusion; vacuum forming; calendering; laminating; slush molding; dip coating; and compression molding. After fabrication, parts may be joined by solvent welding, heat sealing, and even stitching (in the case of flexible sheets).

ALLOYS

Alloys are blends of two or more thermoplastic polymers whose resultant properties are usually intermediates between those of each constituent. Three major alloys are discussed below.

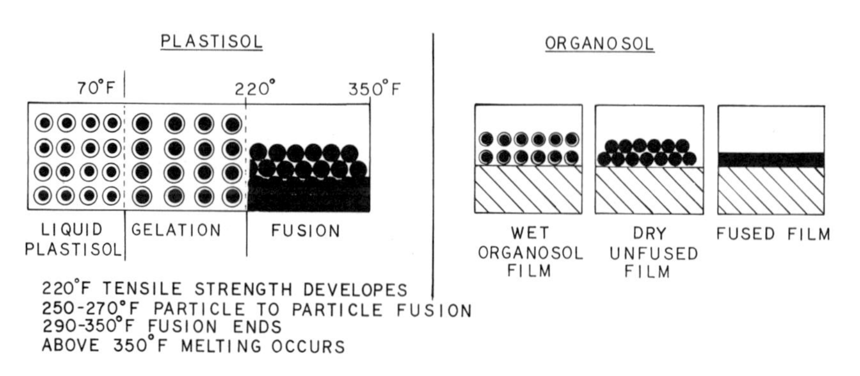

Figure 1–20. There are two types of PVC dispersions, vinyl plastisols and vinyl organosols.

ABS-Polycarbonate Alloy

By alloying ABS with polycarbonate, the resulting material is close in cost to ABS, but displays higher physical properties than ABS. For one such alloy, izod impact is about the same as would be expected for a rubbery, high-impact ABS. However, in contrast to high-impact ABS, the alloy offers high-heat-distortion temperature, high modulus, and high strength. The phenomena of varying izod impact strength with thickness that is characteristic of polycarbonate is reduced in the alloy. Processability of the alloy resembles that of ABS rather than polycarbonate, which is comparatively more difficult to process than ABS.

ABS—PVC Alloy

By alloying ABS with rigid PVC, a self-extinguishing material results that is a happy marriage of the properties of both constituents. The alloy offers good stiffness with higher-heat-distortion temperature than is normal for rigid PVC, while retaining excellent impact and self-extinguishing properties. Applications for these alloys include: electrical plugs; power-tool housings; and television coil yokes.

Acrylic-PVC Alloy

Acrylic–PVC alloys are principally found in sheet form. Toughness, flame resistance, and impact strength are the forte of this alloy. Chemical resistance and elongation of the alloy is superior to that of acrylic, but the familiar translucency of acrylic is absent. However, opaque colors are possible in the alloy, which has found use in thermoformed materials, handling trays, and machine housings.

ELASTOMERS

An elastomer is a rubber-like material that can be stretched to at least twice its original length and return to its original length all at room temperature. A rubber compound will do the same as an elastomer but from 0° to 150° F at any humidity. Plastics for all intent and purposes are rigid and do not stretch.

A wide range of elastomers, both thermoset and thermoplastic types, is available (Table 1–2). Thermoplastic elastomers can be processed by injection molding, extrusion, and blow molding. The material can be reground and reprocessed. The principal types of thermoplastic elastomers are polyester copolymers, styrene-butadiene block copolymers, thermoplastic polyolefins, and polyurethanes. Thermoset elastomers include natural and synthetic rubbers. Rubber is not a plastic and fully automatic molding operations is difficult and expensive to obtain. Rubber does not freeze first at the parting line or around ejector pins. The rubber flows until vulcanization stops the movement and injection pressure is removed. Metal molds for rubber molding are designed for high heat and pressures. The pressures are around 20,000 psi. and the temperatures up to 450° F. The runners are short, round, and as direct as possible to the part.

TABLE 1–2. THIS TABLE ILLUSTRATES A WIDE RANGE OF ELASTOMERS, BOTH THERMOSET AND THERMOPLASTIC.

GUIDE FOR ELASTOMERS

	NR NATURAL RUBBER	IR SYN POLY-ISOPRENE	SBR STYRENE BUTADIENE	IIR BUTYL	EPDM ETHYLENE PROPYLENE	ECO EPICHLOR-HYDRIN	CR CHLORO-PRENE	NBR NITRILE	PU URETHANE	SI SILICONE	CPE CHLORIN-ATED POLY-ETHYLENE	FLUORO ELASTOMER
SPECIFIC GRAVITY	0.93	0.94	0.94	0.92	0.85	1.27	1.23	1.00	1.25	1.1-1.6	1.16	1.40-1.95
TEMP. MAX. SERVICE F°	212	212	225	250	300	275	250	250	250	550	250	600
MIN. SERVICE F°	-60	-60	-60	-50	-60	-50	-40	-60	-65	-160	-40	-40
HARDNESS-SHORE "A"	20-100	20-100	40-100	30-100	30-100	60-90	20-90	30-100	50-95	20-95	40-95	60-90
MAX. TENSILE ROOM T. PSI.	4000	4000	3500	3000	3000	2500	4000	4000	5000	1500	3000	2400
MAX. TENSILE 250°F.	1800	1800	1200	1000	2000	500	1500	700	1800	850	500	300-800
WEATHER RESISTANCE	FAIR	FAIR	FAIR	VERY GOOD	EXCELLENT	EXCELLENT	VERY GOOD	FAIR	EXCELLENT	EXCELLENT	EXCELLENT	EXCELLENT
ELECTRICAL PROPERTIES	EXCELLENT	GOOD	GOOD	GOOD	GOOD	FAIR	GOOD	POOR	GOOD	EXCELLENT	FAIR	EXCELLENT
ADHESION TO METAL	EXCELLENT	EXCELLENT	EXCELLENT	GOOD	FAIR	EXCELLENT	EXCELLENT	GOOD	EXCELLENT	EXCELLENT	GOOD	FAIR
FLAME RESISTANCE	VERY POOR	V. POOR	V. POOR	V.P.	VERY POOR	FAIR	GOOD	POOR	POOR	VERY GOOD	VERY GOOD	EXCELLENT
ADVANTAGES	ABRASION RESISTANCE	SYN. NAT. RUBBER	HEAT RESISTANCE	OZONE RES.	OZONE RES.	GOOD TEMP. AND WEAR	OIL RES.	OIL RES.	OIL AND ABRASION RES.	WIDE TEMP. RANGE	OZONE RES.	RES. TO OIL AND HEAT
USES	DRIVE BELTS	SHOCK MOUNTS	SEALS TIRES	DIAPH-RAGMS	DRIVE BELTS	GASKETS SEALS	BELLOWS	SEALS	GEARS	SEALS GASKETS	CHEM. RES. PARTS	H.T. OIL PARTS

2

Molding Processes

INTRODUCTION

The molding and converting of plastics into end products is a highly specialized field within which more than 15 different processes are used. Any detailed account of the methods used in the fabrication of finished molded articles is beyond the scope of this chapter. However, this chapter outlines the methods and indicates the advantages and limitations of each process from the standpoint of design implications. The selection of any process for a given part must be made on the basis of part geometry, process compatability, and the cost to produce a given number of units.

INJECTION MOLDING

The injection molding process is one of the most versatile production methods in the plastic manufacturing industry. It is a process that is capable of producing molded parts of relatively intricate configuration with good dimensional accuracy. Injection molding is a method of forming or molding objects from powdered or granular types of thermoplastic materials. Thermosetting powdered materials can also be injection molded.

The granular or powdered plastic is fed from a hopper to a heated chamber where it is heat-softened in a flowable condition and is then injected into a cool metal mold by a ram or screw-ram. Pressure is held on the molten plastic material until the mass has hardened enough for removal from the mold.

Injection molding machines are rated in sizes that are determined by the maximum amount of general-purpose polystyrene material that the machine can process. This refers to the capacity of the heating cylinder and is generally stated as the number of ounces of uniformly heated polystyrene that can be shot with one complete stroke of the injection ram or screw-ram. Another important measure of machine capacity is the pounds/hours of plastic that can be uniformly heated by the heating cylinder. Molding machine capacities range from a fraction of an ounce to many hundreds of ounces.

Most injection molding machines are of the horizontal type, with either a mechanical toggle clamp or a full-hydraulic clamp ranging to 400-ton locking pressure. The clamping action holds the mold closed and provides the ejection force to remove the parts from cavities of the die. The injection part of the machine is used to plasticize the material at a rate consistent with the time required to cool a previously molded shot. An exact amount of material for each shot is metered by volume or weight. The shot size, including the sprue and runner system, should not exceed two-thirds the rated shot capacity of the press.

The mold temperature must be maintained below the softening point of the material being molded. It must also be high enough to prevent shrinkage of the plastic, due to hardening of the outside of the mass too quickly. Mold temperatures run from 125 to 300° F. Circulating cold water and electric cartridge heaters are used to maintain the proper temperatures.

The average injection molding pressures ranges from 10,000 to 30,000 psi, depending on the molding compound and mold design.

There are four basic types of injection molding equipment in use today (Fig 2–1). The four types are: (1) conventional injection-molding machine; (2) piston-type preplastifying machine; (3) screw-type preplastifying machine; and (4) reciprocating-screw injection machine.

Conventional Injection-Molding Machine

In this process, the plastic granules or pellets are poured into a machine hopper and fed into the chamber of the heating cylinder. A

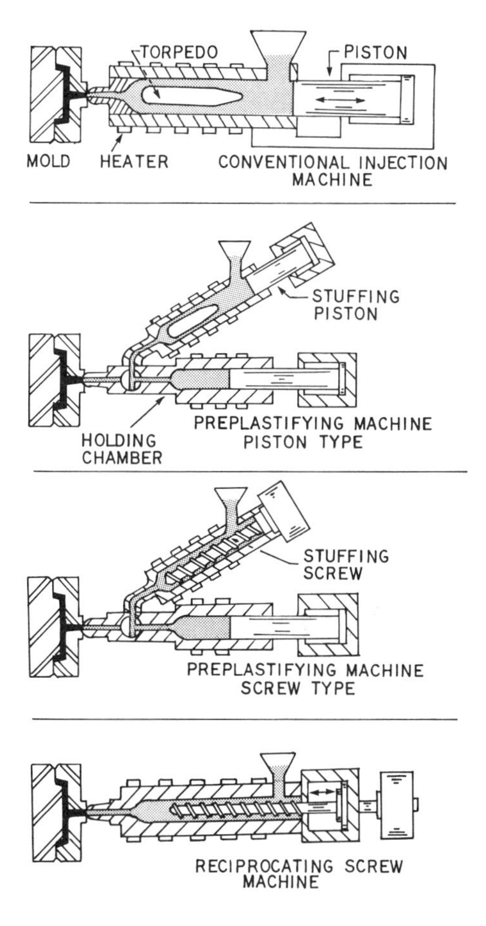

Figure 2–1. The four basic types of injection molding equipment.

plunger then compresses the material, forcing it through progressively hotter zones of the heating cylinder, where it is spread thin by a torpedo. The torpedo is installed in the center of the cylinder in order to accelerate the heating of the center of the plastic mass. The torpedo may also be heated so that the plastic is heated from the inside as well as from the outside.

The material flows from the heating cylinder through a nozzle into the mold. The nozzle is the seal between the cylinder and the mold; it is used to prevent leaking of material caused by the pressure used.

The mold is held shut by the clamp end of the machine. For polystyrene, two to three tons of pressure on the clamp end of the machine is generally used for each inch of projected area of the part and runner system. The conventional plunger machine is the only type of machine that can produce a mottle-colored part. The other type of injection machines mix the plastic material so thoroughly that only one color will be produced.

Piston-Type Preplastifying Machine

This machine employs a torpedo ram heater to preplastify the plastic granules. After the melt stage, the fluid plastic is pushed into a holding chamber until it is ready to be forced into the die. This type of machine produces pieces faster than a conventional machine, because the molding chamber is filled to shot capacity during the cooling time of the part. Due to the fact that the injection plunger is acting on fluid material, no pressure loss is encountered in compacting the granules. This allows for larger parts with more projected area. The remaining features of a piston-type preplastifying machine are identical to the conventional single-plunger injection machine. Figure 2–2 illustrates a piston or plunger preplastifying injection molding machine.

Screw-type Preplastifying Machine

In this injection-molding machine, an extruder is used to plasticize the plastic material. The turning screw feeds the pellets forward to the heated interior surface of the extruder barrel. The molten, plasticized material moves from the extruder into a holding chamber, and from there is forced into the die by the injection plunger. The use of a screw gives the following advantages: (1) better mixing and shear action of the plastic melt; (2) a broader range of stiffer flow and heat-sensitive materials can be run; (3) color changes can be handled in a shorter time, and (4) fewer stresses are obtained in the molded part.

Figure 2–2. A plunger preplastifying injection molding machine. A single cavity mold and parts of a ventilation grill are shown. *(Courtesy HPM Company)*

Reciprocating-Screw Injection Machine

This type of injection molding machine employs a horizontal extruder in place of the heating chamber. The plastic material is moved forward through the extruder barrel by the rotation of a screw. As the material progresses through the heated barrel with the screw, it is changing from the granular condition to the plastic molten state. In the reciprocating screw, the heat delivered to the molding compound is caused by both friction and conduction between the screw and the walls of the barrel of the extruder. As the material moves forward, the screw backs up to a limit switch that determines the volume of material in the front of the extruder barrel. It is at this point that the resemblance to a typical extruder ends. On the injection of the material into the die, the screw moves forward to displace the material in the barrel. In this machine, the screw performs as a ram as well as a screw. After the gate sections in the mold have frozen to prevent backflow, the screw begins to rotate and moves backward for the next cycle. Fig. 2–3 shows a reciprocating-screw injection machine.

There are several advantages to this method of injection molding. It more efficiently plasticizes the heat-sensitive materials and blends colors more rapidly, due to the mixing action of the screw. The material heat is usually lower and the overall cycle time is shorter. A breakdown of the injection molding cycle time is illustrated in Fig. 2–4. A molded part takes 1 min. or 60 sec. to mold. From this is shown the different time in seconds of the machine operations. It should be noted that the actual injection of the material into the mold is only 5 sec., while the plastic part cooling time in the mold is 47 sec.

Injection Molds

An injection mold is usually made in two halves or sections and held together in the closed position by the molding press. The mold is generally made out of tool steel and is provided with channels for cooling, heating, and venting. Ejector pins and other devices may be incorporated.

There are six basic types of injection molds in use today. They are: (1) two-plate mold; (2) three-plate mold; (3) hot-runner mold; (4) insulated hot-runner mold; (5) hot-manifold mold; and (6) stacked mold. Figs. 2–5 and 2–6 illustrate these six basic types of injection molds.

Two-Plate Mold

A two-plate mold consists of two plates with the cavity and cores mounted in either plate. The plates are fastened to the press platens. The moving half of the mold usually contains the ejector mechanism and the runner system. All basic designs for injection molds have this design concept. A two-plate mold is the most logical type of tool to use for parts that require large gates.

Figure 2–3. An injection molded automobile fan shroud made from asbestos filled polypropylene. (A) This shows the fan shroud with the sprue and runner system. (B) A molded fan shroud placed on a shrink or cooling fixture. This is a reciprocating screw injection machine. *(Courtesy HPM Co.)*

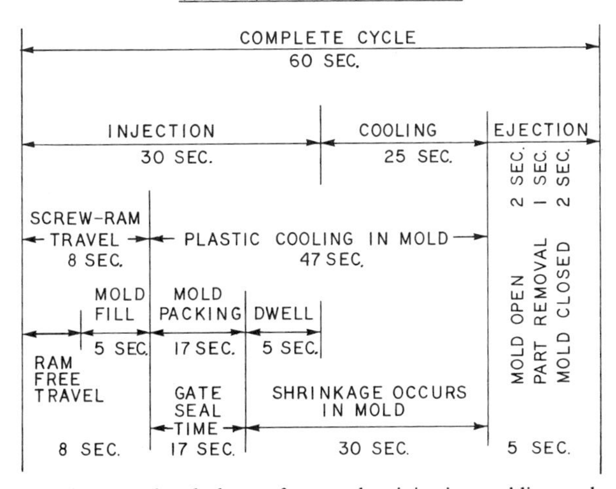

Figure 2–4. A break-down of a complete injection molding cycle.

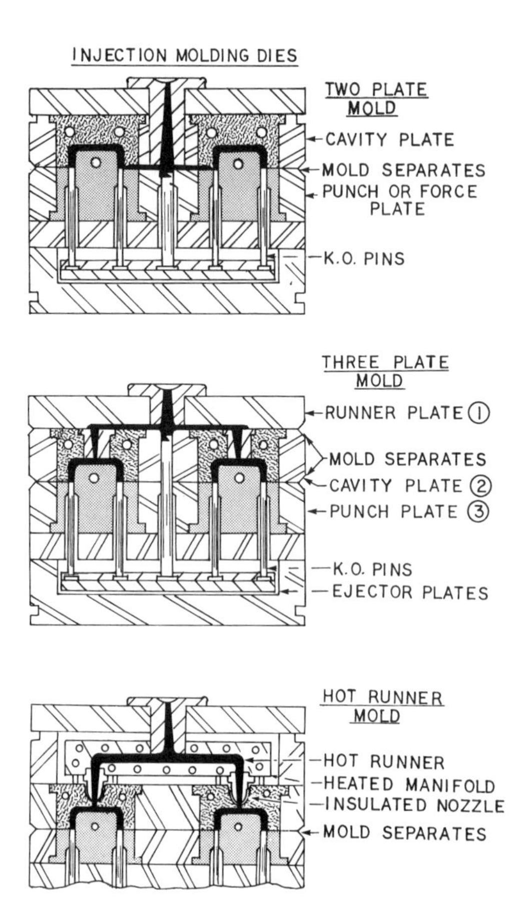

Figure 2–5. This illustrates three of the six basic types of injection molding dies. (1) Two plate injection mold (2) Three plate injection mold. (3) Hot runner mold. See Figure 2–6 for the other three types.

Three-Plate Mold

This type of mold is made up of three plates: (1) the stationary or runner plate is attached to the stationary platen, and usually contains the sprue and half of the runner; (2) the middle plate or cavity plate, which contains half of the runner and gate, is allowed to float when the mold is open; and (3) the movable plate or force plate contains the molded part and the ejector system for the removal of the molded part. When the press starts to open, the middle plate and the movable plate move together, thus releasing the sprue and runner system and degating the molded part. This type of mold design makes it possible to segregate the runner system and the part when the mold opens. The die design makes it possible to use center-pin-point gating.

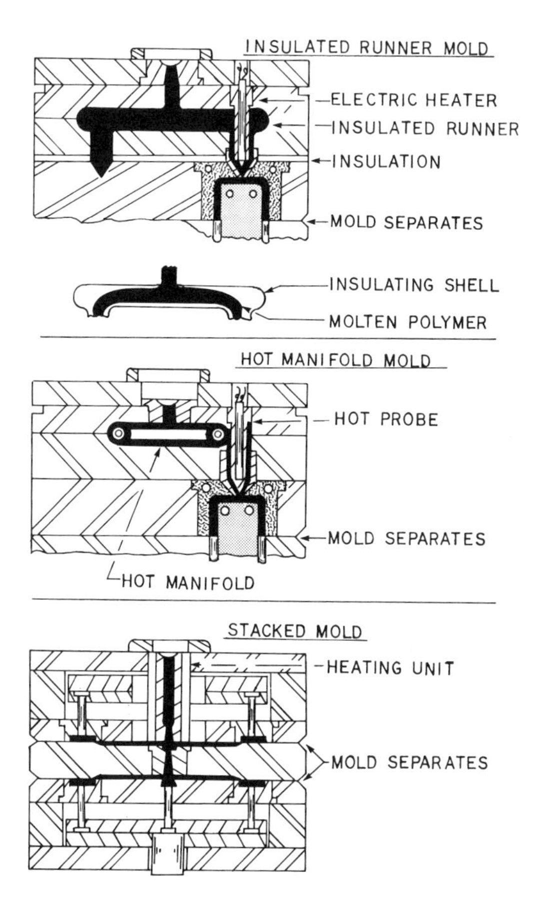

Figure 2–6. This illustrates three of the six basic types of injection molding dies. (1) Insulated runner injection mold. (2) Hot manifold injection mold. (3) Stacked injection mold. See Figure 2–5 for the other three types.

Hot-Runner Mold

In this process of injection molding, the runners are kept hot in order to keep the molten plastic in a fluid state at all times. In effect this is a "runnerless"molding process and is sometimes called the same. In runnerless molds, the runner is contained in a plate of its own. Hot runner molds are similar to three-plate injection molds, except that the runner section of the mold is not opened during the molding cycle. The heated runner plate is insulated from the rest of the cooled mold. Other than the heated plate for the runner, the remainder of the mold is a standard two-plate die.

Runnerless molding has several advantages over conventional sprue runner-type molding. There are no molded side products (gates, runners, or sprues) to be disposed of or reused, and there is no separating of the gate from the part. The cycle time is only as long as is required for the molded part to be cooled and ejected from the mold. In this system, a uniform melt temperature can be attained from the injection cylinder to the mold cavities.

Insulated Hot-Runner Mold

This is a variation of the hot-runner mold. In this type of molding, the outer surface of the material in the runner acts like an insulator for the molten material to pass through. In the insulated mold, the molding material remains molten by retaining its own heat. Sometimes a torpedo and a hot probe are added for more flexibility. This type of mold is ideal for multicavity center-gated parts.

Hot-Manifold

This is a variation of the hot-runner mold. In the hot-manifold die, the runner and not the runner plate is heated. This is done by using an electric-cartridge-insert probe.

Stacked Mold

The stacked injection mold is just what the name implies. A multiple two-plate mold is placed one on top of the other. This construction can also be used with three-plate molds and hot-runner molds. A stacked two-mold construction doubles the output from a single press and reduces the clamping pressure required to one half, as compared to a mold of the same number of cavities in a two-plate mold. This method is sometimes called "two-level molding."

Mold Materials

Special alloy steels generally are used in making injection molds. Cast beryllium-copper is used in cavities and cores with supporting steel around them. Other materials, such as kirksite, aluminum and epoxy have been used in prototype work and for very short-run jobs. Fig. 2–7 shows a large two-cavity, hot-runner injection mold made from

Figure 2–7. A two unit injection mold for the making of battery cases for automotive lead-acid batteries. The battery case is molded from translucent polypropylene copolymer resin. (A) Shows the four sections of the split cavity. (B) The force or punch. (C) The molded battery case.

tool steel. This two-cavity mold is used to make automobile battery cases from polypropylene copolymer resin. Note the four sections of the split cavity. All four sections of the cavity move away from the molded part at the end of the molding cycle. This is necessary because all four sides of the battery container include undercuts in the form of texturing, recesses for retention of battery hold-downs, horizontal ribs, and raised lettering.

JIGS AND SHRINK FIXTURES

In many molded parts, it may be necessary to control the warpage and critical dimensions of the part by placing it on a jig or shrink fixture immediately after it comes from the mold. The shrink fixture holds the molded part in the proper shape until it cools. In order to reduce the time required for a part to be on a fixture, a water quench tank may be used in addition to the fixture. Shrink fixtures are made from metal, plastic, or wood and are usually designed for a specified application.

Screw Injection Molding of Thermosets

The screw injection molding of thermosets is similar to the method of screw injection molding of thermoplastics (Fig. 2–8). In this process, loose thermosetting powder is heated, softened, and densified by a reciprocating screw. Heat is generated by friction from the screw rotation and electric heater bands around the barrel. The barrel temperature ranges from 130 to 240° F. The die temperature varies from 250 to 350° F. The injection pressure ranges from 12,000 to 20,000 psi. As the screw rotates, material travels forward along the flights and is thoroughly preplasticized by a mechanical shearing action (Fig. 2–8a). The material buildup at the end of the screw pushes the screw back. The heated preplasticized material is then forced from the barrel into the mold cavities by the forward movement of the screw, which acts like a ram (Fig. 2–8b). After the material in the die cavities has polymerized, the mold opens and the parts are ejected (Fig. 2–8c).

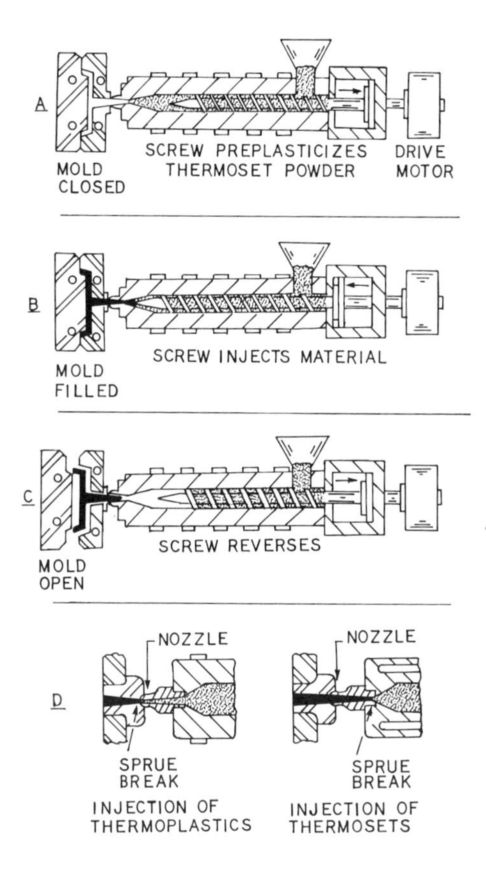

Figure 2–8. This illustrates the screw injection molding of thermosets. (A) The thermosetting plastic powdered material is preplasticized as it travels along the flights of the screw. (B) Under high pressure, the screw is forced forward, injecting the preplasticized material into the mold. (C) After the mold is filled, the screw reverses and begins a new cycle. (D) The "sprue break" in regular injection molding of thermoplastics is at the end of the nozzle. The "sprue break" in thermoset injection molding is at the beginning of the nozzle. The tapered nozzle and sprue openings must be cleared after each shot.

The sprue length should be kept as short as possible. The "sprue break" in regular injection molding of thermoplastics is at the end of the nozzle. The "sprue break" in thermoset injection molding is at the beginning of the nozzle (Fig. 2–8d). The tapered nozzle and sprue openings must be cleared after each shot.

The reciprocating screw is a natural venting device by which gases and moisture vent through unplasticized material in the screw flights and out through a vent hole in the hopper. If the mold is vented properly, degassing and bumping of the mold is not necessary. Full-round runners are recommended for screw injection molding. The frictional heat developed during the injection of the material into the mold causes considerable wear in the runner and gate system. This is particularly true in the use of mineral-filled and glass-filled materials.

If the thickness of a molded part is .125 in. or over, the molding cycle in many cases is faster than those for similar thermoplastic molded parts. It should be remembered that thermosetting materials cure or polymerize in the mold and are discharged hot with no cooling required, while thermoplastic parts must be cooled in the mold to become rigid enough to be removed without distortion.

RAM INJECTION MOLDING OF THERMOSETS

This molding process is sometimes referred to as horizontal transfer molding (Fig. 2–9). The thermosetting molding material, in granular form, is preweighed and preheated before it is placed in the heated injection cylinder. The cylinder or barrel temperature ranges from 175 to 250° F. As the ram moves forward, it compresses the preheated thermosetting material and converts it to a viscous fluid. The pressure on the material may range between 20.000 to 40,000 psi. The viscous fluid is injected into a die that is heated from 250 to 350° F. After the plastic material has polymerized or set, the sprue is automatically removed, and the mold is opened and the part is ejected.

Generally, the mold does not have to be cracked open (bumped) in order to "out-gas" the molding compound that splits off water or other volatile materials when it polymerizes. Usually, these gases can be released by proper venting of the mold. If these gases were locked in the mold, they would cause voids in the molded part. The size and location of the sprue bushing, runners, gates, and cavities must be determined for each type of thermosetting resin that is to be run. This molding process is much faster than standard transfer molding, and it shows a decrease in material loss due to spillage, preform damage, and dirt getting into the powder.

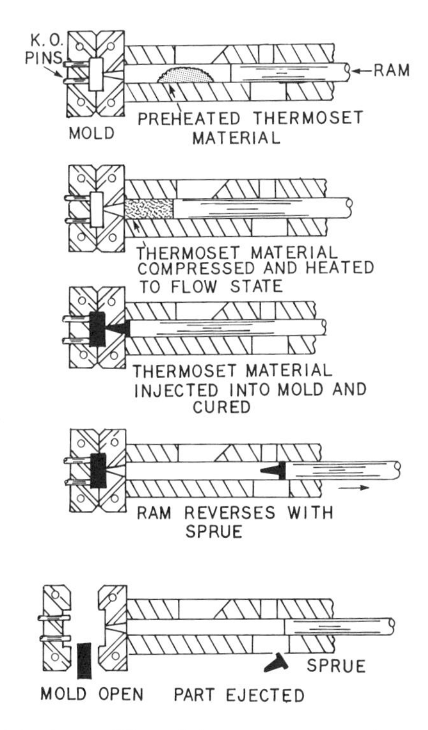

Figure 2–9. Ram injection molding of thermosets or sometimes referred to as horizontal transfer molding.

COMPRESSION MOLDING

Compression molding is a process by which a measured quantity of a plastic material is shaped or formed by heat and pressure applied to a hardened steel mold in which the plastic material has been placed. The steel mold is generally made in two halves that are attached to the upper and lower platens of a press. The steel mold halves are called the force or plunger and the cavity. Heat is applied to the mold or indirectly through the platens. The heat source is steam, circulated hot oil, or electrical resistance heaters.

The molding compound as a powder or preform is placed in the mold cavity, and the mold is closed. The mold closes to apply heat and exert pressure, causing the plastic material to soften, flow, and fill the confines of the cavity. Depending on the characteristics of the plastic material and the design of the mold, the temperature ranges from 280 to 375° F, and the pressure ranges from 2000 to 10,000 psi. The mold remains heated and closed until the plastic material cures or sets. A small clearance, usually 0.003 to 0.006 in., is allowed between the mold halves to permit escape of excess material and volatiles.

A molding pressure of 2000 psi on the total projected area of the part, including lands in the mold, is generally recommended. To this must be added approximately 800 psi for each inch of vertical height of the molded part.

After the material has hardened sufficiently, the mold is opened, and the piece is ejected by knockout pins. The cure time depends on the size and thickness of the part. This may vary from 20 sec. to 10 min. The cure time cannot be predicted accurately, but must be determined by experience or experiment.

Typical compression molded parts include: automotive distributor caps; washing machine agitators; camera cases; handles; knobs; radio and television receiver cabinets; and office machinery housings. Fig. 2–10 illustrates a compression molding press showing the mold, the press operator, and molded terminal block parts. Thermosetting compression-molded parts vary in weight from a few ounces to 30 lbs. or more. This type of molding is also used for some cold-molded products and occasionally for thermoplastic materials. Fig. 2–11 shows a 2000-ton clamp compression press and some of the molded products.

Characteristics of Compression Molding

This type of molding process is ideal for the production of parts of large area and deep draw that have relatively simple shapes. A number of special materials are moldable only by this method.

The cost of finishing compression-molded parts is high when compared to the finishing costs of injection-molded parts, because all compression-molded parts have flash, at the parting line, that must be

Figure 2–10. Compression molding press (1) Force or punch. (2) Cavity. (3) Molded parts. *(Courtesy Westinghouse)*

removed after molding. The relatively slow molding cycles of compression molding require a greater number of cavities for given production requirements, as compared to other molding processes. Compression molded parts require greater tolerances across parting lines. It is sometimes difficult to hold delicate inserts in place during the molding process.

Auxiliary Operations

Compression molding requires many auxiliary operations such as preforming of molding powder, preheating of molding material, and finishing of the molded part.

Preforms. Preforming consists of compressing the molding powder into a pill or tablet of the proper size and weight. Preforming is done at ambient temperature in a machine designed especially for that

Figure 2-11. A 2000-ton compression molding press without any molds in the press. Four molded parts that have been molded in this press are shown. (1) Polyester fiberglass laundry tub. (2) Polyester sisal and asbestos automobile heater housing. (3) Phenolic nose cone for outer space. (4) Phenolic washing machine agitator. *(Courtesy The General Industries Co.)*

purpose. Preforms can be made in almost any size or shape to fit the cavity of a compression mold or the pot of a transfer mold. Preforming of molding powders helps in preheating the material and shortening the molding cycle. It reduces handling and helps obtain a thinner cutoff at the parting line.

Preheating. Thermosetting molding materials are generally preheated to shorten the molding cycle, promote free flow of material in the mold, produce a more uniform part, and improve the physical, chemical, and electrical properties. The four most-common methods or units used to preheat thermosetting molding materials are: (1) electronic or high-frequency units; (2) hot-air circulating units; (3) steam or moist-air ovens; and (4) infrared lamps.

Electronic preheating is a fast uniform method of heating thermosetting plastic molding compounds. It is sometimes called dielectric heating, RF heating (radio-frequency heating), and high-frequency heating. This type of heating process is made by an alternating electrical current of sufficiently high frequency and intensity to cause molecular stresses and friction within the plastic preform. Most thermosetting plastic molding compounds have dielectric loss characteristics sufficiently high to be heated in this manner.

There are several distinct advantages to electronic preheating. Owing to the fact that the pill or preform has been preheated prior to its insertion in the mold, the plastic is much more moldable, fluid, or plastic at the time of the closing of the mold. The result is that lower pressure is required in the molding operation. For this reason, much more delicate pins and mold sections can be incorporated into the mold. It is also felt that thinner wall sections and deeper draws can be designed in the parts.

There are essentially two types of electronic preheat ovens. The first conventional type has the preform stacked on a flat plate. Fig. 2–12A shows diagrammatically the high-frequency field in a flat plate-type preheater. Fig. 2–12B illustrates the heating principle of a roller type electronic preheater. Fig. 2–13 illustrates a flat-plate or drawer type electronic preheater in actual operation. Fig 2–14 is a photograph of a roller-type electronic preheater. The roller-type pre-

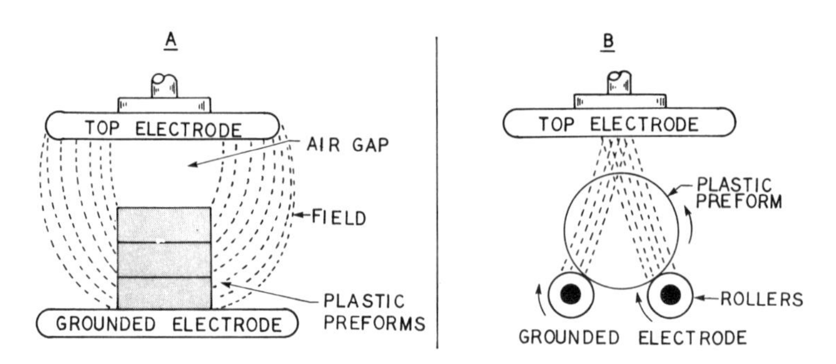

Figure 2–12. (A) This illustrates the heating principle of a flat plate type electronic preheater. (B) This illustrates the heating principle of a roller type electronic preheater.

Figure 2–13. A drawer type electronic preheater. The operator is loading the preheater with melamine preforms. The heated preforms will be placed in the open compression molding die. Note that the die or mold is a "stacked" type. Melamine plastic dishes are being molded.

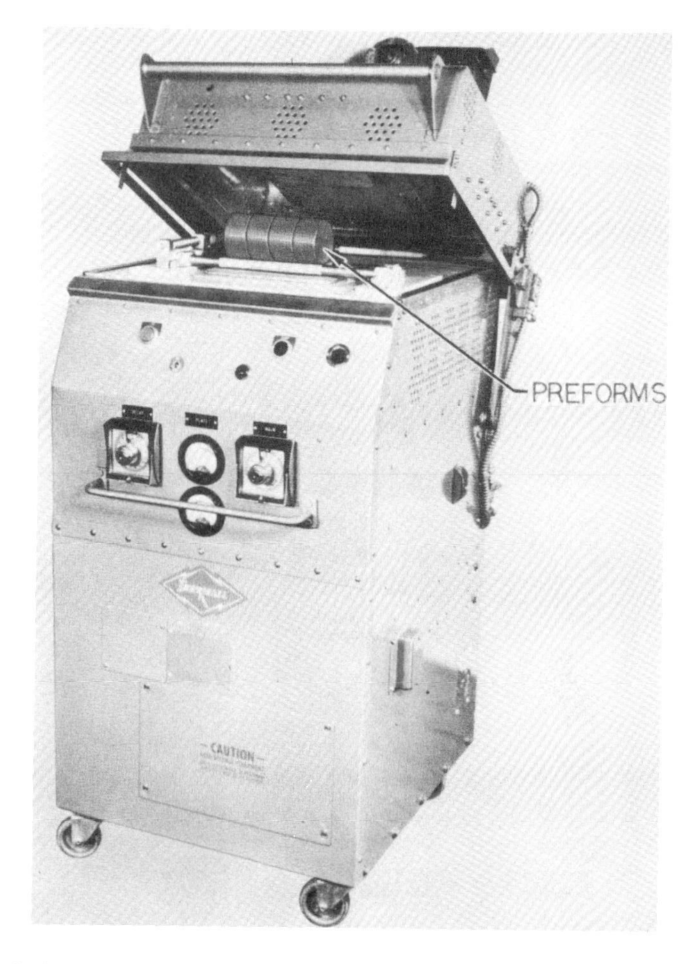

Figure 2–14. A five K.W. roller-type dielectric preheater used to heat thermosetting materials. Note the round preforms are rolled during the preheating period. This produces a much more uniform heat in the preform.

heater is more efficient and gives a more uniform heat to the plastic preform.

Finishing. Plastic molding materials give a faithful reproduction of the surface of the mold in which they are formed. Highly polished mold surfaces impart with true accuracy the corresponding high finish to the molded piece. Compression molded parts have parting lines, or joints, that produce flash or fins. The flash or fins must be economically removed from the molded part. One method of removing flash is by a commercial tumbler shown in Fig. 2–15. Abrasive grit is sprayed over the parts as they are tumbled on an endless belt. Following the removal of flash drilling, tapping, and machining operations are preformed. Buffing or polishing is added to the desired quality and luster of finish.

Figure 2–15. This picture illustrates a mechanical tumblast deflasher. (A) Phenolic compression molded parts being deflashed. (B) Endless rubber belt with large holes for the abrasive return. (Photo by George Freund)

Molds of Dies for Compression Molding

Flash-type molds. Fig. 2–16A illustrates a flash-type mold. This type of mold is not suitable for deep-draw parts, because the molding pressure exerted is not sufficient to make the plastic material flow any great distance. The material flowing across the land area has no other restriction to its movement. This characterizes the name "flash mold." Only flat and shallow parts should be molded in a flash-type mold. The thermosetting plastic molding material should be of the general-purpose type. The impact-type materials are not used, because they are bulky and there is insufficient loading space.

The flash is always horizontal. If the mold is closed too slowly, a heavy flash will result. If the mold is closed too quickly, the density of the molded part will be low and the strength impaired. The advantages of flash-type molds are: (1) lower mold costs; (2) good for small parts; (3) ease of loading inserts; and (4) experimental molds may be made for prototype parts.

Positive-type mold. A positive-type mold is shown in Fig. 2–16B. This type of mold fully confines the molding material, and full mold pressure is exerted at all times. There is insufficient clearance between the cavity and the plunger for the molding material to escape. The full pressure is exerted on the molded part. The travel of the force or plunger is limited only by the amount of plastic molding powder placed in the die cavity. No external pressure pads or stop blocks are used to limit the closed height of the mold. The molding material must be weighed or measured accurately, since there is little chance for escape of excess molding material. The clearance space between the plunger and cavity, sometimes called the "vertical flash ring," is 0.002 to 0.005 in. per side. The positive-type mold is always most desirable where the plastic part must be very dense. It is also used for molding high-impact thermosetting plastic materials.

Landed positive mold. A landed positive mold is illustrated in Fig. 2–16C. This type of compression mold is sometimes referred to as a "landed plunger mold" of "semipositive mold." The loading well is adequate for loading bulky materials. The land is usually .125 to .187

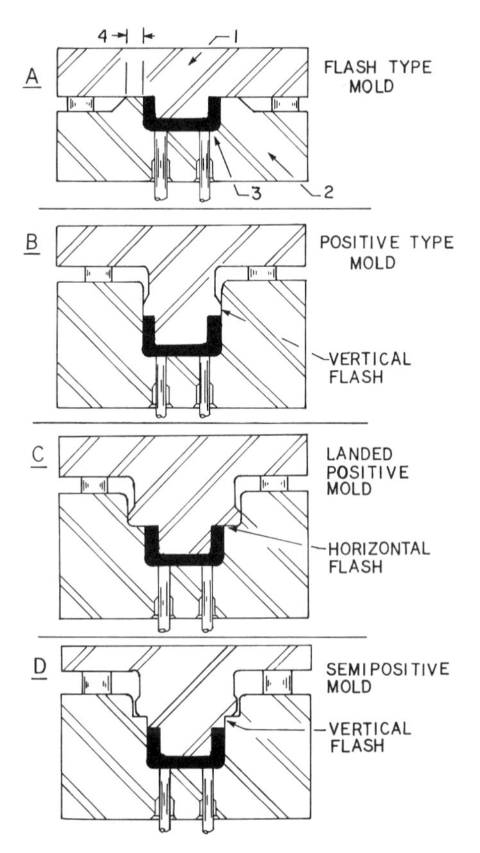

Figure 2–16. (A) A flash type compression mold. (1) Plunger or force. (2) Cavity. (3) Plastic part. (4) Land area. (B)This illustrates a positive type compression mold. (C) This illustrates a landed positive compression mold with horizontal flash. (D) This illustrates a semipositive compression mold with vertical flash.

in. wide. This type of compression mold is not recommended for high-impact materials that may cause heavy flash at the parting line of the die. A modification of the positive mold is known as a "landed positive mold." It is essential that the width of the land should not exceed .375 in. If it does, the applied pressure will be absorbed by the land. This will result in thick flash or fins, and the piece will not be filled out properly. The force or plunger is cut back or relieved about 0.062 in. thus producing a shoulder of this width on its lower end.

This reduced side friction and minimizes the likelihood of the force binding, due to a fin or molding material between it and the side of the mold cavity loading well.

Sometimes three or four perpendicular sprue grooves about 0.031 in. deep are cut into the shoulder of the force plug to provide an escapement in case an excess mold charge is used. When this is done, the mold becomes more of the semipositive type. Flash and positive types of molds should have 0.015 in. average tolerance allowed on the molded part on all vertical dimensions between the two mold halves.

Semipositive mold. Semipositive-type compression molds differ from positive molds, because the force or plunger only telescopes the body of the mold enough to exert positive pressure at the final closing of the mold. Fig. 2–16D shows a semipositive mold. With this type of mold construction, effective pressure on the molding material is assured during the last of the closing cycle, because of the short distance of vertical positive fit between the force and cavity. A slow uniform closing will allow the material to fuse completely before the mold is closed and will increase the effective pressure on the compound. Upon closing the press, positive pressure will be exerted on the part being molded only during the last 0.031 in. This distance generally varies from 0.031 to 0.125 in., depending on the size and shape of the part. This type of mold construction is best suited for quality production molding.

CHROMIUM PLATING OF MOLDS

Hard chromium plating of metal molds provides some good physical properties to mold surfaces. These properties are: (1) low coefficient of friction (a slippery surface); (2) additional hardness; and (3) corrosion protection. Chrome plating will reproduce and accentuate any cracks, pits, or other defects in the mold surface. Therefore, it is necessary to smooth and polish the mold surfaces before plating. Chrome plating is used to cover up defects, weldments, fine cracks, small pits, etc. The minimum limit of plating thickness is 0.0005 in., the average thickness is 0.002 in., and for salvage purposes up to 0.035 in. thickness for wall reduction on large moldings: however, this thick plating must be engineered, since it is far from an all-purpose medium or repair.

A plastic part that has dimensions with tight tolerances may use chrome plating to fill in openings or build up projections in the mold (Fig. 2–17). A matte finish on a mold can be protected by chromium plating. The use of chrome plating is not confined to molds. It is used on runners and gates, injection machine parts, extruder screws and barrels, etc.

MATCHED-METAL MOLDING OF REINFORCED PLASTICS

Almost every known plastic can be and has been reinforced. Thermoplastics as well as thermosets can be made into stronger products by the addition of such fibers as glass, asbestos, sisal, nylon, and other natural and synthetic materials. The term "reinforced plastics" is used almost exclusively to describe glass-fiber-reinforced thermosetting plastics, sometimes abbreviated FRP.

The fiberglass in the form of fibrous filaments, when added to a plastic, serves to increase mechanical strength, stiffness, and fatigue

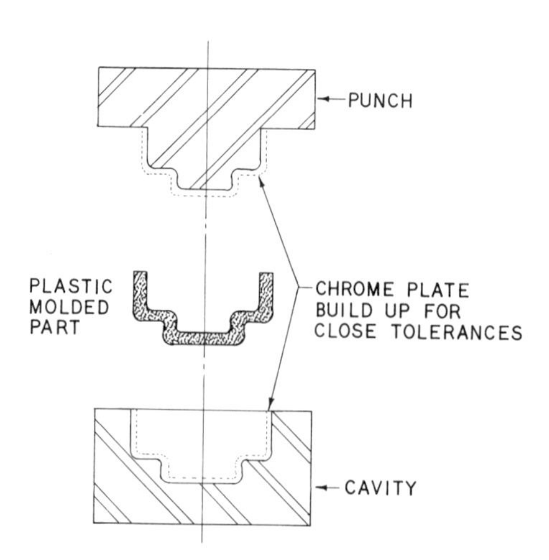

Figure 2–17. Chrome plating can be used to build up mold sections if tight tolerances are required.

impact resistance. The dimensional stability is increased and a wider temperature range is obtained.

This molding process is sometimes known as low-pressure molding and is essentially the same as compression molding, except that lower pressures and temperatures are used. There are three methods in which matched-metal molding may be classified: (1) mat for preform molding; (2) premix compound or bulk molding; and (3) pre-impregnated sheet molding (SMC).

Preform Molding

In this process, a fiberglass preform is made over a perforated metal screen of the same size and shape of the article to be molded. The screen is mounted on a rotating turntable of the preform machine. Chopped strands of glass fiber are blown onto the screen as it slowly revolves. Suction, as much as 6500 cubic ft. of air per min., behind the preform screen, distributes the loose fibers evenly. A measured amount of glass fibers is used in each preform and the fibers are held in place with a resin emulsion binder. The completed preform is then heated to cure the binder so it may be handled. The completed proform, which looks like a thick white blanket, is placed in the cavity or over the male portion of the metal die (Fig. 2–18). A resin mix, comprised of the resin, catalyst, fillers, and pigments, is measured out and poured on the chopped strand fiberglass preform. The press is then closed, and under these conditions, the liquid resin is forced through the preform, forcing out the air before it. The curing of the liquid plastic resin takes place immediately.

Molding temperatures usually range from 225 to 300° F. The molding pressure is 100 to 300 psi. The cure cycle, depending on the thickness, size, and shape of the piece, will cure in from one to five minutes.

Premix Molding

The term "premix" implies that components are mixed mechanically before molding. The premixes are molded similarly to compression

Figure 2–18. The preform molding of reinforced fiberglass liner for a refrigerator. (A) The mold or die. (B) Fiberglass preform. (C) The molded liner.

molding compounds. The molding material is basically glass fibers, mixed with polyester resin, catalysts, fillers, pigments, etc. The Compounds vary in texture from soft putties to clotted fibrous masses. Chopped glass fiber is generally used as the reinforcement, but sisal, asbestos, and some synthetic fibers are also used.

Fig. 2–19 shows a premix polyester compression-molded shower receptor for a bathroom. The die or mold and the compression press that molded the part can be seen in the background. The shower receptor is molded from nylon-fabric-filled polyester premix compound. The molded part is 32 by 48 in. and weighs 60 lb. The only finishing operation necessary is the removal of light flash around the parting line.

Premix material is formed into accurately weighed charges and placed in heated metal molds and subjected to pressures varying from 100 to 1500 psi. Cure temperatures range from 225 to 300° F. the cure time can vary from 30 sec. to five min., depending on wall

Figure 2–19. A compression molded shower receptor, together with the mold and press. The material is a polyester, nylon fiber premix molding compound. (Courtesy The General Industries Co.)

thickness. Polyester premix compounds can be molded by compression or transfer methods. In compression molding, the semipositive mold is recommended. Typical parts molded from this type of molding include trays, boxes, laundry tubs, automotive heater and air-conditioner housings, etc.

Sheet Molding Compound (SMC)

Sheet molding compound (SMC) is a thermosetting polyester resin mixture reinforced with fiberglass strands and formed into a sheet. The sheet can be handled easily, cut to shape, and charged into a compression mold. Figure 2–20 shows a typical sheet molding production line. Polyester resins used in this process may be classified as general-purpose or rigid, flexible, chemical-resistant, flame-resistant, heat-resistant, and low-profile. The low-profile polyester resin system is basically a mixture of a polyester thermosetting resin and a thermoplastic such as acrylic. Low-profile systems are so called because of good dimensional accuracy, low shrinkage, and a very smooth surface on the molded part.

The fiberglass used in SMC can be in lengths from 1/4 to 4 in. The fiberglass in the sheet can be orientated into random-fiber, continuous-fiber, and directional-fiber. The orientation of the fiber pattern in the SMC will give different strengths in the molded part. High glass loadings of 65% can be produced but the standard is 20 to 30%.

Tooling for Matched-Metal Molding

The die or molds used in matched-metal molding of preform, premix, or sheet molding compounds are similar to those used for compres-

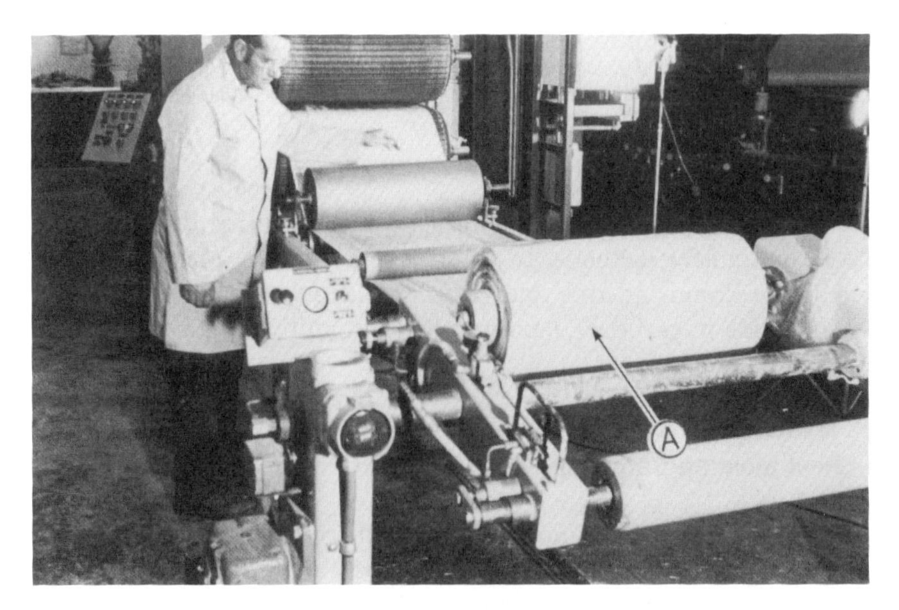

Figure 2–20. This picture shows the end of a sheet molding line. (A) The sheet in roll form is ready for package and delivery to the molding press line. (Courtesy Owens Corning Fiberglas)

sion molding (Fig. 2–21). The steel should be fine grained, not porous, and free from defects. The dies are made of cast iron, cast aluminum, forged steel, or cast steel. Molds are often chromium plated. Matched-metal molds are usually for large parts and often have only one cavity.

TRANSFER MOLDING

This molding process is used mainly for thermosetting resins. The advantage of transfer molding lies in the fact that the mold is closed, as in injection molding, at the time the plastic material enters the die. The parting lines that might give trouble in compression molding are held to a minimum in finishing in the transfer-molding process.

Delicate inserts can be molded and not be subject to movement as compared to compression molding. It is also possible to mold more intricate parts with side cores, uneven wall sections, small holes, etc. Longer core pins can be used to produce side holes and very deep holes unobtainable with compression molding.

There are three basic types of transfer molding: (1) pot or sprue-type transfer molding; (2) plunger transfer molding; and (3) screw transfer molding.

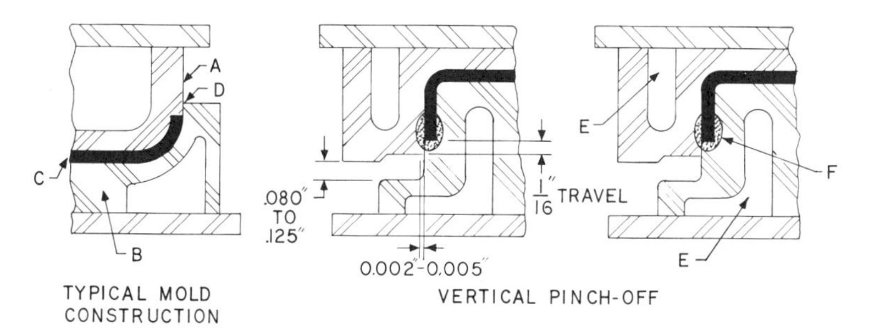

Figure 2–21. Matched-metal molding die for reinforced plastics. (A) Force or punch. (B) Cavity. (C) Molded part. (D) Pinch-off area. (E) Cored steam chest. (F) Flame hardened pinch-off area.

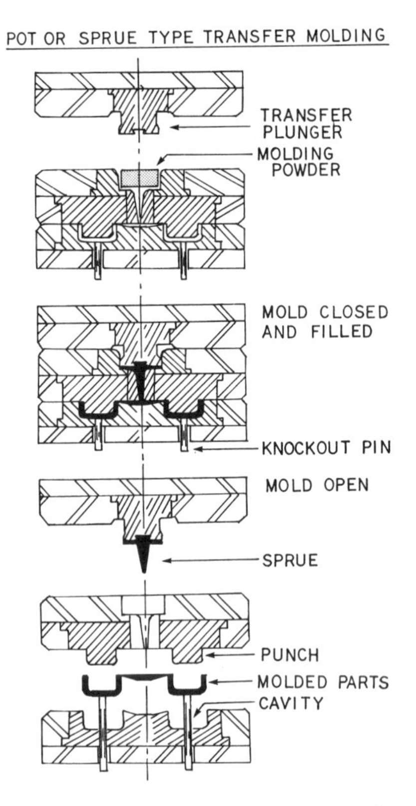

Figure 2–22. This illustrates the pot or sprue type transfer molding of thermosets.

Pot- or Sprue-Type Transfer Molding

The thermosetting molding material, which may be preheated, is placed in an open pot at the top of a closed mold (Fig. 2–22. In this pot, sometimes called a loading well, the thermosetting material is plasticized into a viscous mass by heat and pressure and is then forced into the mold cavities by the pressure. Molding temperatures of 300 to 350° F and pressures of 6000 psi are used. The area of the pot is approximately 15% larger than the projected area of all the cavities and runners in the mold. Following the curing cycle, during which the material is polymerized, the press is opened and the molded parts are ejected.

Plunger Transfer Molding

In this molding process, an auxiliary ram is installed on the transfer press to operate a plunger that forces the material into the mold cavity (Fig. 2–23 and 2–24). This process differs from the pot-type molds in that the plunger or force is pushed to the parting line of the die. The plunger enters the loading chamber and transfers the molding compound to the cavities. The plunger backs off, the mold opens, and the knockout pins eject the parts from the mold. Plunger transfer

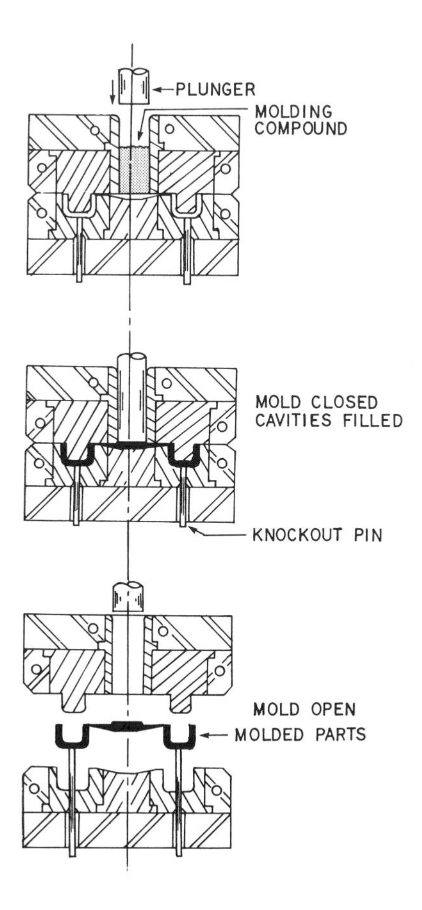

Figure 2–23. This explains the plunger transfer molding of thermosets.

Figure 2–24. This shows four handles for electric toasters that have been plunger transfer molded from phenolic material. Note the large flat runner system. *(Courtesy HPM Co.)*

molding develops more frictional heat than the pot-type transfer molding and reduces the molding cycle time.

Fig. 2–25 shows a molded shot taken from a 20-unit plunger transfer mold. The part is a decorative urea formaldehyde utensil handle. Note that a decorative overlay sheet of the printed design was laid in the mold before the transfer process was started. The molding cycle was approximately one minute.

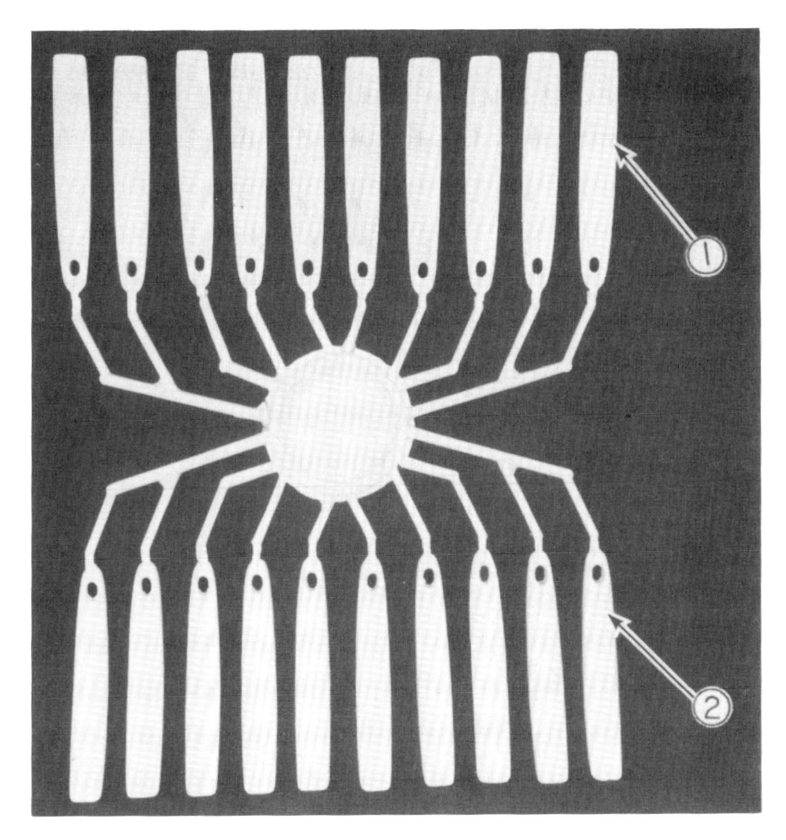

Figure 2–25. A 20-unit plunger type transfer molded shot of kitchen utensil handles. The material is urea formaldehyde. (1) Shows the decorative side of the handles made with a melamine resin impregnated foil overlay. (2) Shows the plain back of the handles.

Screw Transfer Molding

In this screw-transfer-molding process, the reciprocating screw is used to preheat or preplasticize loose molding powder (Fig. 2–26). The molding compound is hopper-fed to the reciprocating screw barrel. As the screw rotates, material travels forward along the flights and is thoroughly preplasticized by a mechanical shearing action (Fig. 2–26a). The plasticized material builds up to the end of the screw and forces it to move backward a preset distance (Fig. 2–26b). The heated preplasticized material is then forced from the barrel into the transfer pot by the screw, which acts like a ram (Fig. 2–26c). The transfer ram then advances to inject the material through the runner system, as would be the case in any transfer press (Fig. 2–26d). The mold is opened and the cycle is completed as the part is ejected (Fig. 2–26e).

This type of molding has been used with general-purpose and reinforced phenolics, melamines, filled alkyds, diallyl phthalates, epoxies, and glass-reinforced polyesters. With the development of transfer-moldable, soft, flowing materials such as the epoxies, diallyl phthalates, and silicones, it is now possible to transfer mold or encapsulate extremely fragile insert components and assemblies.

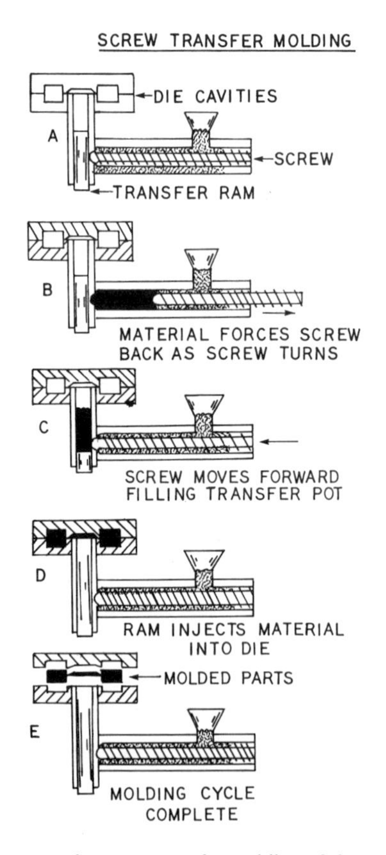

Figure 2–26. This illustrates the screw transfer molding of thermosetting plastic materials.

Reaction Injection Molding (RIM)

Polyurethane molded parts made by the RIM (Reaction Injection Molding) process are produced by mechanically or stream impingement mixing together large amounts of two highly reactive liquid components. The mixed material is injected into a mold where it cures within a very short time (Fig. 2–27, and Fig. 2–28). The liquid material flows to the lowest point in the mold, so tilting the mold provides the easiest method of venting in order to remove gases.

Polyurethane can be produced in many varieties, from flexible to extremely rigid, and from solid to cellular. This is accomplished by the wide choice of different polyols and isocyanates in combination with crosslinking agents. Polyurethane moldings can be produced in practically any desired size and shape and with cycle times which are comparable with those of thermoplastic injection molding. RIM technology, so far, has been developed exclusively using polyurethane materials. Accurate metering of the RIM components is an absolute necessity, because just a slight imbalance in the ratio of components can give entirely different properties to the part.

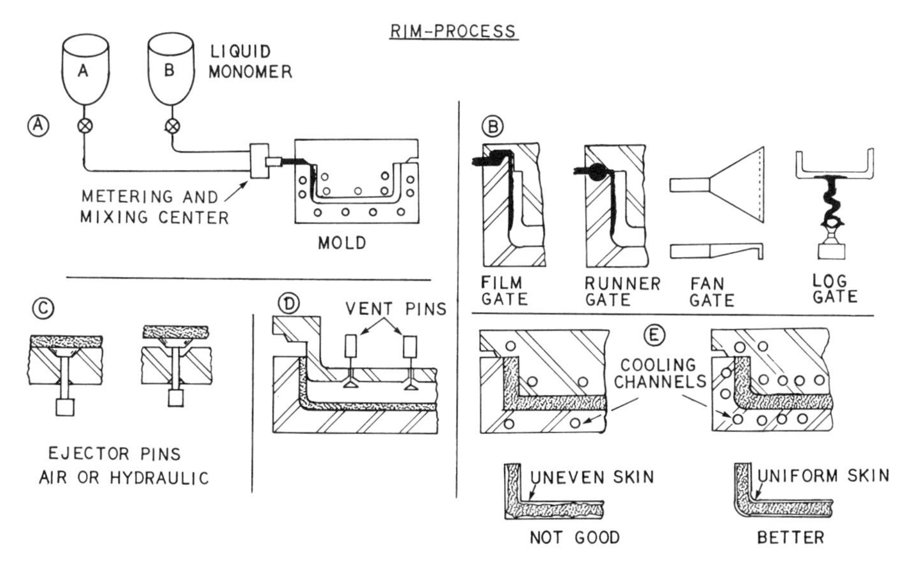

Figure 2–27. Outline drawings of the essential parts to the RIM process. (A) Molding operation. (B) Types of gates. (C) Ejector pins. (D) Vent pins. (E) Adequate cooling channels.

Figure 2–28. This picture illustrates a typical RIM set-up. (A) Metering equipment. (B) Tilting mold platen or carrier. (C) Control panel. (D) Mold. *(Courtesy Cincinnati Milacron)*

MOLDS

RIM molds are designed for injection pressure of 100 psi. All molds should include facilities for heating and cooling, and provide for effective, uniform temperature control. The temperature of the mold affects the skin thickness and can have an effect on the ultimate properties of the finished product. High temperatures result in thin skins. The molds used in the RIM process can be made from electroformed and vapor-formed nickel shells embedded in epoxy, machined steel, aluminum, kirksite, and plated cast steel. Epoxy molds are not recommended for production tools. Good heat-conducting materials should be used for RIM molds. Uniform temperature control of the mold must be maintained at all times. The best medium for controlling mold temperature has proven to be tempered water.

A summary of materials for molds.

Steel–Good wear, large complicated molds
Aluminum–Less expensive
Zinc alloys–Kirksite, inexpensive
Nickel–Electroless nickel plating. Good for grain design. Small molds
Epoxy–Prototype only
Copper alloy–No good. Stick problem.
Cast iron–No good. Porosity.

Mold runner should be designed to change a turbulent material flow to a more laminar flow. The flow in the mix head is, of necessity, extremely turbulent, and unless the flow becomes laminar as it enters the mold cavity, air entrapment is a certainity. It is important that all molds be lubricated after each molding cycle. Flow problems can be studied by spreading a thin film of pigment paste along the film-gate inlet.

The air in the mold must have an escape route, ideally located at the highest point in the mold. The parting line should be through the highest point of the cavity to insure complete venting. Venting slots should be milled in the parting surface at the highest point in the cavity to obtain controlled venting.

Wall sections up to two inches are possible but the thicker the section the longer the mold cycle time. Optimum section thickness is .375 to .500 in. Large bosses or those with a very deep draw should be avoided. Generous radii are preferable to sharp corners.

Scrapless Forming

Scrapless forming is a solid phase forming process (Fig. 2–29). The process starts with a square blank of thermoplastic material. The blank can be a single layer sheet or a coextruded multilayer sheet. The heated blank is forged into a preform by a compression type mold. In the making of the preform the outer periphery of the final molded article is made. The preform is then immediately vacuumed, pressure, and plug assist formed in a mold to the shape of the finished article. During or after the heating of the blank it is lubri-

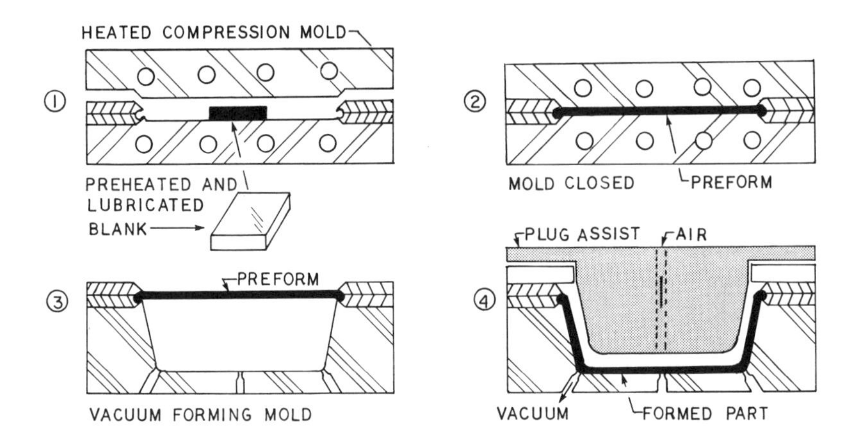

Figure 2–29. This illustrates the process of scrapless forming. (1) A heated plastic blank is placed in a compression die. (2) The heated blank is forged into a preform. (3) The preform is indexed to a vacuum forming mold. (4) The preform is then vacuum formed into a finished article.

cated. The blank is lubricated in order to reduce warpage, reduce pressure in forging, and to promote "plug-flow." Plug-flow is the ability of the coextruded composite to flow evenly in the molding or forming process. The lubricated blank is heated in an infrared oven to a temperature above the softening point of the plastic but below the melting point.

This type of forming is suited to handle materials such as polypropylene and polyethylene, which is difficult to thermoform. Also coextruded materials, that are used to make food containers requiring barriers to oxygen, water vapor, and light, are used in this process.

Monaforming

Monaforming is a process that combines extrusion and compression molding with thermoforming to make open mouthed containers (Fig. 2–30). A reciprocating extruder is mounted vertically over the compression cavity. The extruder meters a precise amount of melted plastic material (melt-slug) into a compression mold. The compres-

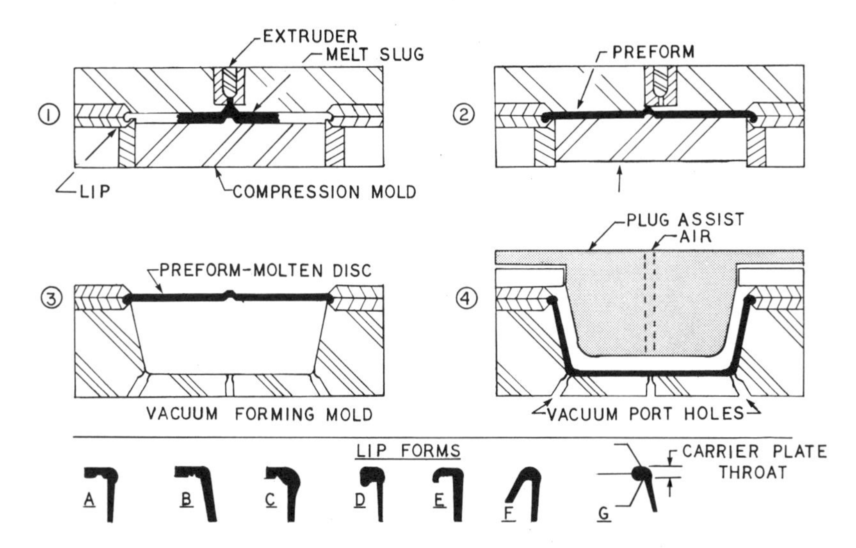

Figure 2–30. This illustrates the Monaforming process of producing plastic open mouthed containers. The lip form A, B, and C are heat seal types. The special purpose types are D, E, and F. The drinking edge is G. *(Courtesy Hayssen Mfg. Co.)*

sion mold closes making a preform or a molten disc. The range of preform thickness used is between 1.2 mm (.047 in.) and 2.00 mm (.078 in.). A lip or rim is created during the compression process. After the preform is made it is indexed to the forming station. Here the molten disc is shaped to the confines of the mold by air pressure, plug assistance, and vacuum. The finished container or part is indexed to the ejection station where it is removed. The process has three stations: (1) make the preform, (2) form the part, (3) eject the part. It has accurate definition and wide selection of lip geometry. The process can produce wide mouth plastic containers without any offal or scrap. Polystyrene, ABS, SAN, and acrylics have been used in making containers.

Extrusion Blow Molding

In extrusion blow molding, a heat-softened thermoplastic tube called a "parison" is extruded in the rough shape of the article to be made.

This shaped parison is inserted in a female mold and air is blown into the plastic. The plastic stretches and takes the shape of the female mold (Fig. 2–31 and 2–32A).

In the extrusion process, the blown parts must be formed from a tubular length of softened plastic. This is done by pinching the tube at two points, then inflating the pinched-off cylinder inside a hollow mold. The part is ejected from the mold with the pinched-off parison material at the bottom of the article and the top of the neck section. The neck section is trimmed off by a machine. The bottom part of the article, called a "tail", is easily removed manually.

Figure 2–31. An extrusion blow molding machine showing: (1) The extruder. (2) The plastic parison. (3) The two halves of the die. *(Courtesy The Dow Chemical Co.)*

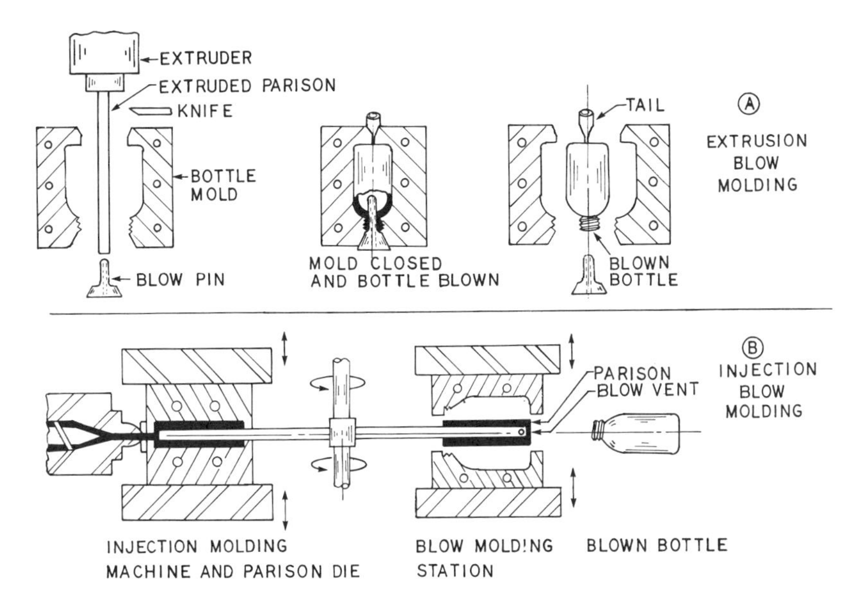

Figure 2–32 A and B. Schematic diagrams of extrusion blow molding (A) and injection blow molding (B).

Injection Blow Molding

The injection-blow molding process involves the injection molding of a parison over a mandrel and its subsequent transferral into a blow mold. Fig. 2–32B provides a schematic illustration of these basic steps. The injection-molding machine molds a thermoplastic parison around a mandrel. The injection die opens, and the parison is indexed to the blow-molding station. Here the blow mold closes, and internal air pressure through the "blow vent" forces the thermoplastic parison to the confines of the bottle mold.

An advantage of the injection-blow molding process is that it can produce a bottle or hollow article that is completely finished.

Material movement within a blow mold is a stretching rather than a flowing action. This reduces the problems of weld and flow lines, mold erosion, trapped gas, etc. Blow molds are only cavities. This eliminates matching force plugs and cavities. Undercuts on blow-molded items can be stripped more easily than injection-molded pieces. This eliminates the need for expensive sliding inserts, levers, cams, etc. Blow molds are made from cast aluminum alloys, zinc alloys, steel alloys, and beryllium copper. It is possible to blow mold any thermoplastic, and essentially all have been blow molded on an experimental basis.

Thermoforming

This is a process of forming a thermoplastic sheet into a three-dimensional shape. Thermoforming is made possible by the ability of the thermoplastic sheet to be softened and reshaped when heated and to retain the new shape when cooled.

Basically, all thermoforming includes movement of the heated sheet by vacuum, air pressure, mechanical drawing, or a combination of the three. There are several types or names given to the basic techniques for the thermoforming of thermoplastic sheets (Fig. 2–33).

Almost all thermoforming operations follow a set cycle regardless of the end product: heating of the thermoplastic sheet or film; forming the article; cooling; and trimming. In order to be certain of uni-

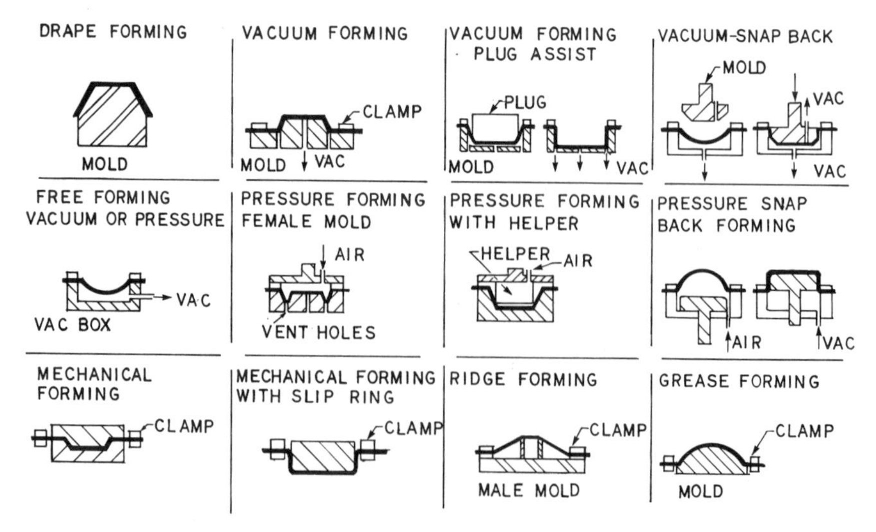

Figure 2–33. The twelve basic techniques for the thermoforming of plastic sheet is shown in this illustration.

form stretch while forming, the plastic sheet must be uniformly heated. The entire plastic sheet should be gradually and evenly brought up to the forming range. There are three types of heat that can be used; radiant heating; convection heating; and conduction heating.

One of the biggest advantages of thermoforming is the variety of materials that can be used for molds. These include wood, plaster, epoxy thermosetting plastics, and metals. Wood and plaster are usually limited to prototype work. Plastic molds are used for short runs and limited production. Metal molds are used for production runs. The major difference between plastic and metal for production molds is the difference in thermal conductivity of the two materials.

The thermoplastic sheet or film materials that are used in thermoforming are: acrylics; ABS; polystyrene; vinyls; cellulose acetate; and cellulose acetate butyrate. Molded articles made by thermoforming include door liners, tote boxes, signs and displays, contour maps, and aircraft domes. Fig. 2–34 illustrates one process of making a refrigerator door liner. The door liner was thermoformed out of a polystyrene sheet in a metal die.

MOLDING EXPANDABLE POLYSTYRENE FOAM

Expandable polystyrene foam is an expanded, closed-cell, rigid, plastic material. It starts as tiny beads or spherical particles of polystyrene impregnated with an expanding agent. These particles expand when exposed to heat, forming a rigid honeycomb structure of individual air cells. When exposed to heat, without restraint against expansion, these beads "puff" from a bulk density of 35 lbs. per cubic ft. to a low of one lb. per cubic ft.

These beads may be molded indirectly in a closed mold, or they may be pre-expanded by heating and then molded. Also, the beads may be extruded into thick sheets. Fig. 2–35 illustrates the steps used in molding polystyrene beads. In Fig. 2–35A, the small expandable polystyrene beads have been placed in the split-cavity mold. Live steam is passed into the steam chamber and enters the cavity through the .031 in. diameter holes. The heat of the steam causes the beads to expand and fill the confines of the mold. In Fig. 2–35B, the beads are

Figure 2-34. A refrigerator door liner. The door liner was thermoformed out of a polystyrene sheet in a metal die. *(Courtesy Dow Chemical Co.)*

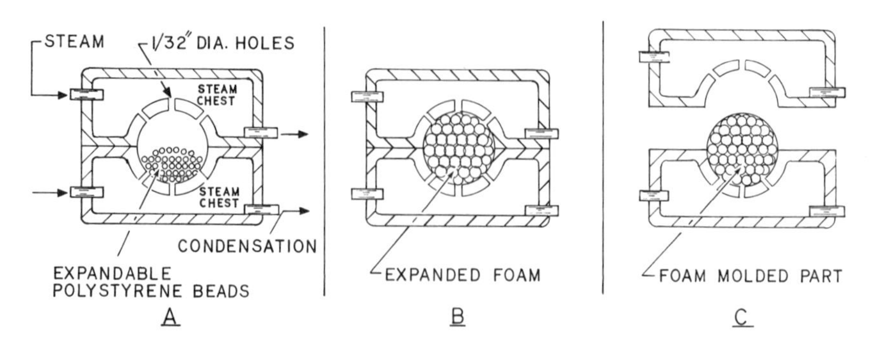

Figure 2-35. This illustrates the construction of a steam chest mold to expand polystyrene beads.

fully expanded. After the mold and the molded part have been cooled with water, the mold is opened and the part is removed (Fig. 2–35C).

Proper venting of the mold is essential for satisfactory expansion of the beads. Water forms in the cavity by condensation of the entering steam. In the expansion process, the water is squeezed out of the cavity by the expanding mass. The water generally escapes at the parting line of the die. Holes in the mold for steam to enter must be no larger than .031 in. in diameter. Generally, one hole for each two sq. in. of surface area is satisfactory. To insure even heating and cooling of the molded part, the cavity walls should be of uniform thickness. Aluminum and brass are preferred for mold construction materials. The steam pressures recommended for this process are in the range of 10 to 30 psi, read at the pressure regulator. Fig. 2–36 shows a typical bead-foam mold used to make a large rectangular polystyrene block.

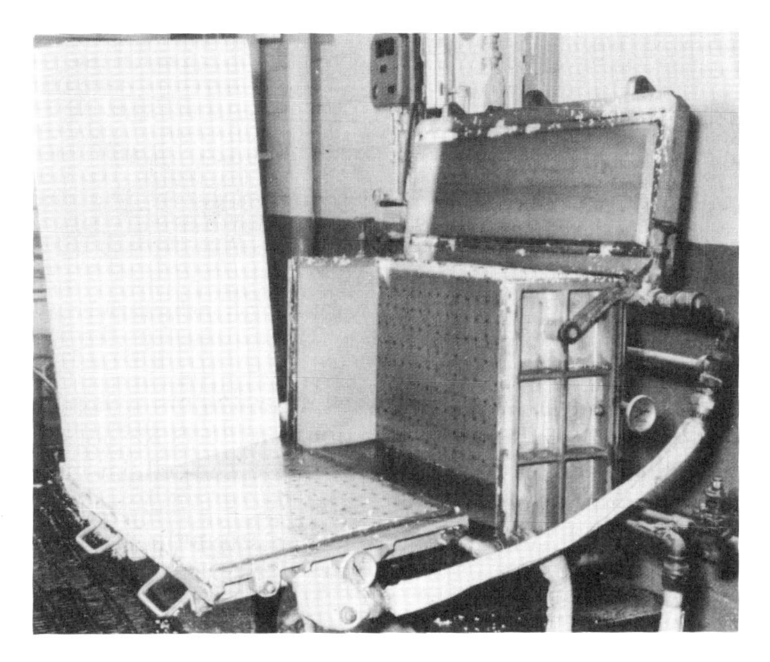

Figure 2–36. A polystyrene bead foam mold and a molded rectangular block of foam. *(Courtesy Dow Chemical Co.)*

Figure 2–37. A group of picnic coolers and containers made from expandable polystyrene beads. *(Courtesy Dow Chemical Co.)*

Expandable polystyrene foam is used in such items as shipping and packaging containers, thermal-insulation containers for hot and cold liquids, toys and novelties, and display racks. The cellular material is molded to almost any conceivable shape. Fig. 2–37 shows a group of picnic coolers and containers made from expandable polystyrene beads.

3

Mold Design and Processing

Although the molding of plastic materials follows somewhat the laws of hydraulics, the rules that govern hydraulics cannot be applied. Flow (in the molding of plastic) will be considered as the forced movement of a molten plastic material through the runner, gate, and into the part being molded during the fabrication process. Many plastics soften gradually over a wide range of temperatures. This gives the most familiar indications by which a plastic is judged.

Solid plastic polymers may be divided into three general types: amorphous; crystalline; and cross-linked. Thermoplastic materials are either amorphous or crystalline, and thermosetting plastic materials are crosslinked (see Chapter 1). The plastic material is usually in the form of powder, beads, flakes, or lumps.

The mold cavity in injection molding is actually a shrink fixture during the cooling or cure stage. The plastic material entering the mold cavity should flow as shown in Fig. 3–1a. As the molten plastic enters the cavity, it tends to "skin" as flow progresses. Additional material passing through the gate continues as a molten core between the two outer chilled layers. The chilled layer should be kept as thin as possible until the mold is filled. This can be accomplished by increasing the mold surface temperature, increasing the plastic material temperature, or increasing the speed at which the mold is filled. The plastic material entering the mold cavity should not jet into the mold, as shown in Fig. 3–1a. The "jetting and worming" will produce a turbulent flow of plastic resin, which in turn will give weld lines and poor fill rate.

WELD LINES

A weld line is sometimes called a knit line. If the plastic flow in a mold is split by an obstruction such as a pin, insert, corner, or slot, a weld line will usually result when the flow fronts meet. (Fig. 3–1b). Weld lines are particularly noticeable in transparent and translucent materials. Not only do they mar the looks of the piece, but they are a point of potential failure. This problem is most common to thermoset and thermoplastic materials used in injection and transfer molding. Weld lines can appear in compression molded parts, but they are rare. Weld lines can be prevented by designing the mold so as to permit the material to move with maximum freedom. Fig. 3–2 shows an ABS injection molded automobile arm rest. The weld lines in the arm rest are very prominent, due to the metallic flakes that are used for coloring.

RUNNERS

In injection and transfer molding, the runner is a channel that connects the sprue with the cavity gate. The name also applies to the

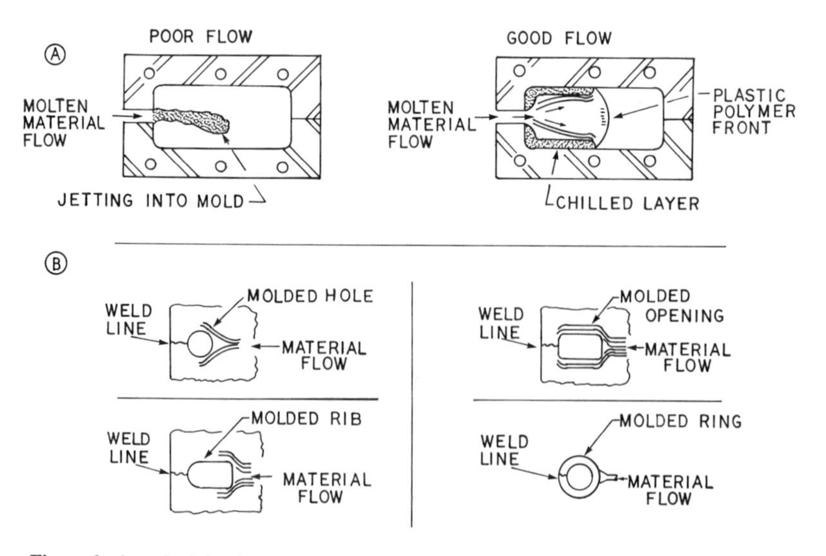

Figure 3–1a. In injection molding the molten plastic material should enter the cavity with an even flow and not jet into the mold. Fig. 3-lb. The flow patterns of molded plastics around holes, ribs, bosses, and opposite gates should always be noted and kept in mind.

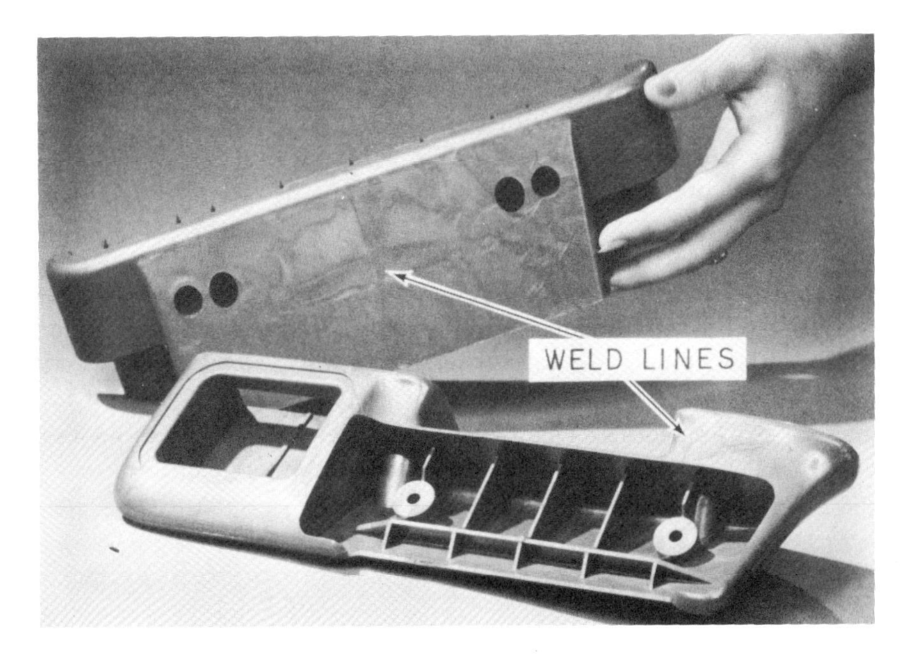

Figure 3–2. An automobile arm rest injection molded from metallic colored ABS. The weld lines are very prominent due to the metallic flakes in the color pigment. *(Courtesty Dow Chemical Co.)*

mold area and to the material itself that remains in the channel after the molding cycle. Runners can be machined on either half of the mold. They should be on the ejection or moving half of the mold in order to increase the tendency of the molded article to remain in that particular half when the mold is open.

Friction between the molten plastic material and the metal runner is very high. Consequently, runners should be highly polished to keep this friction to a minimum and to help reduce the molding pressure that is required to fill the cavities. There are generally five types of runners in use (Fig. 3–3). They are full round, half round, quarter round, trapezoidal and modified trapezoidal. The preference is given to the round runner, because it has the least amount of friction surface in relation to its volume. This type of runner allows the plastic mass in the center to move more rapidly. It also gives the least pressure drop and is the easiest to eject from the mold. Trapezoidal runners are satisfactory for injection molding, but should have positive ejection provisions, such as knockout pins.

The motion or movement of the molten plastic material through the runner can be classified as either a laminar or turbulent flow. Laminar flow or streamlined flow of thermoplastic material in a mold or runner system is achieved by solidification of the molten plastic layer in contact with the metal surface, thus providing an insulating tube through which more material flows. This type of flow is desirable for molded parts that require a minimum of stresses, weld marks, etc. The turbulent flow is generally brought about by small runners and gates. The small gates and runners turn the flow of material into a chaos of swirling eddies. This type of flow is not always desirable, because it has a tendency to increase stresses, sink marks, etc.

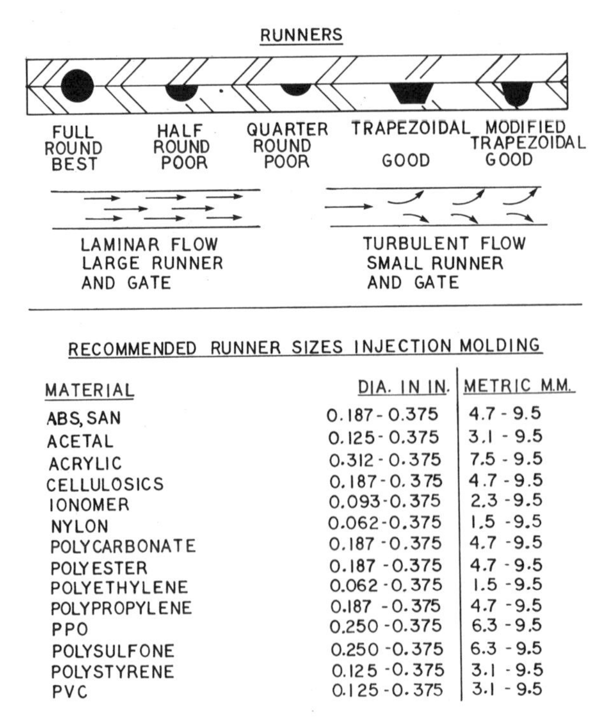

RECOMMENDED RUNNER SIZES INJECTION MOLDING

MATERIAL	DIA. IN IN.	METRIC M.M.
ABS, SAN	0.187 - 0.375	4.7 - 9.5
ACETAL	0.125 - 0.375	3.1 - 9.5
ACRYLIC	0.312 - 0.375	7.5 - 9.5
CELLULOSICS	0.187 - 0.375	4.7 - 9.5
IONOMER	0.093 - 0.375	2.3 - 9.5
NYLON	0.062 - 0.375	1.5 - 9.5
POLYCARBONATE	0.187 - 0.375	4.7 - 9.5
POLYESTER	0.187 - 0.375	4.7 - 9.5
POLYETHYLENE	0.062 - 0.375	1.5 - 9.5
POLYPROPYLENE	0.187 - 0.375	4.7 - 9.5
PPO	0.250 - 0.375	6.3 - 9.5
POLYSULFONE	0.250 - 0.375	6.3 - 9.5
POLYSTYRENE	0.125 - 0.375	3.1 - 9.5
PVC	0.125 - 0.375	3.1 - 9.5

Figure 3–3. In injection molding round runners and laminar flow of material in the runners is recommended.

Figures 3–4. A 16-cavity injection molded shot of caster wheels. A perfect balanced runner system. Note all cavities are equal distance from the sprue. A three-plate mold was used. Material is polystyrene. *(Courtesy General Industries Co.)*

Runners should be kept as small as possible in order to reduce the cooling time and the projected area of the mold. The pressure drop on the flow of a plastic material gets less and less as the cross section gets smaller. A molten plastic material does not flow across a runner or die surface like water might flow. It lays back against the cold mold surface and continues to flow through the middle. Cold slug wells should be provided at the ends of runners and at 90° turns. A cold slug well is a space or pocket made in the runner system, in order to catch material that is too cold to be molded.

A balanced runner system is very desirable. If it is at all possible, the cavities in injection molding should be equally distant from the sprue. This will allow the same length of runner flow. Fig. 3–4 illustrates a perfect balanced mold. All of the 16 molded caster wheels are an equal distance from the sprue. A three-plate injection mold was used in making the caster wheels. As the mold opens, the caster wheels and gates are separated and both drop out of the mold automatically. Fig. 3–5 shows an injection-molded shot taken from a comb mold. The material has not completely filled the cavities. In the plastic industry, this is called a "short shot." Note that the cavities closest to the sprue are the most completely filled.

GATES

In injection and transfer molding, the gate is considered that portion of the molded piece that allows the molten resin to flow from the runner or from the sprue into the cavity. The gate may be the same size and shape as the runner, but generally it is much smaller. There is an optimum gate size. It should be large enough for suitable fill rate and small enough to seal off and prevent backflow or packing.

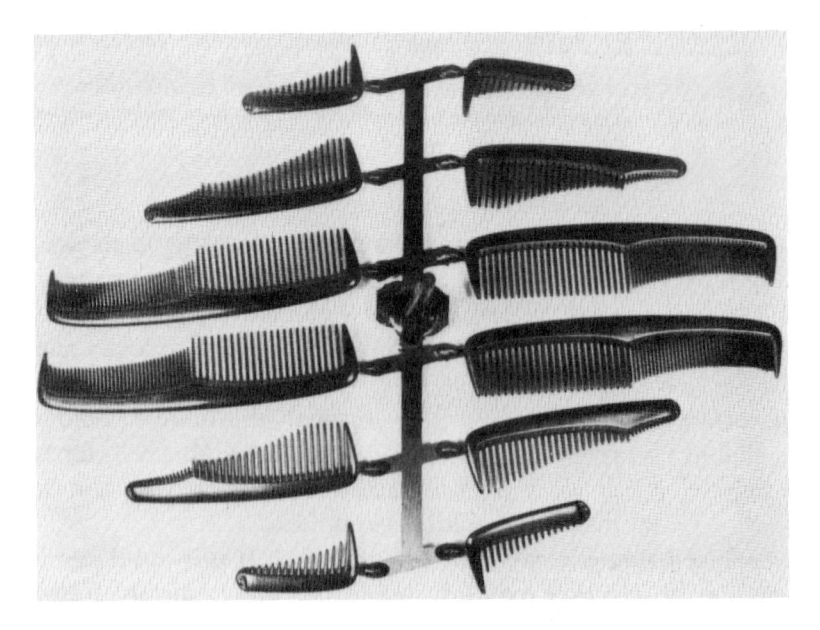

Figure 3–5. An injection molded shot from a comb mold. The shot did not fill out completely. This is called a "short shot" in the plastic industry.

If possible, gates should be located for economical removal and finishing of the molded part. No definite rule can be given for the depth of a gate, as the shape, size, thickness, and details of the molded part each have an influence in determining the size of a gate. A good practice is to start with a gate approximately 0.025 in. in depth and then gradually open the gate by removing metal from the mold until a good part is obtained. This is done while varying the pressure and temperature on the molten plastic material in the injection machine. Round gates are recommended wherever applicable, and they should be so located that the plastic melt leaving the gate impinges against a mold surface to build up a smooth flow of material into the cavity, thus preventing "jetting and worming." Generally, the gate thickness is 40 to 60% of the part thickness. There are several types of gates used today (Fig. 3–6).

Standard Gate

This type of gate is used more often than any other, since it is adaptable to most injection-molded parts. The main advantage of this gate is its ease of removal, which minimizes the cost of finishing. This type of gate can easily be removed by a trim fixture or by a pair of hand nippers. No extra finishing is required on the molded part.

Ring Gate

This ring-type is best suited for hollow cylindrical parts such as pencil barrels. The material enters around the core pin and flows down evenly around the pin. The easy material flow prevents the trapping of air and eliminates weld marks that would be present if a standard-type gate were used. The removal of the ring from the molded piece generally increases the cost of the article.

Pinpoint Submarine Gate

If the gate diameter is small (approximately 0.020 in.), it is called a pinpoint gate. A submarine gate is made to carry the plastic material

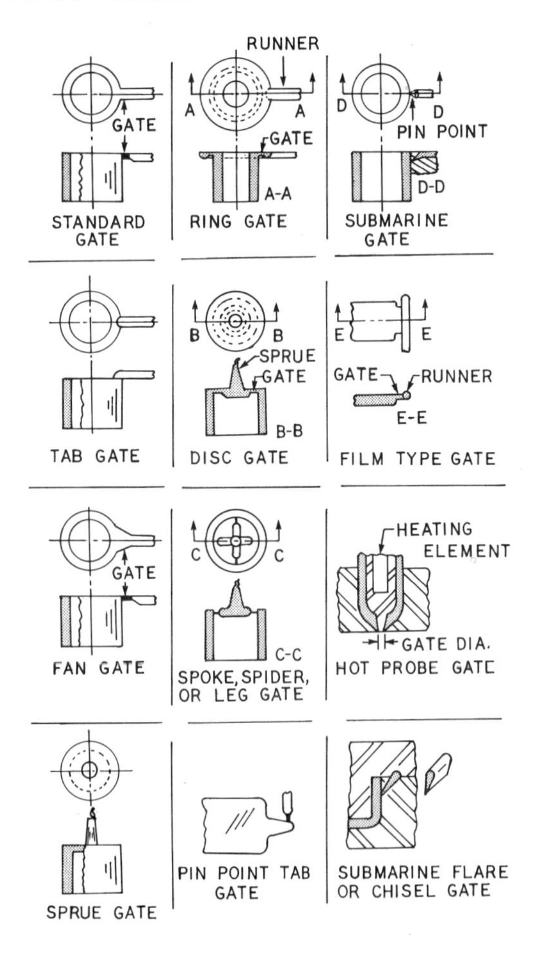

Figure 3–6. This illustrates the many types of gates used in the plastic industry.

down below the parting line of the mold into the cavity. The combination of the two is called a pinpoint submarine gate. This type of gate permits automatic ejection from the runner system. It is used only on small parts.

Tab Gate

This type of gate extends the runner system into the molded part. Generally, an extra operation is required to remove the tab from the molded part after molding.

Disc Gate

The disc gate is the reverse or opposite of the ring gate. The material flows around the center core to the periphery. With this type of gate, good material flow virtually eliminates the occurrence of weld lines and trapped air, if the mold is properly vented. This gate is used mostly on flat, annular parts such as bezels, etc.

Film-Type Gate

A film-type gate is used for large parts that are being molded out of metallic-colored plastic material. This helps to eliminate the unsightly weld lines caused in metallic molding materials.

Fan Gate

A fan-type gate is relatively wide with a thin edge. It is used for such articles as boxes, covers, or other parts having flat, thin sections. The fan gate helps to spread the material, thus improving and minimizing the possibility of flow lines.

Spoke, Spider, or Leg Gate

This type gate may produce more weld lines in the molded part, but the part will be stronger than the weld lines produced in a standard gated part of the same size and shape.

Hot-Probe Gate

This may also be called an insulated runner gate, and is used in runnerless molding. In this type of molding, the molten plastic material is delivered to the mold through heated runners, thus minimizing finishing and scrap costs.

Sprue Gate

This type of gate is used for a large piece in a single cavity mold. Gating at the center allows for an even flow of material into the cavity and eliminates the possibility of trapped air and weld lines.

Pinpoint Tab Gate

In this design, the pinpoint gate is attached to a tab that is joined to the molded part.

Submarine Flare Gate or Chisel Gate

This type of gate is very similar to the pinpoint submarine gate, except that it is larger and has the shape of a chisel edge. It is used on large molded parts and permits automatic ejection from the runner system.

Gates and Runners for Injection Molded Thermosets

The runner and gate system for injection molded thermosets should be designed to avoid excessive frictional heat build-up in the plastic material. The runner system should be full round and sharp corners at runner bends should be avoided (Fig. 3–7). Most molds have re-

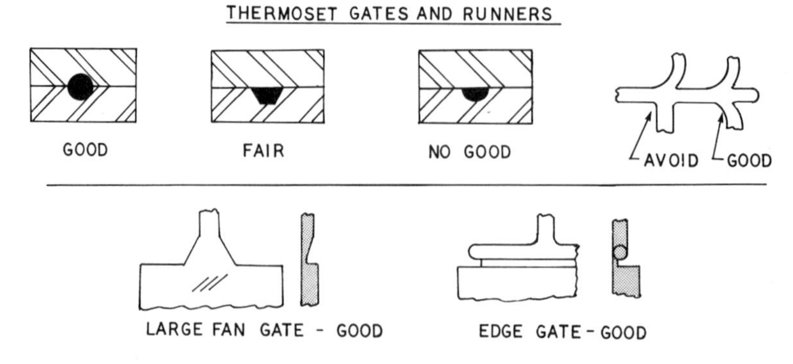

Figure 3–7. This illustrates the best gate and runner system used in thermoset injection molding.

placeable gates and runner sections. They can be easily replaced if they wear in production. The molds should be chrome plated to reduce wear and eliminate sticking.

SPRUE PULLERS

A sprue puller is a device used to pull or draw a molded sprue out of the sprue bushing. It is generally a straight round pin with the end machined in the form of an undercut. There are six methods in general use for providing an anchor to pull the sprue (Fig. 3–8). The sprue lock pin is located where the small depression at the mold entrance meets the runners. It is fastened to the knockout or ejector mechanism and runs through the movable part of the mold in direct line with the mold entrance.

VENTING OF MOLDS

All molds contain air that must be removed or displaced as the mold is being filled with a plastic material. The compression or air in a

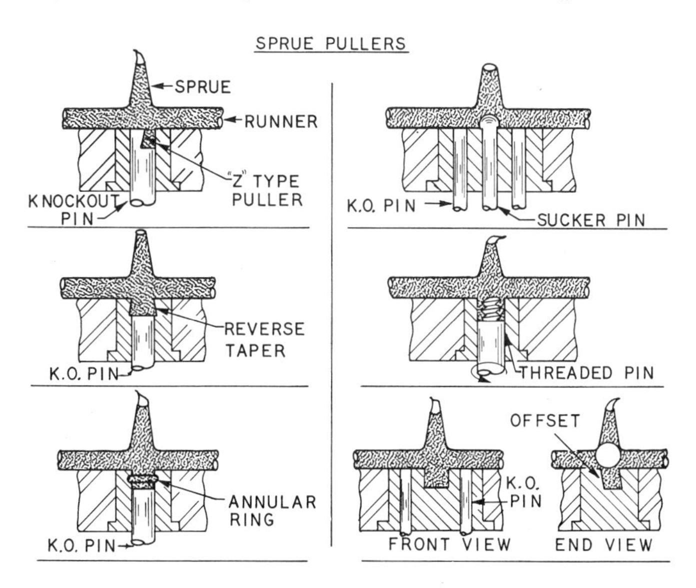

Figure 3–8. Sprue pullers in plastic injection molding are used to pull or draw a molded sprue out of the sprue bushing.

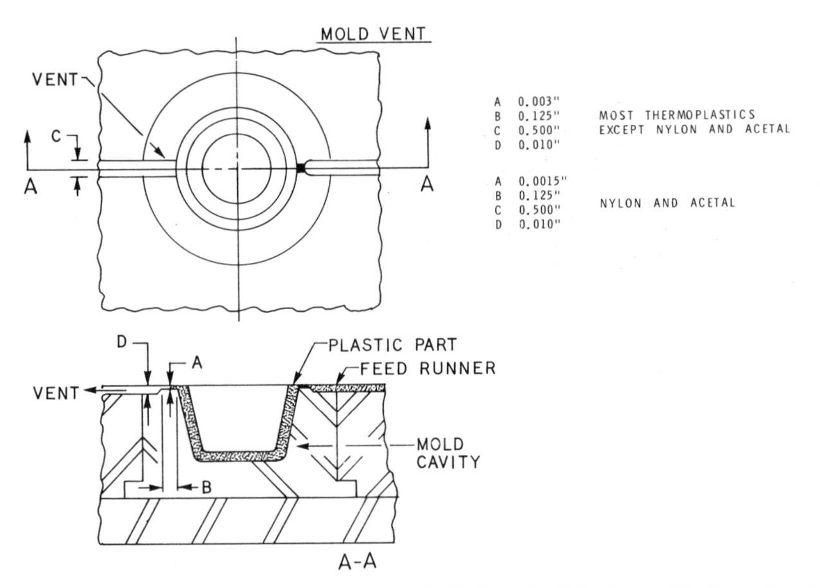

Figure 3–9. This drawing illustrates one method of venting injection molds. Note that this is for thermoplastic materials.

mold creates heat, and enough heat will tend to burn the plastic material. Generally, every mold is vented opposite every gate and on the same half of the mold as the runners. Vents are sometimes placed in a mold at the point that fills out last.

Deep recesses in a mold can be vented by using pins with adequate clearance to allow passage of air. Loose-fitting knockout pins may be used for venting. Also, scratch marks on mold surfaces may be sufficient. In some cases, it is necessary to pull a vacuum on the mold to help remove the entrapped air. Fig. 3–9 shows approximate dimensions used for venting injection molds. Vents for transfer molds should be from 0.002 to 0.005 in. deep and from .125 to .250 in. wide.

DESIGN FOR FLOW AND SHAPE

In molding with any plastic material, parts should be designed with ample curves, except at the parting lines of the mold cavity section. If the material, as it is being molded, does not sweep across the con-

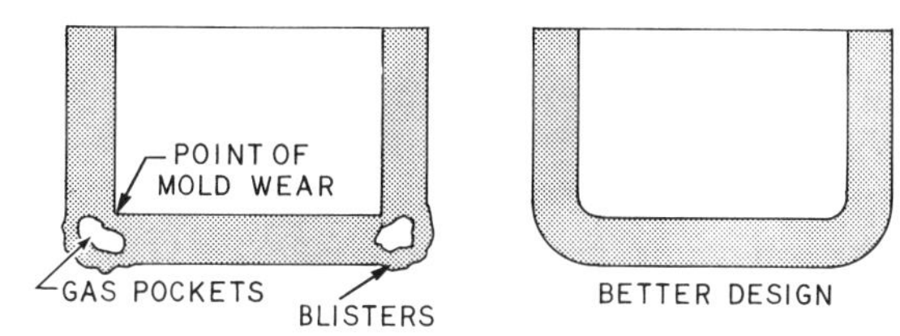

Figure 3–10. Streamlining of the plastic part will help to prevent gas pockets.

fined areas of the mold, gas pockets or voids may develop. This results in blisters or sink marks on the surface nearest the pocket. With thermosetting materials, these gas pockets may be caused by trapped gases produced during chemical cross-linking of the material (Fig. 3–10 and 3–11). With thermoplastics, these voids may be the result of "case hardening" of the melt as it momentarily hesitates and cools in the corner.

The strength of a plastic part depends largely on good design. When plastic materials flow around protruding sections of a mold, they knit or weld on the other side. Good design calls for consideration of this flow route in the part. With the thermosetting materials, this knit or weld line may be a weak point, due to the fact that the

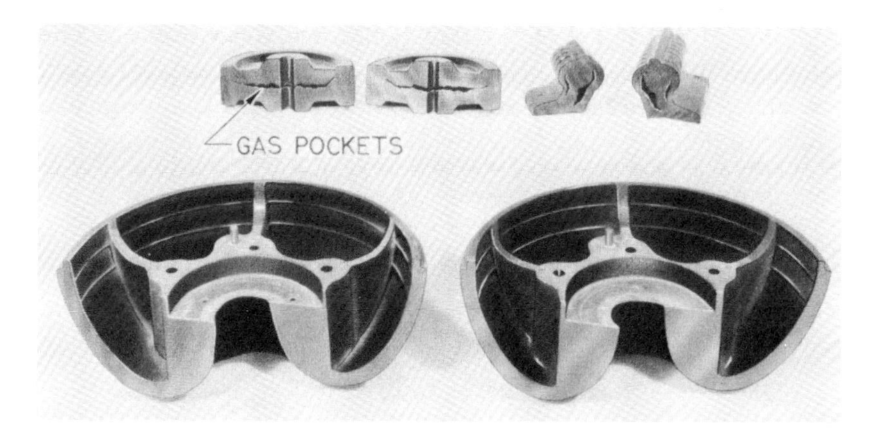

Figure 3–11. If possible avoid thick and thin or uneven wall thicknesses in compression molded phenolic molded parts as gas pockets may form.

plastic material has approached the last stages of polymerization before the two streams meet or weld on the opposite side. Thus, they do not bond well. The thermoplastic materials tend to cool as they fill the mold cavity, with the result that the weld or knit line will be weaker than the adjoining material.

DESIGN OF RIBS AND BOSSES

Bosses may be defined as protruding studs on a part that assist in the assembling of the plastic part with another piece (Figs. 3–12 and 3–13). Because they are frequently the anchoring member between the plastic part and mating part, they are subject to strains and stresses not found in other areas of the part.

Ribs (Fig. 3–14) may be defined as long protrusions on the part, which may be used to decorate or strengthen the part or, if properly placed, to help prevent it from warping. Bosses and ribs normally present unusual flow problems. The boss or rib is made in the cavity section or in the plunger section of the mold, but not by a combination of the two mold sections. If the boss or rib is made by either the cavity or plunger of the mold, it is preferable to locate these projections in the corners or sides of the part (Fig. 3–15a). Such a location allows easy removal of condensation products, in the case of compression molding, and trapped air in the case of transfer and injection molding. If the rib is to be extremely thin and away from the sides of the part, filling may be helped by adding intermittent heavy sections on which a vent pin can be placed in the mold to allow trapped air or condensation to escape (Fig. 3–15b).

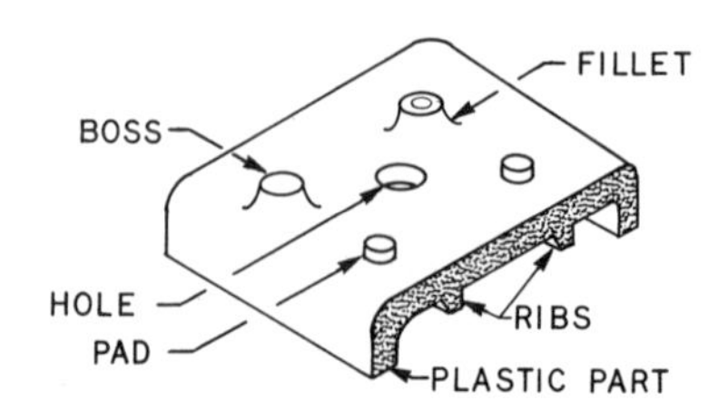

Figure 3–12. This drawing explains or shows what a boss, hole, pad, fillet, and ribs are on a molded plastic part.

Figure 3–13. This picture illustrates a section of a portable radio case showing an inside view. (1) A pad. (2) A boss. (3) A boss with rib supports. (4) A boss at a wall. Note all ribs, bosses, and pads are radiused at their junction with the main wall.

Ribs placed on a part need not result in a non-uniform wall thickness. If the rib is hollowed out from the back, added strength will still be present, and a slight saving in material will result (Fig. 3–15c). If long thin bosses are required, ribs should be placed on the side to improve flow and add strength and density to the rib (Fig. 3–15d). If a part requires deep slots or grooves, these will be made by thin sections of metal in the mold. Wherever possible, these slots should be widened to provide mold strength (Fig. 3–15e). If the slots cannot be widened at intervals, the depth of the slot should never be more than three times its thickness.

Ribs must be designed in the right proportions if they are to fill out in the molding. They should be rounded and located in the corners of the part when possible. Rather than one heavy rib, it is better to use a series of ribs. Fig. 3–16 illustrates the rib design and proportions used in most thermoplastic materials. It will be noted that all dimen-

Figure 3–14. This picture illustrates the use of ribs in the making of an automobile arm rest.

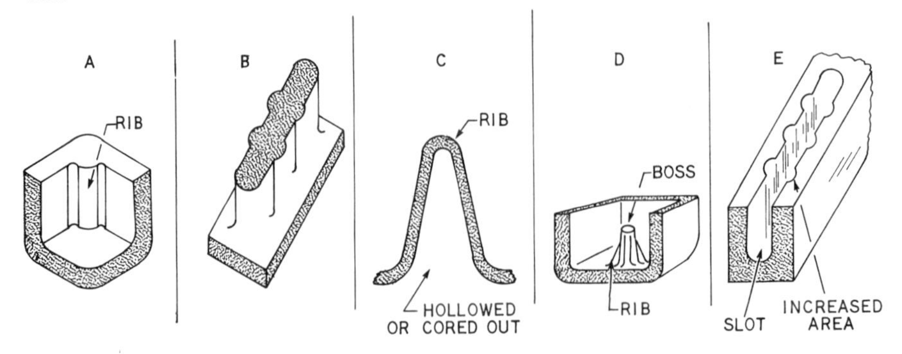

Figure 3–15. This illustrates some fundamental design factors for ribs. (A) A rib should be located in the corner or side of a molded part. This will lower the mold cost and allow easier filling of the plastic part. (B) Bosses and fillets will help to fill out ribs. (C) A uniform wall thickness can be maintained in a rib if it is hollowed or cored out from the back. (D) Ribs on the side of long bosses will aid in filling them. (E) Increasing the area of slots at intervals in the molded parts will add strength to the mold.

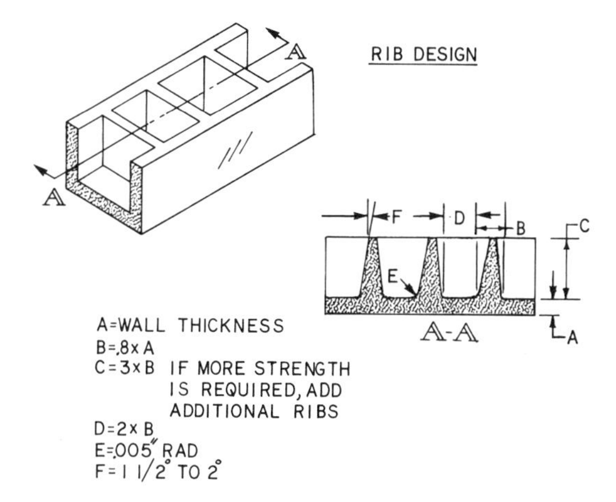

Figure 3–16. This drawing illustrates the rib design and proportions used in most thermoplastic materials. *(Courtesy Dow Chemical Co.)*

sions are a function of the wall thickness. The "B" dimension is the distance over the radius. If the designer goes over 80% of the wall thickness on the "B" dimension, sink marks will start to appear. Most parts are developed by using 60% of dimension "B". Recommended proportion for ribs molded with thermosetting materials should be as indicated in Fig. 3–17.

If cloth-filled phenolic or similar material is required and thin ribs are desired, sometimes the cloth filler can be forced into the end of the rib by adding bosses or fillets to the ribs. In effect, this procedure thickens the wall, allowing a freer flow of the material. If deep, thin ribs are molded with thermosetting and thermoplastic materials, knockout pins are often placed on the rib. The clearance between the knockout pin and the rest of the mold section will allow trapped air to escape and the rib to fill out with material.

A large rib placed on a part should not result in a non-uniform wall thickness and cause a sink mark area. It is better to make many smaller ribs out of the large rib (Fig. 3–18a). Sometimes, it may be better to design two narrow ribs instead of one large rib, or better still, one long rib. This will result in a slight savings in material cost

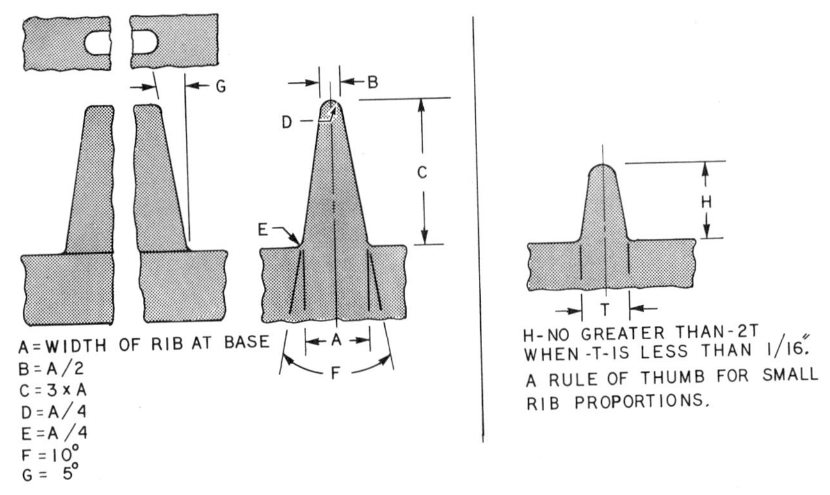

Figure 3–17. Recommended proportions for ribs molded with thermosetting materials.

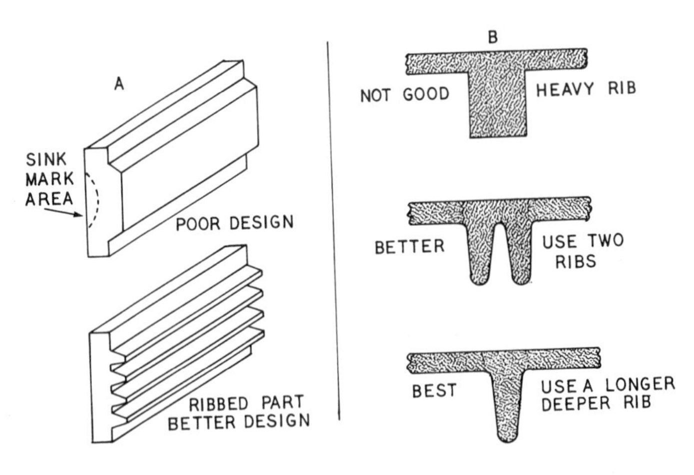

Figure 3–18. (A) In order to eliminate sink marks it is better to have many small ribs instead of one large rib as illustrated in the drawing. (B) Good rib design calls for narrow ribs instead of one large heavy rib.

(Fig. 3–18b). Fig. 3–19 illustrates an excellent use of ribs in the handles of grass shears that were injection molded from acetal resin. A boss at the wall is shown in Fig. 3–20. It will be noted that most of the dimensions are a function of the wall thickness. Any attached member to the wall should not be over 80% of the wall thickness. Over this amount, sink marks will occur in most thermoplastic materials. A boss away from the wall is shown in Fig. 3–21. It will be

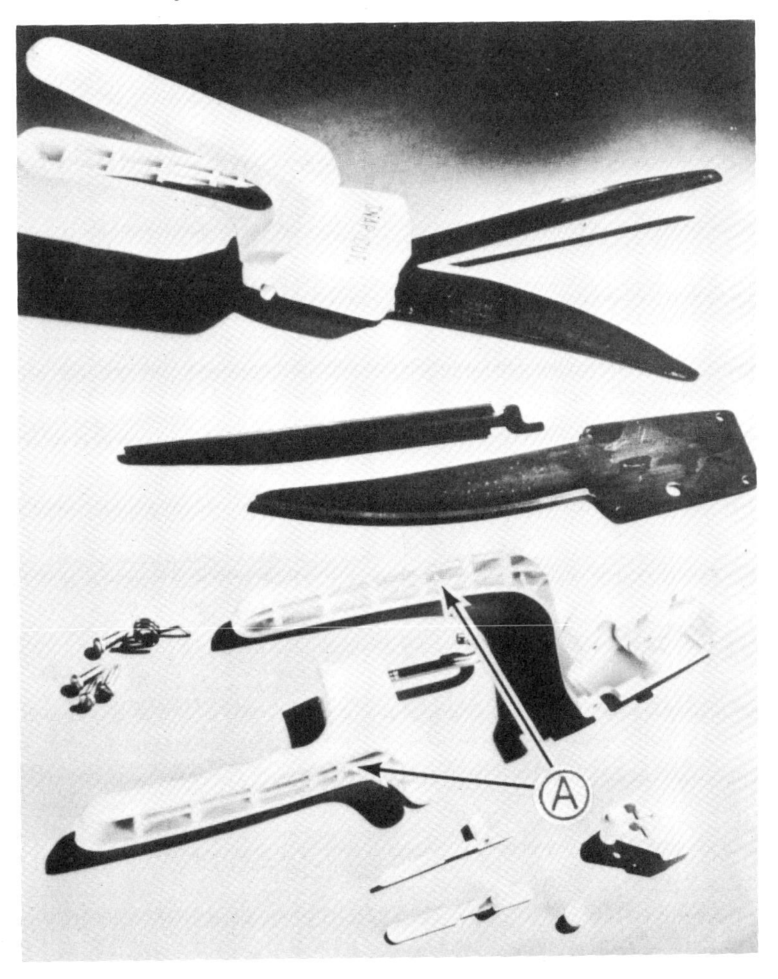

Figure 3–19. Grass shears with handles injection molded of Celcon acetal copolymer. Note (A) cored-out ribbed sections in the handles. *(Courtesy Celanese)*

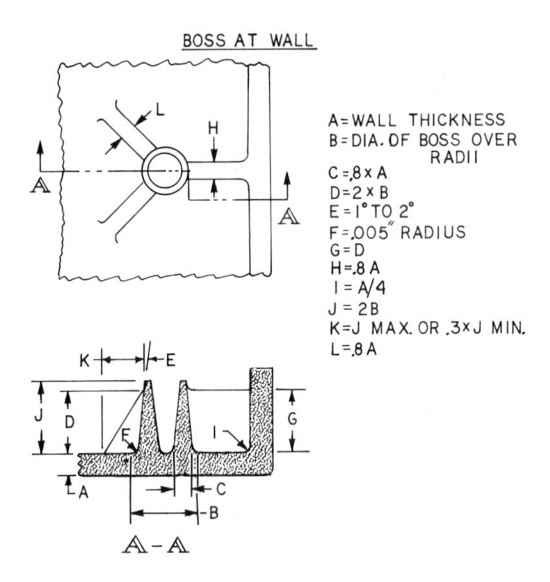

Figure 3–20. This drawing illustrates design and proportions of a boss at a wall. It is used in most thermoplastic materials. *(Courtesy Dow Chemical Co.)*

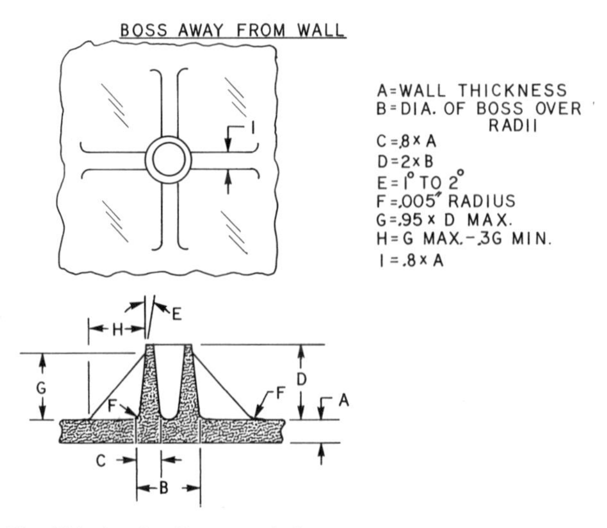

Figures 3–21. This drawing illustrates design and proportions of a boss away from the wall. It is used in most thermoplastic materials. *(Courtesy Dow Chemical Co.)*

noted that this type of boss is very similar to the boss at the wall. An outside boss (Fig. 3–22) is used in attaching or holding a part to an assembly. The height and the width of the boss should be the same dimensions, with any thick sections cored out from the back. If possible, bosses too near a corner, as illustrated in Fig. 3–23a, should be avoided. Bosses too near corners or walls require thin wear sections of mold material. The designing of bosses with cross sections other than round ones should be avoided. Square or oval holes are difficult to machine into mold steel and will result in a more expensive mold (Fig. 3–23b). If it becomes necessary to design a square or rectangular boss, the corners should have a radius of at least 0.015 in. If close tolerances are required on bosses and projections, it may be necessary to develop these dimensions. This can be accomplished by gradually removing metal from the mold a little at a time until the desired dimensions are obtained (Fig. 3–23c).

RIM DESIGN

A rim design is an undercut in the part and is used to give added strength and support (Fig. 3–24). A mold needed to produce this type of design is expensive and requires retractable sections in the

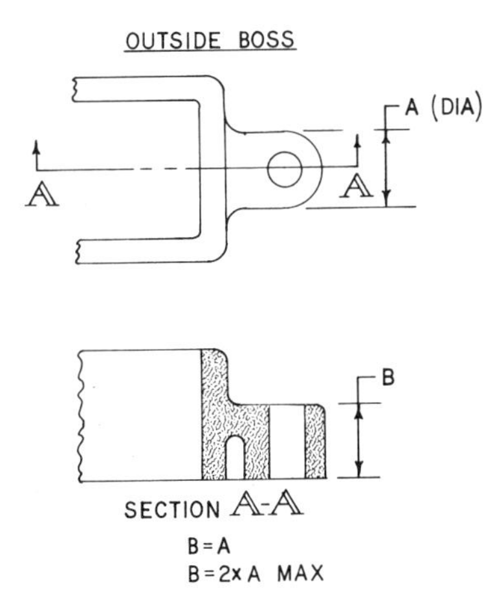

Figure 3–22. Recommended proportions for a boss on the outside of the molded part.

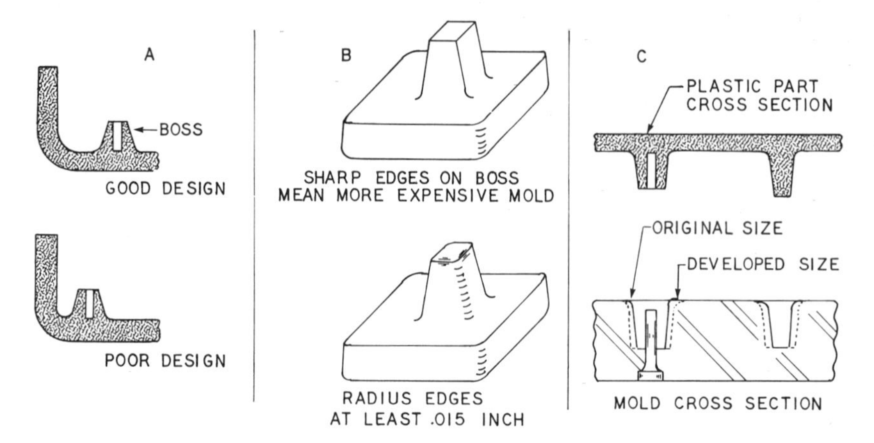

Figure 3–23. (A) Avoid, if possible, bosses too near an outside wall. This will cause weak sections in the mold. (B) Sharp edges on bosses should be avoided. (C) Close tolerances on bosses and projections can be obtained by development. This is done by designing the mold undersize, then removing metal from the mold until the correct dimensions are obtained on the molded part.

mold, as the part cannot be stripped out normally. Loose sections in the mold must be used and the mold cost is approximatley 40% more than a conventional mold. The same or almost equivalent strength can be achieved by using gussets.

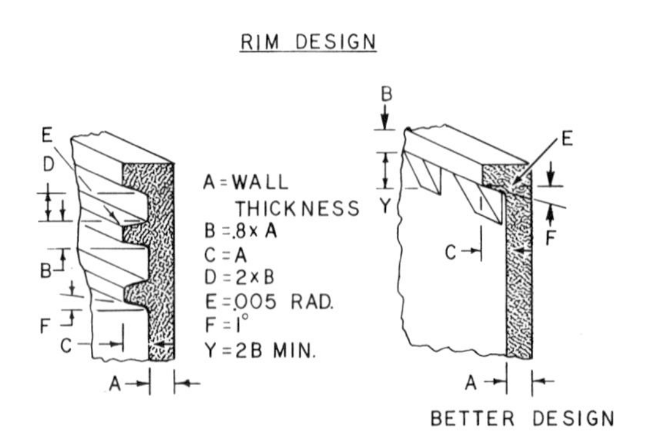

Figure 3–24. This drawing illustrates design and proportions of a rim. A better part can be obtained by using gussets. This design is used in most thermoplastic materials. *(Courtesy Dow Chemical Co.)*

GUSSETS

Gussets are supporting arrangements for an edge (Fig. 3–25). All dimensions of the gusset are a function of the wall thickness. For additional support, two or more gussets do the job better than increasing the height or width of one.

RADII

Surfaces of all intersections should be rounded. Large radii should be positioned at a junction of two or more surfaces to insure proper material flow and relieve any possibility of stress concentration. Any sharp corner that is subjected to loading and twisting will tend to be weak and break. All of the stressing within the part will be at the sharp corner. Corner radii should be a minimum of 1/4 of the part thickness, as shown in Fig. 3–26a.

Flow of a material at a corner presents no problem if the corner is rounded as shown in Fig. 3–26b. If the corner is square, the plastic material lodges in the corner and impedes even flow, thus causing flow marks and non-uniformity.

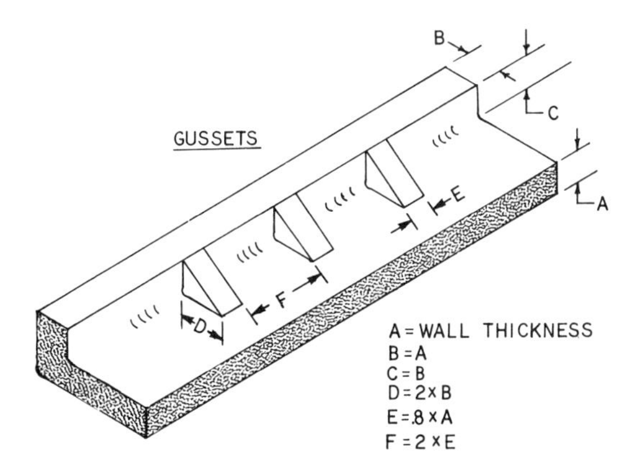

Figure 3–25. This drawing illustrates design and proportions of gussets. Gussets are supporting arrangements for an edge. This design is used in most thermoplastic materials. *(Courtesy Dow Chemical Co.)*

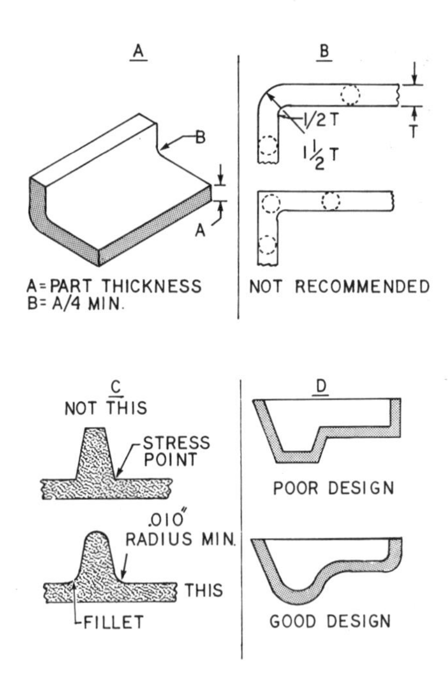

Figure 3–26. (A) Corner radii should be a minimum of 1/4 of the part thickness for most thermoplastic materials. (B) Plastic corner flow should present no problem if the corner is rounded as shown in the above drawing. (C) Fillets should be placed at the junction of bosses and ribs with the main body of the part. (D) Curves and fillets in a molded part prevent stress concentrations, add strength, and help eliminate warpage.

FILLETS

Fillets (or radii) are used at the base of ribs or bosses to facilitate the flow of plastic material and to eliminate sharp corners, thus reducing stress concentrations in the molded part. All plastic parts requiring bosses should be provided with fillets at the junction of the boss with the main body of the plastic part. Radii of these fillets should be at least 0.010 in. and preferably 0.030 in. (Fig. 3–26c). The addition of a fillet increases the strength of the mold and the molded part. Fillets generally reduce the cost of the mold, the molded part is more stream-lined, and the corners of the molded part are easier to keep clean of dust. Fig. 3–26d illustrates a well-designed part that has adequate curves and fillets to prevent stress concentrations and warpage.

4

Mold Design for Part Requirements

PARTING LINES

Parting lines (sometimes called cutoff lines) may be described as those lines made by the juncture of the male and female die and loose mold sections. In designing the mold, parting lines should be kept in one plane, if possible, and located along an edge of the molded part, rather than over the center of a flat surface. The parting line should never be placed partially down the side of the part, because this would result in high labor cost of buffing out the parting line flash from the part. The parting line should be around the section of the part having the largest cross-sectional area. In compression-molded parts, the location of the parting line is most important, because heavy flash usually occurs on these molded pieces and must be removed by filing, tumbling, or tumblast.

Parting lines, as illustrated in Figs. 4–1 and 4–2, are typical of molded parts that have undercuts. In order to mold an undercut in a part, such as illustrated in Fig. 4–1, it is necessary to develop split sections in the mold. These sections, of course, are open or taken out of the main cavity before the part, or as the part, is removed. The proper location of cutoff and parting lines may save many dollars in finishing and rejections. For example, suppose a cylindrical or cup-shaped item is to be molded. The parting line should never occur along the side of the cylinder, but should be placed at the top of the

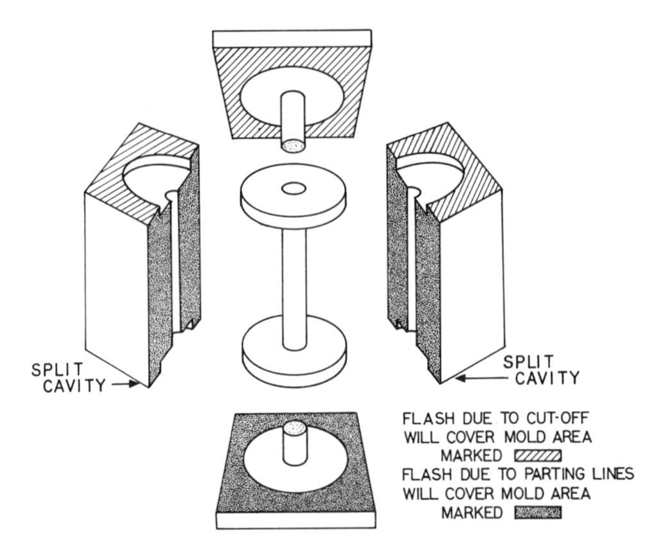

Figure 4–1. One method of parting mold sections when compression molding a spool. Flash due to cutoff and parting lines will affect the dimensions of the part. Note that the mold does not have sufficient loading space for the molding material. The drawing shows only those areas affected by cutoff and parting line flash.

part (Fig. 4–3a). If placed part way down the side, a buffing operation will be required to remove the objectional appearance of the parting line. If placed at the top, no buffing should be required, since the top edge will disguise the parting line.

If a compression-molded part is to be tumbled after molding, avoid designing a part having a thin wall at the parting line. Unlike a thick wall, a thin wall can break or chip during tumbling (Fig. 4–3b). If a compression molded part is to be filed in order to finish the part after molding, the part should be designed so that the parting line is along a straight surface. Dips and recesses in the wall should be kept to a minimum (Fig. 4–3c).

The thermosetting materials are the only plastics molded in volume that are taken out of the mold with flash, which must be removed from the part. Normally the flash can be removed satisfactorily by tumbling. Flash produced when molding with the cloth-filled materials can not be removed effectively by this inexpensive method. If flash can be removed by tumbling, it is advisable to

Figure 4–2. A plastic molded spool. The parting lines of the die are shown by heavy white lines.

bear in mind that some tiny bits of it will always remain, even after tumbling. Thermoplastic parts are not tumbled to remove flash, as they have very little or no flash when they are removed from the mold. The injection mold that produces thermoplastic parts is closed before the plastic is injected into it. This eliminates the possibility of flash. Thermoplastic parts may be tumbled to remove sharp edges and corners or to finish by tumble-buffing.

Compression-molded plastic parts can be molded with either a horizontal or vertical flash line (Fig. 4–4a). The designer of a proposed part should study the piece to determine which one of the two flash lines will be the most economical to clean or to remove. Flash

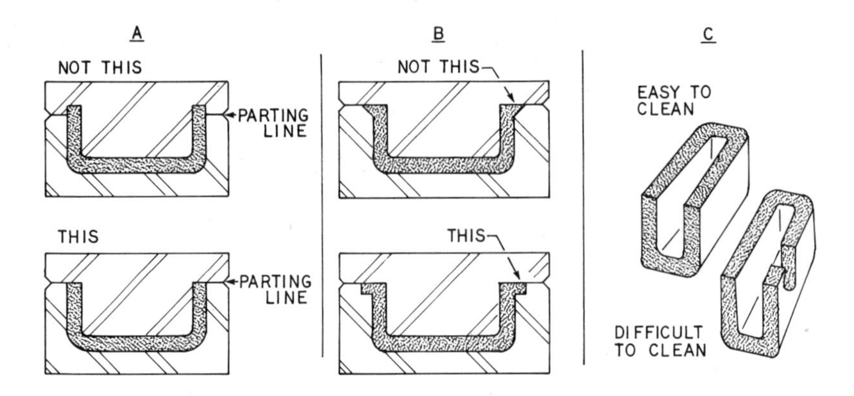

Figure 4–3. (A) Whenever possible, parting lines or cutoff lines should be located at the top of the part to facilitate finishing operations. (B) Avoid designing a part having a thin sharp wall at the parting line. It will break or chip very easily during finishing operations. (C) If filing is required to remove flash at the cutoff and parting lines, the lines should be located where they will be easy to clean.

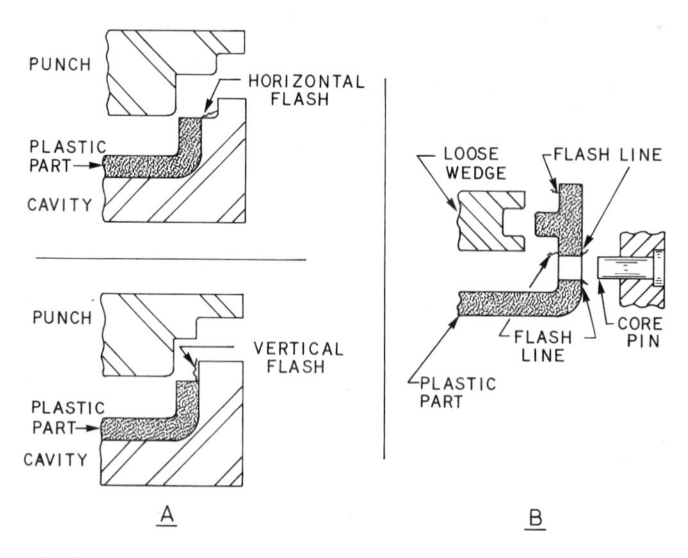

Figure 4–4. A compression-molded part will have a vertical or a horizontal flash line as shown in (A). Also flash lines will be made by loose wedges and side core pins as shown in (B).

lines will also show at the junction of loose wedges and side core pins. These areas must be treated as parting lines and sometimes require a cleaning operation (Fig. 4–4b).

If a compression-molded part has decorative ribs on the outside walls, the ribs should not extend to the top of the parting line (Fig. 4–5a). The ribs make an irregular parting line that will be difficult to clean. It is much better to have the decorative ribs stop immediately below the parting line, so that the cleaning or removing of the flash line is done on a straight surface (Fig. 4–5b). If the design of the part has many irregular and difficult parting lines, it will be wise to consider molding the article by injection or transfer molding.

In compression-molded parts such as knobs, bottle caps, handles, and any other part with a rounded section where the parting line of the die must be placed, it is advisable to use one of the designs illustrated in Fig. 4–6. A beaded parting line is used where the appearance is important and the adjoining surface should be protected during the finishing operation. A beaded parting line also helps in disguising mold misalignment. This is sometimes known as placing a parting line on a ridge. A peaked parting parting line is located above

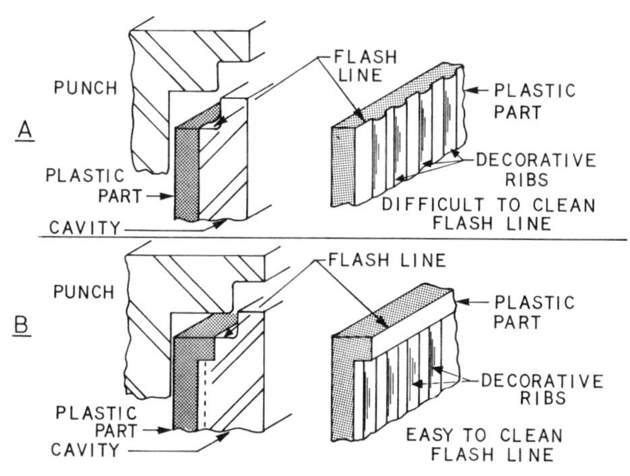

Figure 4–5. Decorative ribs on a compression-molded part should not run to the top of the parting line. The ribs will be difficult to clean as indicated in (A). A better design is shown in (B).

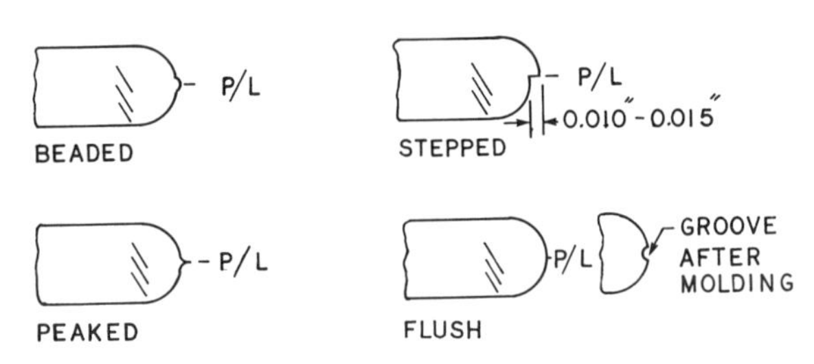

Figure 4–6. Compression molded parts that have parting lines on rounded sections may use any of the above type designs.

the surrounding surfaces of the part, and the flash can be cleaned without marring the surfaces. A stepped parting line will also allow clean and easy finishing of the molded part; the flash will be horizontal and easy to remove mechanically. In the event that a bead is not permissible, a flush parting line may be necessary. The flash may be removed by adding a decorative effect made by grooving the part at the parting line. This method, however, is practical only on round parts that may be spindled by turning the part rapidly on the end of a motor shaft and holding a cutting tool against it. The grooved line may be disguised by painting it after undercutting. Fig. 4–7 illustrates two acceptable compression-molded phenolic knobs that have different types of parting lines. A cross-section of each knob is shown. The top knob was molded with a flush parting line and then grooved. The bottom knob was molded with a peaked parting line.

The cost of finishing or removing the flash from a compression molded part is generally a large percentage of the direct labor cost of the molded piece. Simple straight parting lines should be designed into the part if at all possible. The designer of a plastic part will have no difficulty with flash lines if they are located where they may be removed easily and inexpensively and where they may be disguised or at least unobjectionable.

Parting lines for injection molds are shown in Fig. 4–8. The standard-type parting line that is flat and square at the top of a part is

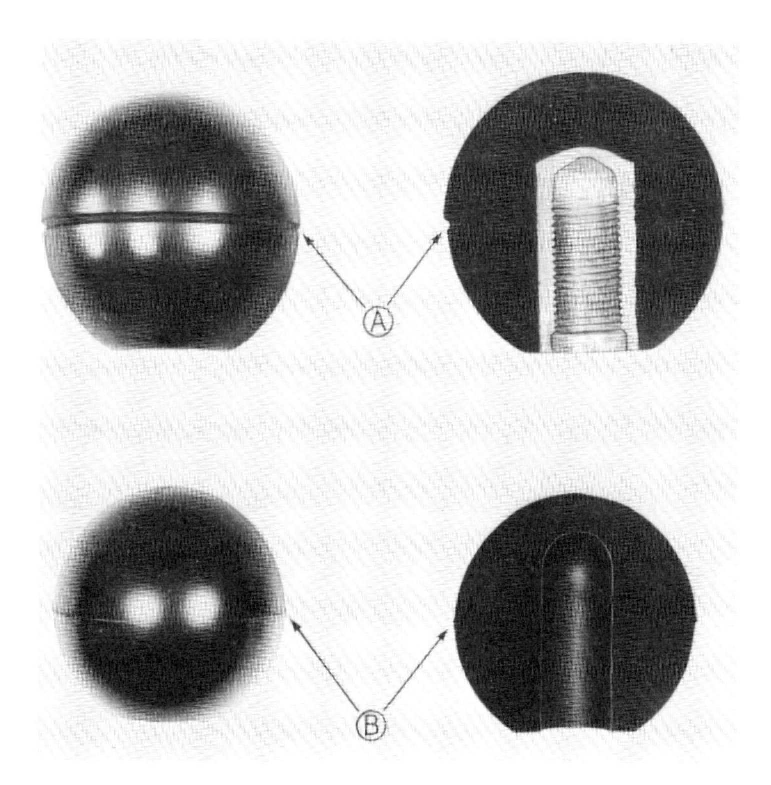

Figure 4–7. Two compression-molded phenolic knobs that have different types of parting lines. (A) This knob had a flush parting line and then was grooved. (B) This knob has a peaked parting line.

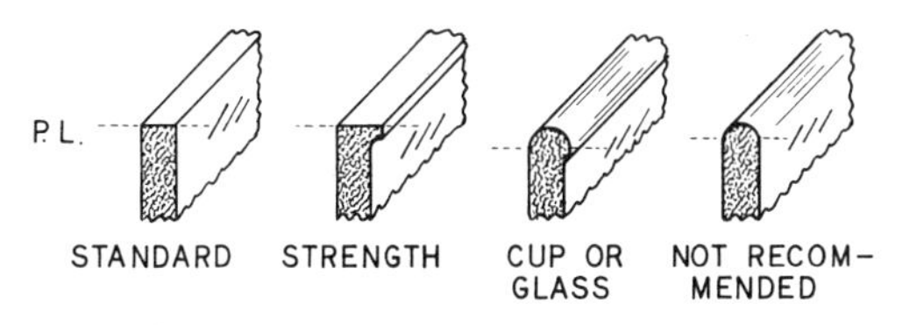

Figure 4–8. Recommended parting line designs for injection molding.

most often designed. If strength is needed, a slight edge may be added. If a drinking-glass or cup-type is desired, this design may be used. The parting of a die on a radius is not recommended.

PARTING LINE MISMATCH

Repeated opening and closing of molds will cause them to wear. Wear between the plunger and cavity pins and guide pin bushings causes the misalignment. The excessive wear at mold parting lines can create a mismatch, on a molded part, that appears greatly exaggerated in Fig. 4–9. A mismatch at a parting line of a few thousandths of an inch may appear to look many times larger. Tolerances for misalignment of cavity and plunger should total 0.006 in. for the average mold. The misalignment tolerance of 0.006 in. is the maximum misalignment that may be expected during the normal life of the mold. When parts are first produced from the mold, the tolerance due to

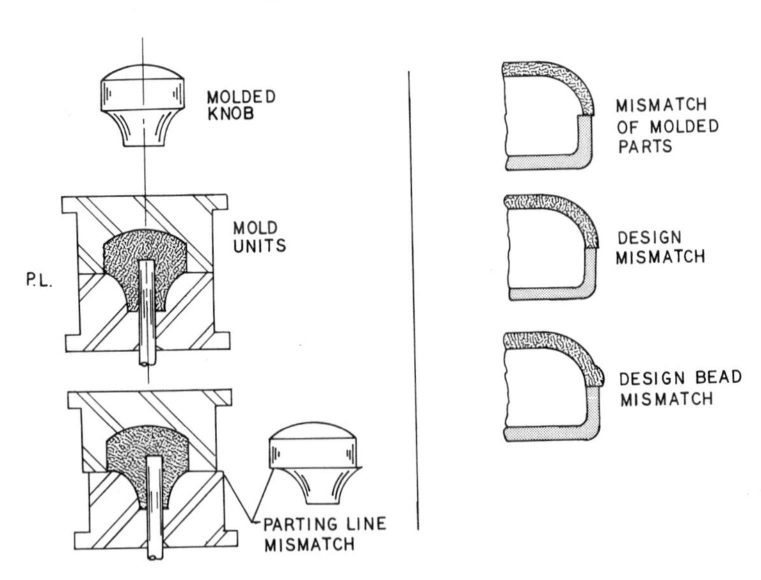

Figure 4–9. Excessive wear at a mold parting line can create a mismatch on the molded part. Mismatch can be used in the design of two molded parts.

misalignment will be much less. A greater tolerance should be allowed for molds producing items as large as TV cabinets.

PARTING LINE LIP DESIGN

In order to increase the rigidity of such items as food containers molded from the less rigid thermoplastics, e.g., polyethylene, polypropylene, EVA, polyvinyl chloride etc., it is necessary to design the lip of the container to make it stiff. Fig. 4–10 shows a number of ways that this may be accomplished while still retaining a well-designed parting line.

TOLERANCES

Dimensional tolerances in plastic molded articles will be considered as allowable variations, plus and minus, from a nominal or mean dimension, as used or set by the plastic industry. Every critical dimension should show the nominal dimension plus acceptable high and low limits. The overall tolerances should be the general rule, with uncompromising tolerances specified only where essential.

Variables to be considered when stating tolerances are: (1) mold maker's tolerances; (2) plastic-material shrinkage variances; and (3) molding process techniques. Dimensional tolerances are affected by several production and tooling variables such as the number and size of cavities and the degree of control to the molding operation. Overall tolerances for a part should be shown in inches per inch, not in fixed values.

Holding extreme accuracy of dimensions in molded parts is expen-

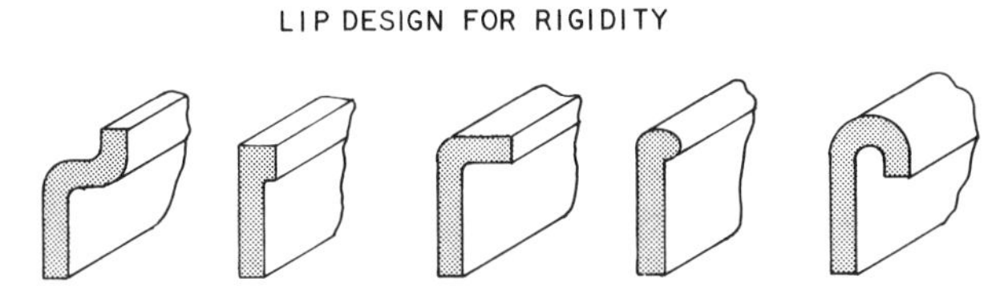

Figure 4–10. The lip on a plastic container can be made more rigid by using one of the above designs.

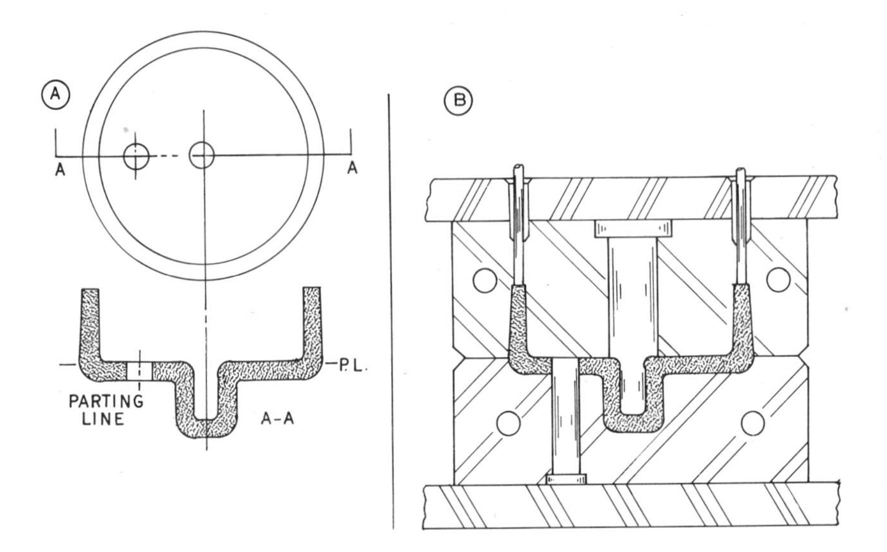

Figure 4–11. (A) A cross-section of a round molded part which has one molded through-hole. The parting line of the die is indicated. (B) A cross-section showing the molded part and the die that made the plastic part. Note that the parting line is placed near the bottom of the piece.

sive. Extremely close tolerances increases the initial cost of the mold. The designer should note that dimensional changes due to temperature variations alone can be three to four times as great as the specified tolerances. Parts should not be measured immediately after molding; they should be allowed to reach equilibrium under the stated environment. As the number of cavities in a mold increases, so must the tolerances on critical dimensions.

STANDARD TOLERANCES ON MOLDED ARTICLES

The table on standard tolerances (Table 4–1) was prepared by the Custom Molders of The Society of Plastics Industry. This table is to be used only as a guide. The dimensions are based on a hypothetical molded article with a cross-section shown in Fig. 4–11a. Fig. 4–11b explains the cross-section of the molded article along with a mold

TABLE 4–1. STANDARD MOLDING TOLERANCES FOR POLYCARBONATE PLASTIC MOLDING MATERIAL.

STANDARDS AND PRACTICES OF PLASTICS CUSTOM MOLDERS

Engineering and Technical Standards

POLYCARBONATE

NOTE: The Commercial values shown below represent common production tolerances at the most economical level. The Fine values represent closer tolerances that can be held but at a greater cost.

Drawing Code	Dimensions (Inches)	Plus or Minus in Thousands of an Inch: 1 2 3 4 5 6 7 8 9 10 11 12 13 14 15 16 17 18 19 20 21 22 23 24 25 26 27 28
A = Diameter (see Note #1)	0.000	Fine, Commercial
	0.500	
	1.000	
	2.000	
B = Depth (see Note #3)	3.000	
	4.000	
C = Height (see Note #3)	5.000	
	6.000	

Drawing Code	Dimensions (Inches)	Comm. ±	Fine ±
	6.000 to 12.000 for each additional inch add (inches)	.003	.0015
D=Bottom Wall (see Note #3)		.003	.002
E = Side Wall (see Note #4)		.003	.002
F = Hole Size Diameter (see Note #1)	0.000 to 0.125	.002	.001
	0.125 to 0.250	.002	.0015
	0.250 to 0.500	.003	.002
	0.500 & Over	.003	.002
G = Hole Size Depth (see Note#5)	0.000 to 0.250	.002	.002
	0.250 to 0.500	.003	.002
	0.500 to 1.000	.004	.003
Draft Allowance per side (see Note #5)		1°	½°
Flatness (see Note #4)	0.000 to 3.000	.005	.003
	3.000 to 6.000	.007	.004
Thread Size (class)	Internal	1B	2B
	External	1A	2A
Concentricity (see Note #4)	(T.I.R.)	.005	.003
Fillets, Ribs, Corners (see Note #6)		.015	.015
Surface Finish	(see Note #7)		
Color Stability	(see Note #7)		

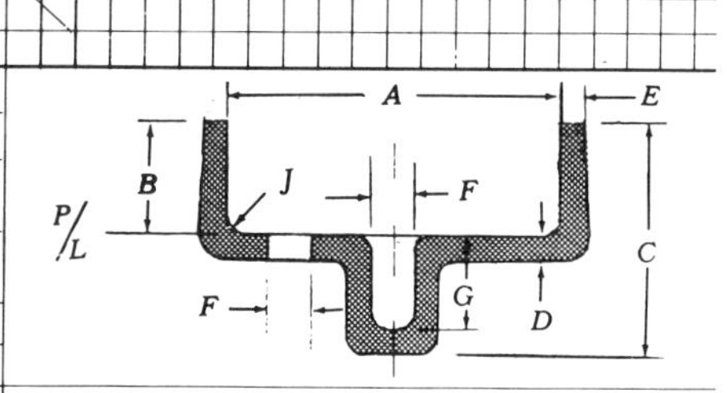

REFERENCE NOTES

1 – These tolerances do not include allowance for aging characteristics of material.

2 – Tolerances based on ⅛" wall section.

3 – Parting line must be taken into consideration.

4 – Part design should maintain a wall thickness as nearly constant as possible. Complete uniformity in this dimension is impossible to achieve.

5 – Care must be taken that the ratio of the depth of a cored hole to its diameter does not reach a point that will result in excessive pin damage.

6 – These values should be increased whenever compatible with desired design and good molding technique.

7 – Customer-Molder understanding necessary prior to tooling.

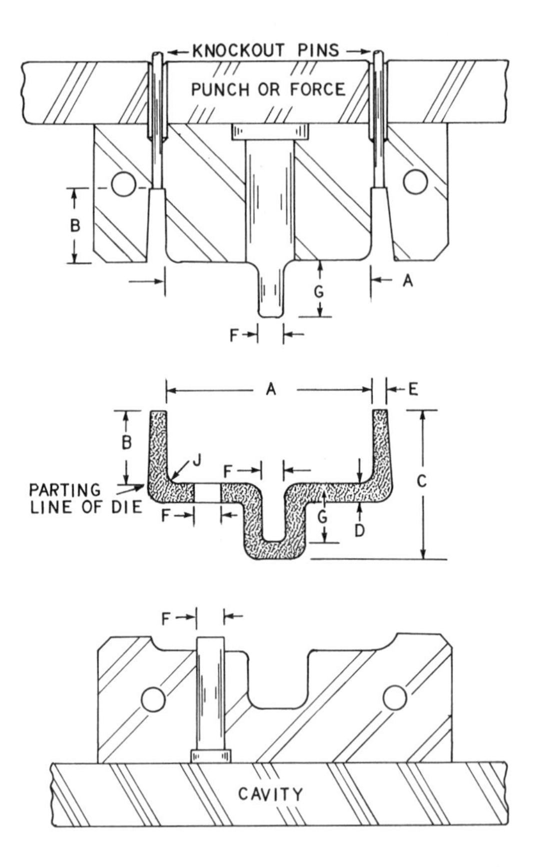

Figure 4–12. This drawing shows each dimension on the part and the corresponding dimension of the die.

that would make the plastic part. Fig. 4–12 illustrates the mold dimensions and part dimensions. The following example shows the reader how to use the table. A round polycarbonate injection molded part is considered in the Table 4–1. Fine tolerances represent the narrowest possible limits of variation obtainable under close supervision and control of production. Commercial tolerances will be that which can be held under average conditions of manufacture. Table 4–2 indicates dimensions given the molded part and tolerances for these dimensions taken from Table 4–1. Additional tables on all

TABLE 4–2. TOLERANCES FOR A POLYCARBONATE MOLDED PART.

Dimensions of the Molded Part	Tolerances (in.) Plus or Minus	
	Fine	*Commercial*
A. 5 in. (made by punch only)	.005	.008
B. 2 in. (made by punch only)	.003	005
C. 4 in. (made by cavity and punch)	.005	.007
D. 0.100 in. (made by cavity and punch)	.002	.003
E. 0.100 in. (made by punch only)	.002	.003
F. 0.250 in. (made by mold pin in cavity or punch)	.0015	.002
G. 0.750 in. (made by punch only)	.003	.004
J. Fillets, Ribs, Corners	.015	.015
Draft allowance per side	1/2 degree	1 degree
Flatness - 5 in.	.004	.007
Concentricity (T. I. R., True Inside Radius)	.003	.005

types of molding materials can be obtained from plastic material companies.

WALL THICKNESS

A non-uniform wall thickness will cause more trouble than any other problem in part design. A thick section will cool last and sink away from the mold, causing a "sink mark." Heavy sections mean long cycle times. Fundamentals of design with any material require that the wall sections be of adequate thickness for the application of the part and be shaped for adequate strength.

Wall thickness should be as uniform as possible to eliminate internal stresses, part distortion, and cracking (Fig. 4–13a). If different wall thickness in the part cannot be eliminated, the wall intersections should be blended gradually (Fig. 4–13b). A good rule to remember is that the thicker the wall, the longer the part will have to stay in the mold in order to cure or to cool properly. This rule is important when molding by compression or injection.

Tables 4–3 and 4–4 give preferred minimum, average, and maximum wall thicknesses for the thermoplastics and thermosetting plastics.

TABLE 4–3. SUGGESTED WALL THICKNESSES FOR THERMOPLASTIC MOLDING MATERIALS.

Thermoplastic Materials	*Minimum (Inches)*	*Average (Inches)*	*Maximum (Inches)*
Acetal	.015	.062	.125
ABS	.030	.090	.125
Acrylic	.025	.093	.250
Cellulosics	.025	.075	.187
FEP fluoroplastic	.010	.035	.500
Nylon	.015	.062	.125
Polycarbonate	.040	.093	.375
Polyester T. P.	.025	.062	.500
Polyethylene (L.D.)	.020	.062	.250
Polyethylene (H.D.)	.035	.062	.250
Ethylene vinyl acetate	.020	.062	.125
Polypropylene	.025	.080	.300
Polysulfone	.040	.100	.375
Noryl (modified PPO)	.030	.080	.375
Polystyrene	.030	.062	.250
SAN	.030	.062	.250
PVC-Rigid	.040	.093	.375
Polyurethane	.025	.500	1.500
Surlyn (ionomer)	.025	.062	.750

TABLE 4–4. SUGGESTED WALL THICKNESS FOR THERMOSETTING MOLDING MATERIALS.

Thermosetting Materials	*Minimum Thickness*	*Average Thickness*	*Maximum Thickness*
Alkyd - glass filled	.040	.125	.500
Alkyd - mineral filled	.040	.187	.375
Diallyl phthalate	.040	.187	.375
Epoxy glass	.030	.125	1.000
Melamine - cellulose filled	.035	.100	.187
Urea - cellulose filled	.035	.100	.187
Phenolic - general purpose	.050	.125	1.000
Phenolic - flock filled	.050	.125	1.000
Phenolic - glass filled	.030	.093	.750
Phenolic - fabric filled	.062	.187	.375
Phenolic - mineral filled	.125	.187	1.000
Silicone glass	.050	.125	.250
Polyester premix	.040	.070	1.000

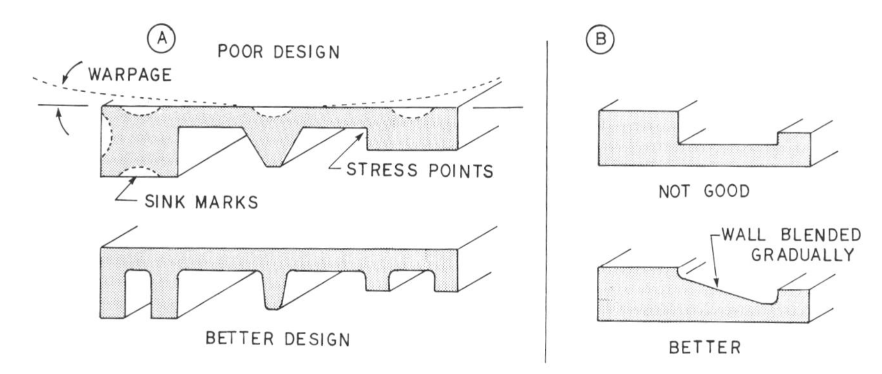

Figure 4–13. (A) Wall thickness in molded parts should be as uniform as possible to eliminate warpage, sink marks, and internal stresses. (B) If different wall thicknesses in a part cannot be eliminated they should be blended gradually.

VARIANCES IN WALL THICKNESS

As was pointed out, wall thicknesses should not be too heavy, nor should they vary greatly. Walls should not vary more than a ratio of three to one. This is true if the part is molded by either the compression or injection method and should be avoided with transfer molding as well.

When parts are made by compression or transfer molding, the parts will remain in the mold while they are being cured. A longer cure is required for a heavy section than for a thin one. In the process of curing a heavy section of a thermosetting plastic part, the thin section may become so overheated that it will discolor, especially with urea and melamine materials.

Parts that are molded by the injection method remain in the mold until they become hard and cool. A longer time is required to cool and harden a heavy section than a thin section. If the heavy section can be reduced in thickness, the mold can produce more and better parts. Whenever a heavy section is encountered in a plastics design, that heavy section should be redesigned (Fig. 4–14). A uniform section will speed up production and give a more satisfactory part, regardless of whether thermoplastics or thermosetting materials are used.

If parts are designed with thick and thin sections, sink marks will be evident on the thick sections. This is true with any type of molding. If translucent or transparent colored materials are used, there will be a variance in depth of color with even the slightest variance in wall thickness. Many times, a part design will call for a thick, heavy, solid section such as are found in knobs or handles. The heavy thick sections are difficult to mold and have long cure cycles. Long cure cycles may be eliminated by molding and assembling two individual moldings to produce the desired part (Fig. 4–15).

TAPER

A taper is a slight draft angle in a mold wall designed to facilitate the removal of the molded part from the mold. Plastic parts designed to be produced by the molding processes must have taper or draft on all surfaces perpendicular to the parting lines of the die. Draft should be provided, both inside and outside (Fig. 4–16a). Plastic materials tend to shrink tightly around core pins and the plunger of the mold. In order to remove the molded part, adequate taper must be used. Parts actually have been molded with a back taper, sometimes called backdraft, or undercut.

The degree of taper will vary according to the molding process,

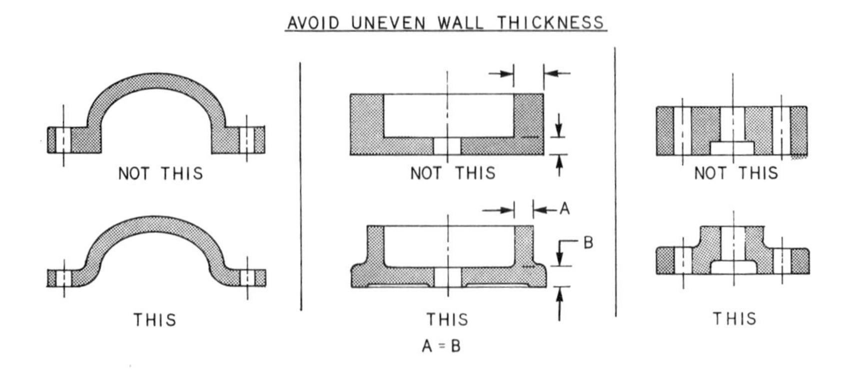

Figure 4–14. If a plastic part is designed with uneven wall thickness, it should be redesigned as shown in parts above.

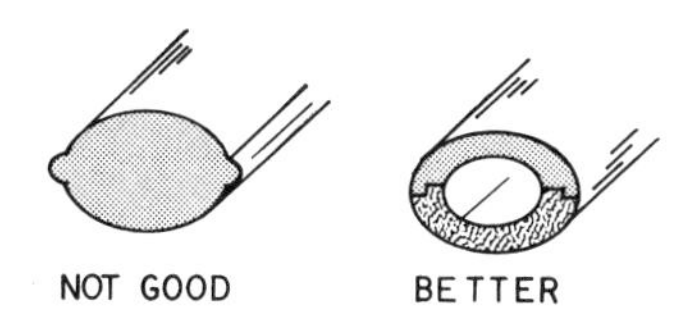

Figure 4-15. Thick, heavy, solid sections such as are found in knobs and handles should be redesigned into two individual moldings.

wall thickness of the part, and the molding material being used. There are no precise calculations or formulas for taper. A designer should get or use the most liberal taper or draft that the part will permit. A minimum of 1/2 degree taper per side is generally adequate, although more taper is desirable for certain high-production parts. Most phenolic materials can be molded with a minimum taper of 1/2 degree, but a 1 degree draft will give better results. Plastics parts molded from thermoplastics require tapers ranging from 1/2 to 3 degrees.

If a taper is placed on only one side of the part, it will be more difficult to mold than if it occurs on both the inside and the outside. If taper or draft is on the inside of the part only (Fig. 4-16b), the part may stay in the cavity as the mold opens. In Fig. 4-16c, the

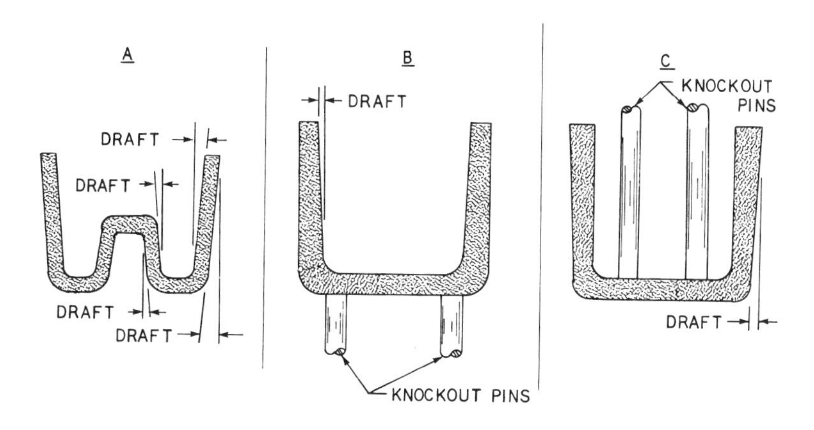

Figure 4-16. (A) Because plastic materials shrink during molding, draft is generally required both inside and outside of the part. (B) When draft is placed in the plunger only, the part will stay in the cavity. Knockout pins will then be required in the cavity. (C) If draft is placed on the outside of the part only, the part will stay in the plunger. Knockout pins will then be required in the plunger.

molded part has draft on the outside. This will cause the part to stay on the plunger as the mold opens. Knockout pins in the plunger will be required to remove the part. A stripper plate (Fig. 4–17a) can be used to remove the part from the plunger. In complete automation of molded parts, it is necessary to have knockout pins on both the cavity and plunger of the mold. This is called "positive ejection" (Fig. 4–17b).

Bosses and ribs should have a draft of 5 degree per side if they are molded by the bottom of the cavity or bottom of the plunger alone. If the boss or rib can be molded by the side of both the cavity and the plunger (Fig. 4–17c), only one side need have a draft of 5 degrees. The other side takes the draft of the part itself.

EJECTOR MECHANISMS

The ejector mechanism releases the molded plastic part from the metal mold. Ejector systems are a component part of the mold and are activated during the opening of the press platens. The design of the ejector is determined primarily by part geometry. The most com-

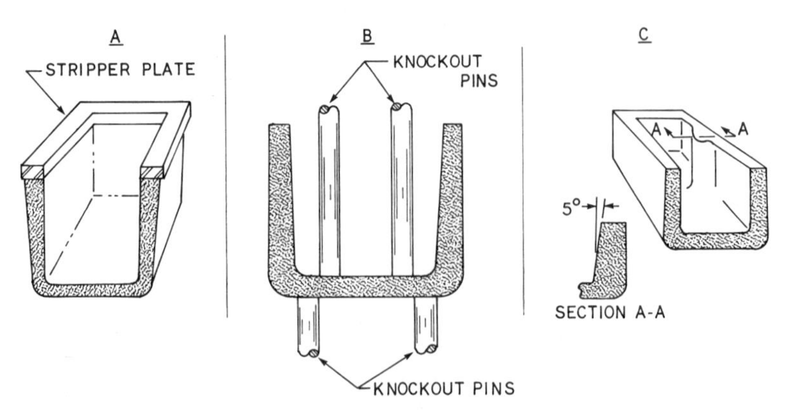

Figure 4–17. (A) A stripper plate instead of knockout pins may also be used to remove the part from the plunger. (B) In complete automation of molded parts it is necessary to have knockout pins on both the cavity and the plunger. This is called positive ejection. (C) When a boss or rib is molded by both the cavity and plunger, i.e., on the side of the part, then one side should have a draft of 5 °. The other side takes the taper of the part itself.

monly used ejecting systems are ejector or knockout pins, stripper plates or rings, and compressed air.

Ejector pins should be located at the thickest possible sections, preferably directly over bosses or ribs, as shown in Fig. 4–18. Knockout pins are generally recessed approximately 0.015 in. into the plastic part, in order not to have any surface extend above the wall. Sometimes knockout pins are also used to mold holes or as a mold pin. A blade-type knockout pin is not recommended, because it is difficult to machine a slot in a mold for a blade, and the mold maintenance is costly. It is good practice to place a circular boss on a narrow rib so the knockout pin will have an adequate surface to push against.

Knockout pins may be located on the runners as well as the molded part (Fig. 4–19). Because knockout pins move upward or downward in the cavity or plunger when they eject parts from the mold, wear between the knockout pin and its adjacent mold section will occur. The small space between pin and mold section may fill with an irregular cylinder of flash extending above the knockout depression. This flash may be removed by tumbling, but a slight ridge will always be left around the depression. Many parts are designed to stay in the cavity section when the mold opens. Normally, knockout

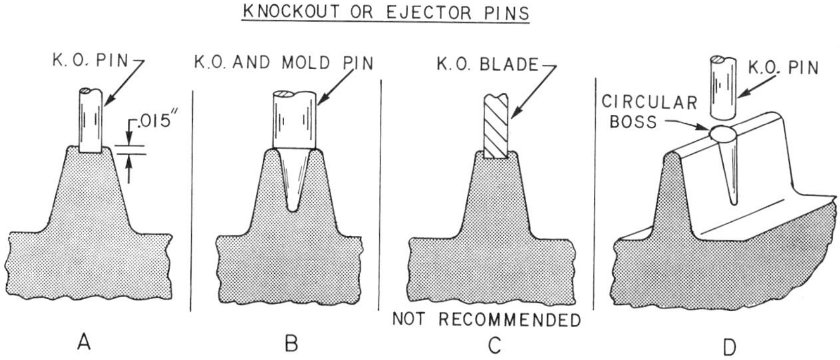

KNOCKOUT OR EJECTOR PINS

K. O. PIN

.015″

A

K.O. AND MOLD PIN

B

K. O. BLADE

NOT RECOMMENDED

C

K.O. PIN

CIRCULAR BOSS

D

Figure 4–18. (A) Knockout pins should be recessed into the plastic surface approximately 0.015 in. (B) Knockout pins are sometimes used as mold pins. (C) Blade type knockout pins are not recommended. They are expensive and difficult to maintain. (D) All knockout pins should have an adequate surface to push against.

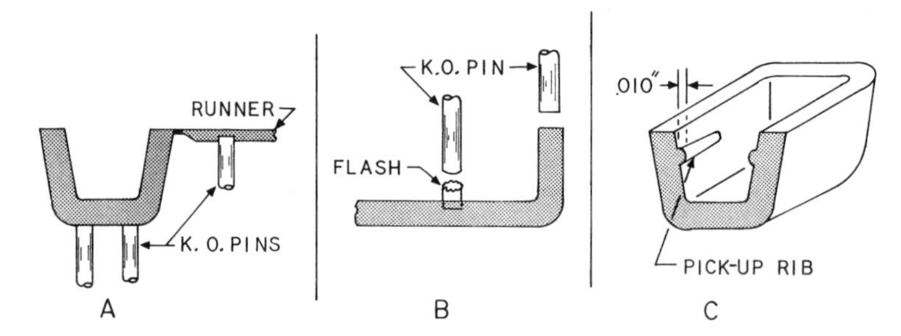

Figure 4–19. (A) Knockout pins are sometimes placed on runners. (B) Wear between the knockout pin and the mold will cause flash to occur around the pin. (C) Pick-up ribs are frequently used to make the molded part remain in the cavity or on the punch.

pins in this case are provided in the cavity. If such a design will not allow knockout marks on the outside of the part, pick-up ribs may be placed on the inside of the part. Pick-up ribs measuring 0.010 in. in radius will be of adequate size to remove the part from the cavity.

If knockout marks are objectionable, a stripper plate may be used to remove the part from the mold. When the part is designed to stay on the plunger as the mold opens, the stripper plate is placed on the plunger. To remove the part, the stripper plate is moved off the plunger, thus pushing the part before it (Fig. 4–20). A stripper plate exerts a more even pressure on the molding than ejector pins and results in less distortion. Ejector sleeves are preferred when the molded

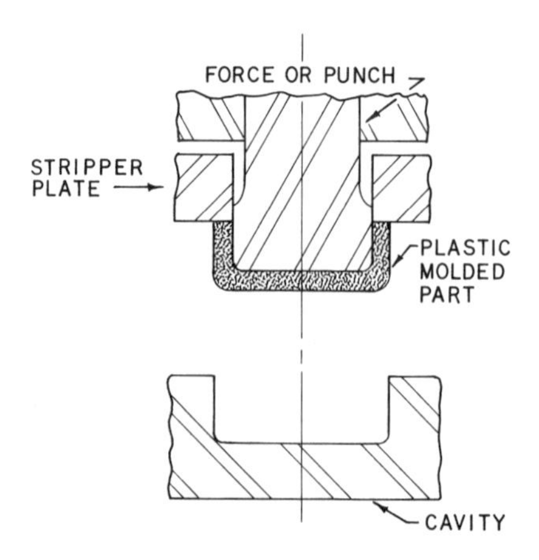

Figure 4–20. Stripper plates exert a more even pressure on the molding than ejector pins and result in less distortion.

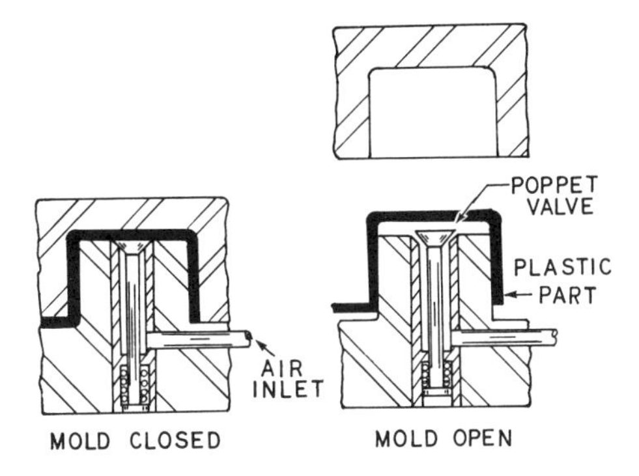

Figure 4–21. Molded parts that have thin walls and deep draws can be ejected from the mold by using air poppet valves.

parts have to be stripped off of one or more cores. Ejector sleeves are subjected to severe stress and wear. Scoring of both cavity and core may take place. The outside diameter of the sleeve should be held to 0.001 to 0.002 in. smaller than the hole in the cavity.

Deep drawn parts can often be ejected by air. Also, air poppet valves are used on parts with thin walls and deep draws. The valves should seat properly to seal air and to prevent molten plastic from flashing under them (Fig. 4–21).

5

Molded Holes and Undercuts

MOLDED HOLES

Molded holes or openings in plastic parts are used for a variety of purposes. Holes are provided to allow assembly with other parts, to decorate the part and give it more eye appeal, or are functional, such as ventilators or louvers.

Holes or openings can be round, square, rectangular, elliptical, etc. (Fig. 5–1); however, for the purpose of this chapter, round holes are discussed. Molded holes may be classified as blind, through, step, recess step, and intersecting (Fig. 5–2).

HOLES PARALLEL TO THE DRAW

A hole parallel to the draw is a hole whose axis is parallel to the movement of the mold as it opens and closes. Holes may also be molded at right angles to the draw and at oblique angles (Fig. 5–3). Oblique-angle holes should be avoided if possible, because it is very difficult and expensive to make a mold to operate at oblique angles, as they require split dies and retractible core pins.

The main point to bear in mind regarding any molded hole is that the hole is made by a pin that is inserted into the mold. The pin is subject to breakage and wear. Holes may extend part way through the molded piece, in which case the steel pin making the hole is sup-

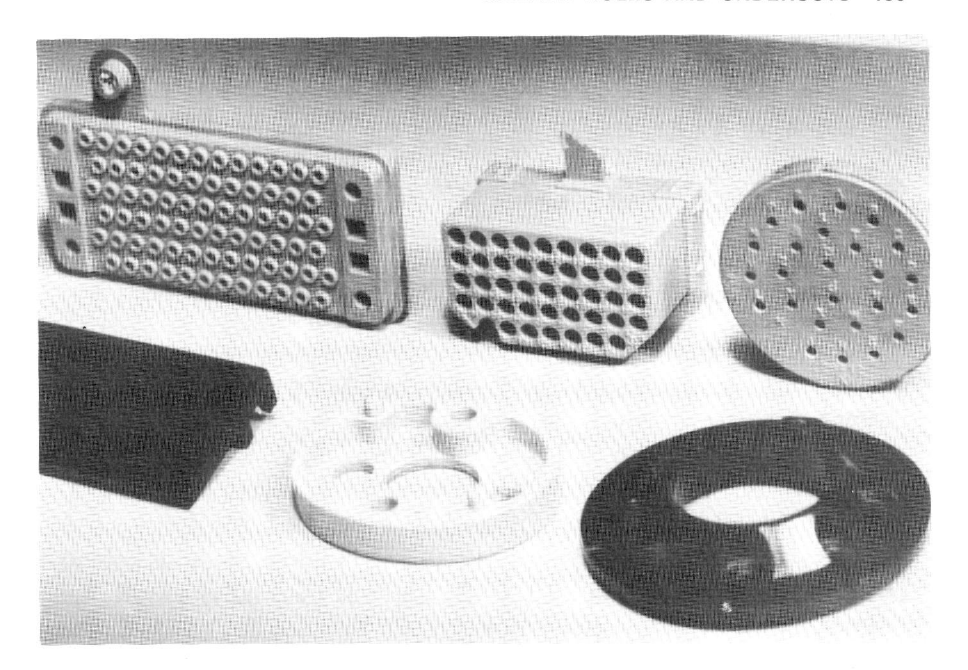

Figure 5–1. A group of transfer molded electronic parts. Note the many different sizes and shapes of molded holes.

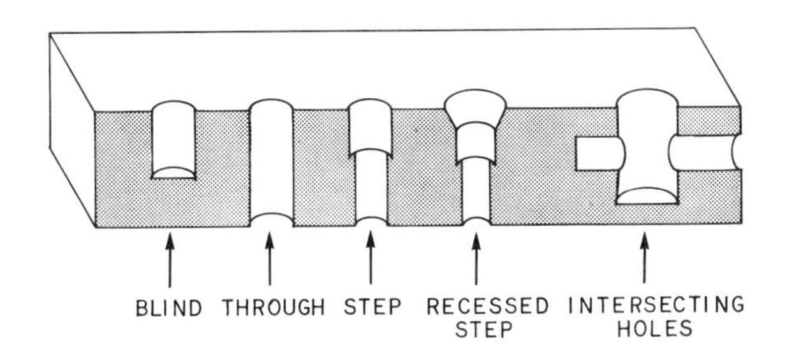

Figure 5–2. Molded holes in plastic parts may be classified as shown in the above drawing.

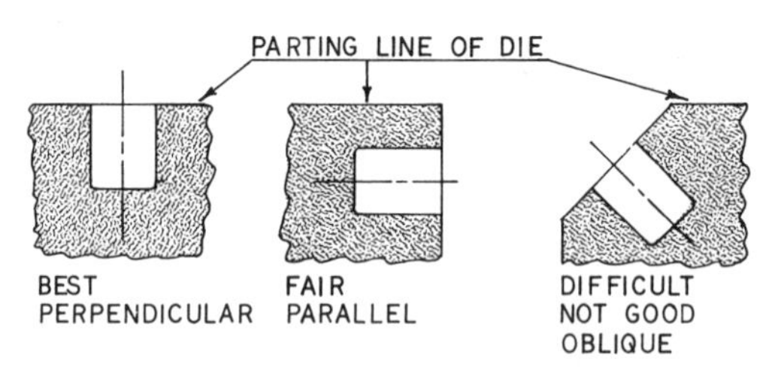

Figure 5–3. Holes in plastic parts may be molded perpendicular to the draw, parallel to the draw, and at oblique angles. Oblique angles are not recommended.

ported only at one end (Fig. 5–4a). Holes may extend entirely through the piece, in which case the pin may or may not be supported at both ends (Fig. 5–4b and 5–4c). The strength of the pin making the hole is influenced by the ratio of its mean diameter to its length called its "slenderness ratio." Fig. 5–5 lists the slenderness ratios for the average holes, using the various types of molding methods, when these holes are molded parallel to the draw. It is usually possible to follow the depth-to-diameter ratios given in Fig. 5–5 for molded side holes.

In some cases, small-diameter, supported at one end and molded by the compression method, these holes should be no longer than their diameter (Fig. 5–6). If small holes are molded by transfer or injection, the slenderness ratio given in Fig. 5–5 should apply.

Through holes, made with a pin supported at both ends (Fig. 5–4b), are not always as practical from a molding standpoint as are

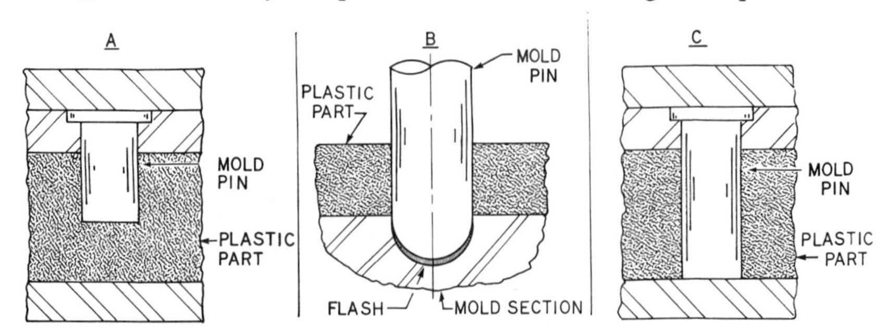

Figure 5–4. (A) Molded hole made by pin supported only at one end. (B) Hole molded through a part by a pin supported at both ends. This is considered poor design in compression molding. (C) Hole molded through a part by the pin supported at one end. This method is most frequently used in molding holes.

LENGTH OF HOLES PARALLEL TO THE DRAW

MOLD PIN
PLASTIC PART
A
B

	ONE PIN
COMPRESSION	A = 2×B
TRANSFER	A = 6×B
INJECTION	A = 6×B

	TWO PINS
COMPRESSION	A = 6×B
TRANSFER	A = 15×B
INJECTION	A = 15×B

Figure 5–5. This drawing lists the slenderness pin ratios for average molded holes, using the various types of molding methods.

holes made by butting the pins. In the case of compression and transfer molding, the recess for a pin supported at both ends usually contains a small particle of flash prior to a new molding cycle. If the pin entering the recess butts against this hard flash, the pin may break. If a hole made by a butting pin is to be used in assembling the part with some other part, namely a bolt or shaft, provision should be made for the misalignment of cavity and plunger and the possibility of the

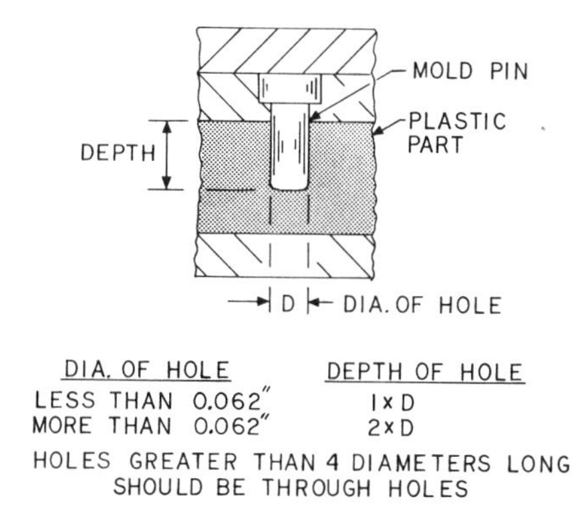

Figure 5–6. Small diameter holes made by compression molding should use the proportions in the above drawing.

pin's bending. If either misalignment or bending occur, the companion part may not go through the hole. Therefore, the pins used in molding butt holes should differ in radius by at least 0.015 in. (Fig. 5–7a).

In order to compensate for the misalignment of cavity and plunger sections, good practice calls for a molded hole to be made by using two different diameter pins. Some designers may prefer to telescope the pin, thus making the hole shown in Fig. 5–7b. It must be pointed out that in the case of compression moldings, flash may occur where the pins have been butted together. Some of the flash may remain on the pin having the recess. When the mold closes, this may cause the pins to bend and break.

Deep holes may be molded by building the holes up in steps, as shown in Fig. 5–7c. The slenderness ratio, however, should be considered as resulting from the ratio of the mean diameter to the total length and should not exceed the slenderness ratios given in Fig. 5–5. Long holes of small diameter are difficult to mold. Sometimes deep holes are molded for a short length or spotted, and then drilled after molding. Fig. 5–8 is a picture of a fuse holder for a high-voltage transformer. It is injection molded from fiberglass-filled nylon. Note the very complicated molded holes and side coring that was necessary to mold this part.

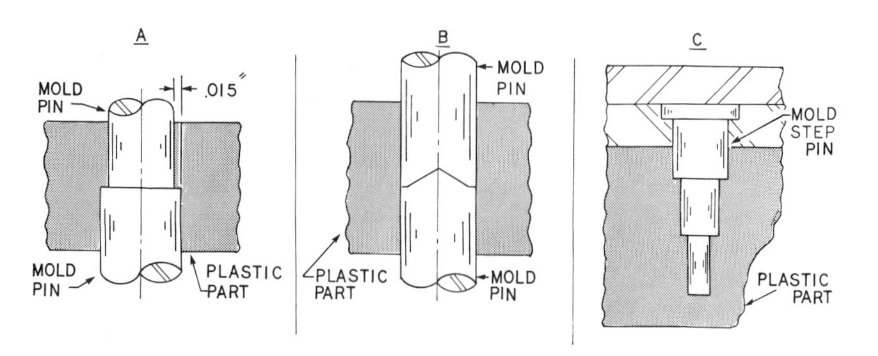

Figure 5–7. (A) A hole molded by the mold pins butted together. Good design requires pins of different diameters. (B) A hole molded by two pins telescoping together. (C) A deeper hole can be molded with a step pin.

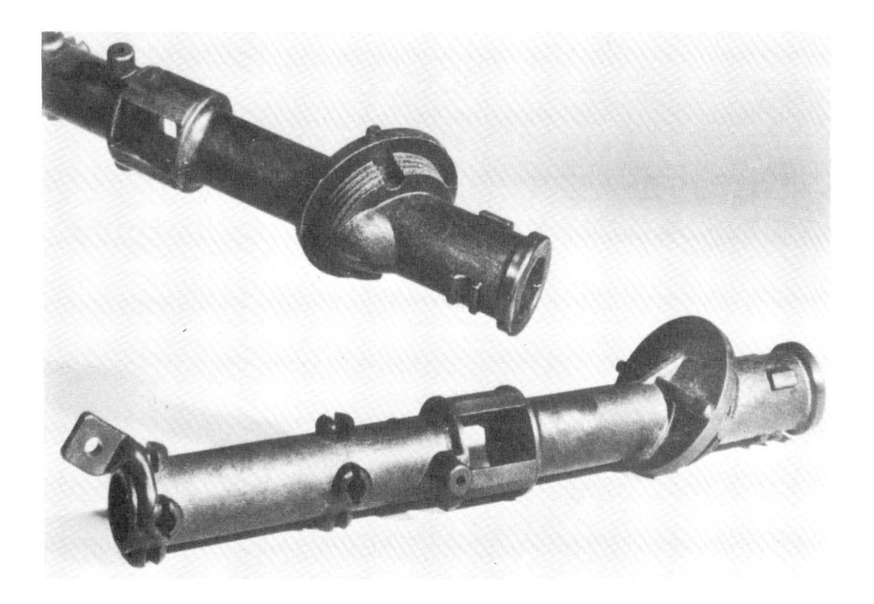

Figure 5–8. A fuse holder for a high-voltage transformer. The holder was injection molded from fiberglass-filled nylon. Note the complicated molded holes.

NEARNESS OF HOLES TO EACH OTHER AND SIDE WALL

Because both the thermoplastic and thermosetting plastics flow and knit inside the mold, strains are set up within the part. The flow of the compound around a pin making a hole and the welding of that compound on the other side of the pin are causes for strain lines.

As has been mentioned previously, wall thickness should be kept fairly uniform. Not only will non-uniform wall thickness cause inequalities in curing time, but the part will be more susceptible to warping at very thin wall sections when these sections are joined to much thicker wall sections. The recommended distances between holes and the distance of the hole from the side wall are presented in Table 5–1. If the holes differ in diameter, the distance between the holes may vary proportionately from that shown in the table. Fig. 5–9 demonstrates an excellently designed injection molded part with many holes. The part is a fire-extinguisher valve injection molded

TABLE 5–1. HOLE PROPORTIONS.

Recommended Distances Between Holes and Distances of Hole From Side Wall		
Hole Diameter In Inch	*Minimum Distance From Edge In Inch*	*Minimum Distance From Each Other In Inch*
0.062	0.093	0.140
0.093	0.109	0.187
0.125	0.156	0.250
0.187	0.218	0.312
0.250	0.250	0.437
0.312	0.312	0.562
0.375	0.343	0.875
0.500	0.437	0.875

Figure 5–9. An injection molded fire extinguisher valve made from Noryl. Note the many cored holes and female threaded sections.

from Noryl. It should be noted that all of the holes are parallel to each other or at right angles to each other. There are no oblique-angle holes in reference to the parting line.

As explained previously, transfer and injection molds require gates. If holes are molded, knit lines will occur on the opposite side of the holes from the gate and between the hole and the side wall. The distance between the hole and the side wall should be at least 0.125 in., if maximum mechanical strength is required.

In molding with most thermoplastic materials, the material between the wall of a hole and exterior wall of a part should be at least the thickness of the hole diameter (Fig. 5–10). It should never be less than 1/4 of the material thickness. The interior wall between holes should be at least one thickness of the hole diameter, and never less than 0.125 in.

If two holes are molded in a thermosetting material from opposite sides of the part, they should be no closer than 0.125 in., if the possibility of cracking between the sharp edges of the holes is to be avoided. If the thermoplastic materials are used, this distance may be as short as 0.062 in.

The bottom wall thickness, of a hole that is not molded through, should be at least 1/6 the diameter of the hole (Fig. 5–11a). If the wall at the bottom of the hole is less than 1/6 the diameter of the hole, the bottom will tend to bulge after molding (Fig. 5–11a). A bet-

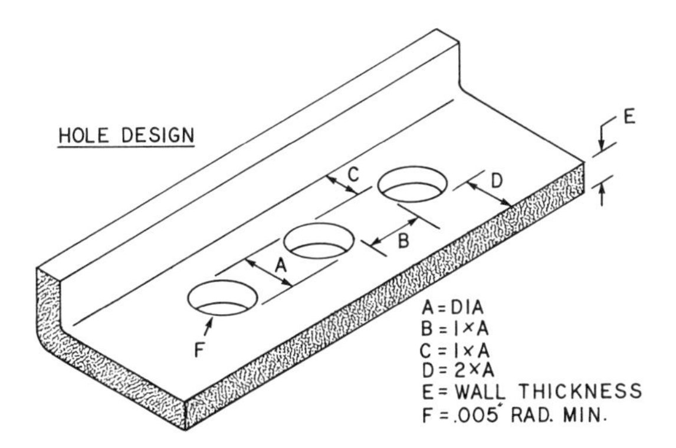

Figure 5–10. This drawing illustrates the relationship of the distance of molded holes from each other and from the side walls.

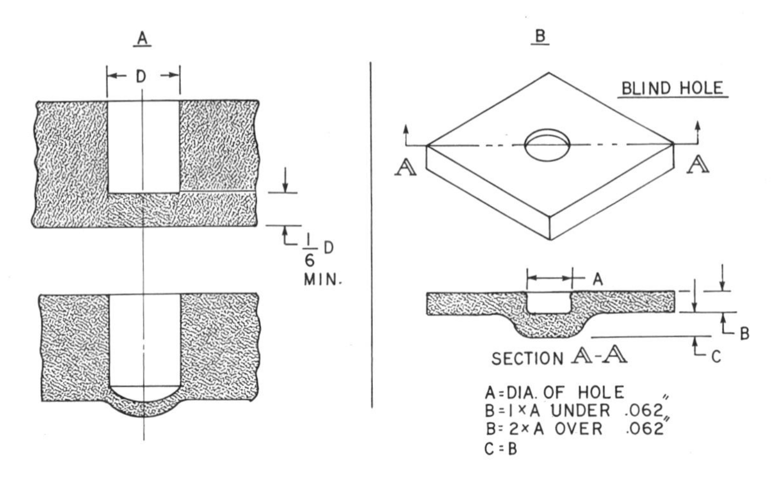

Figure 5–11. (A) Holes molded only part way through should have ample thickness at the bottom. (B) The wall thickness should be kept uniform for small blind holes.

ter design is shown in Fig. 5–11b. It will be noted that the wall thickness is uniform throughout, and there are no sharp corners for stress concentrations to develop.

MOLDING HOLES NOT PARALLEL TO THE DRAW

Whenever the axis of a hole is not parallel to the draw, either the pin making the hole must be removed from the part before the part is extracted from the mold, or the pin must be removed with the part as it is extracted and then taken from the part. Usually, molded holes not parallel to the draw require more complicated molds or more direct molding labor than do holes parallel to the draw, and thus, higher mold and parts costs result. Therefore, holes entering the side of the part should be avoided if possible.

Certain designs of side holes, however, will lend themselves to low-cost molding. Fig. 5–12a illustrates a side hole that can be molded without the necessity of the removal of the side pin, either before or after the part is extracted. This design may be impractical for some applications, because either one or both the upper and lower surfaces of the cavity and plunger sections of the mold making the hole must

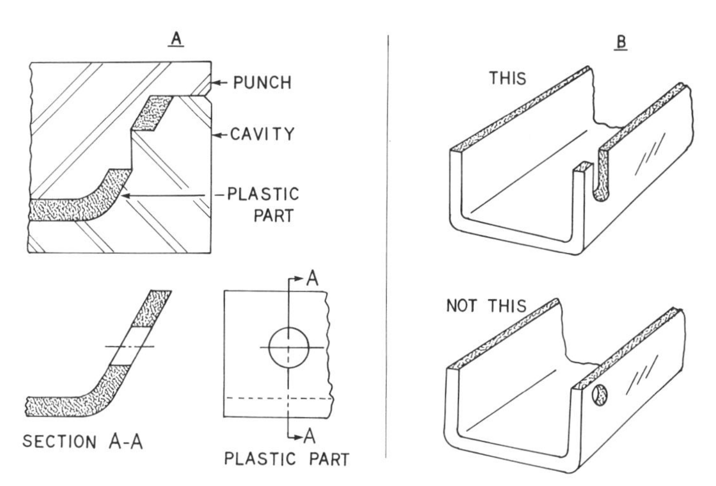

Figure 5–12. (A) Side holes or openings may be molded without the use of pull pins or loose lugs in certain applications by using a "kiss-off" type of die. (B) Extending side openings to the top of the part will lower costs.

slide along each other in order to obtain a straight draw of the part from the mold.

Another design point that is often overlooked by engineers is the low cost of molding that can be effected if a side hole is extended to the top of the part, thus facilitating a straight draw (Fig. 5–12b). This procedure is not practical in many designs, but it will lower price on parts and mold costs if it can be used. Molded-in side holes are troublesome to produce, because extra provision is required to actuate the core pins from the side. Side holes may be molded automatically by using the cam action in the mold or with hydraulic or pneumatic actuators. In compression molding, drilling after the part is molded is usually simpler.

DRILLED AND TAPPED HOLES

When holes parallel to the draw are too slender to be molded, it becomes necessary to drill these holes after the part has been molded (Fig. 5–13a). At times, it may be more economical to drill a side hole than to mold that hole. Good design calls for the spot location of a hole to be drilled (Fig. 5–13b). This spot acts as a guide for the drill entering the plastic. Spots should be made only for holes that are to be drilled parallel to the draw. A spot for a hole to be drilled perpendicular to the draw would be an undercut in the part and should not be used (Fig. 5–13b).

If through holes are drilled in the molded part it is good practice to note on the drawing that the hole may chip on the edge where the drill exits. It is best to avoid designing parts so that drilling must be done on an angled surface. Drills may break the surface or "walk" over the entering surface. Drilled holes should be so designed that the drill enters the part perpendicularly. Molded holes sometimes produce undesirable weld lines (Fig. 5–14). To overcome the weld lines the holes are molded two-thirds of the way through the wall then drilled the remainder to the distance. Frequently, molded holes are tapped. Holes to be tapped after molding or holes for self-tapping screws or drive screws should be countersunk to allow the tap or screw to find its way in and to prevent chipping at the entering end (Fig. 5–15).

Many molded plastic parts have holes drilled in the part after molding. It is generally less expensive to build a drill jig to drill holes in a molded part than it is to incorporate elaborate retractable core

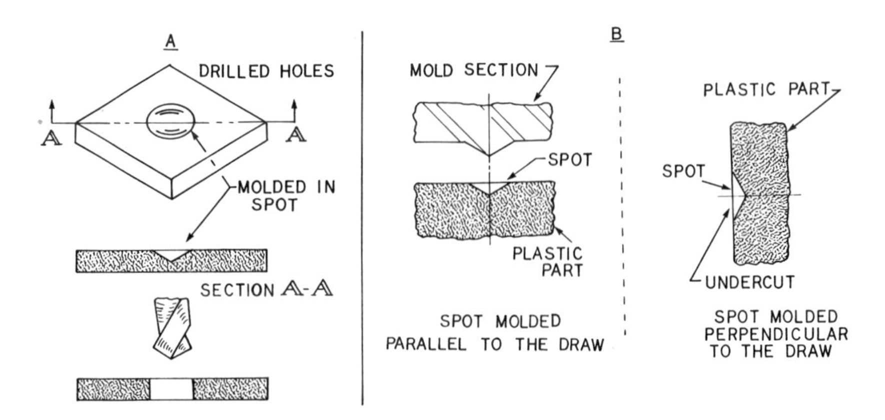

Figure 5–13. (A) It may be more economical to drill a hole than to mold it in the plastic part. Drilled holes reduce mold costs, eliminate weld lines, and reduce mold maintenance. (B) Good design calls for a spot location of a hole that is to be drilled. A spot should never be located perpendicular to the draw as this will constitute an undercut.

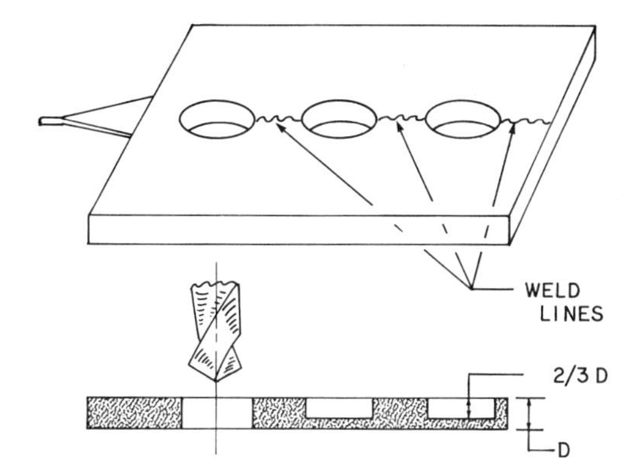

Figure 5–14. Weld lines created by molded holes can be eliminated by molding the holes two thirds of the way through and then drilling the remainder of the way.

pins in the die. Fig. 5–16 shows an electronic instrument case injection molded from high-impact polystyrene. The holes on the side of the case were drilled after molding. This was necessary because of frequent changes in the assembly of the part, in order to make the case more versatile for other uses.

A designer should not call for a perfect chamfer or radius at the open end of a hole, because this will call for extreme precision in the die. If a conical-head screw is not to extend above the surface of the plastic part, the hole for the screw should be designed with a 0.015 in. of vertical depth to allow for the variations in screws.

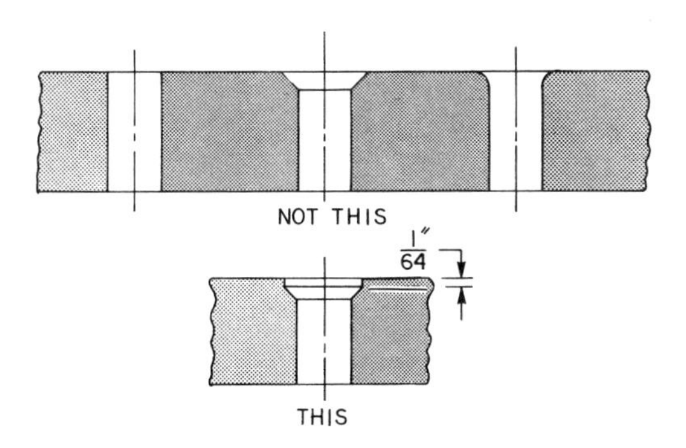

Figure 5–15. Holes to be tapped or used with self-tapping screws should be countersunk.

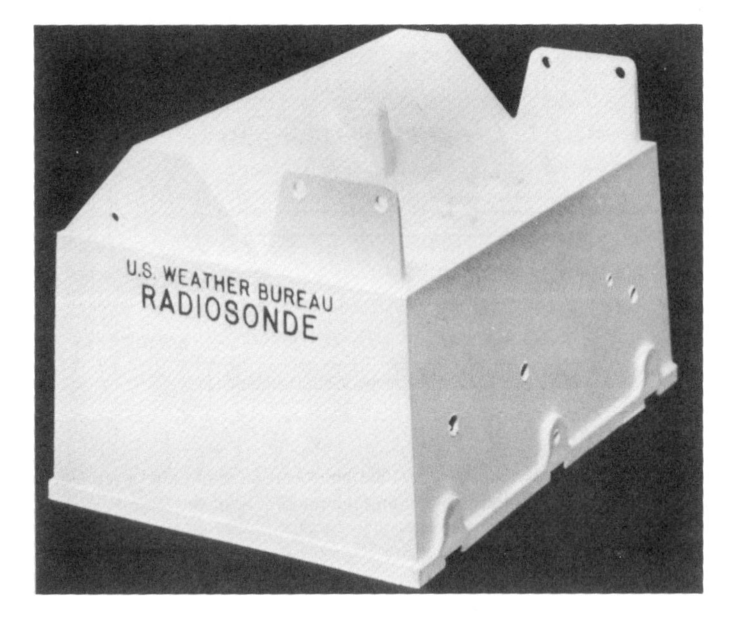

Figure 5–16. An electronic case injection molded from high impact polystyrene. The holes on the side of the case are drilled after molding. This was necessary because of frequent changes in the assembly of the part in order to make it more versatile.

Any molded hole that requires extremely tight tolerances may necessitate size development. This means that the metal mold pin is made oversize and then gradually reduced in size by removing metal from the pin until the exact dimensions are obtained in the molded part (Fig. 5–17a). Elongated holes may be used with plastic materials that have uncontrollable shrinkages such as the polyolefins (Fig. 5–17b).

UNDERCUTS

An undercut is an indentation or projection on a molded part which makes ejection from the simple two-part mold almost impossible. There is a very fine distinction between undesirable and impossible undercuts. Undercuts can be classified as internal undercut, external

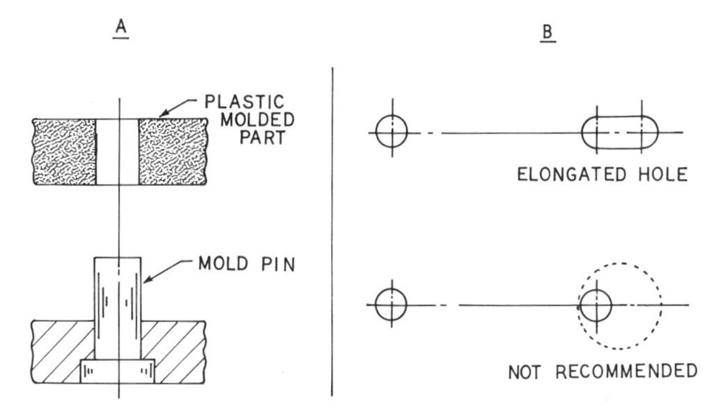

Figure 5–17. (A) Molded through holes with tight tolerances may require size development of the mold pin. The mold pin is made oversize and then reduced. (B) Elongated holes may be used for plastic materials that have uncontrollable shrinkages.

undercut, circular undercut and an undercut on the side wall of a part formed by a core pin (Fig. 5–18).

Undercuts are frequently necessary in a molded plastic part design. However, these should be avoided whenever possible, as they increase mold costs and parts prices and lengthen the molding cycle. Undercuts may be molded by means of split-mold cavity sections and by means of movable side cores that must be drawn away from

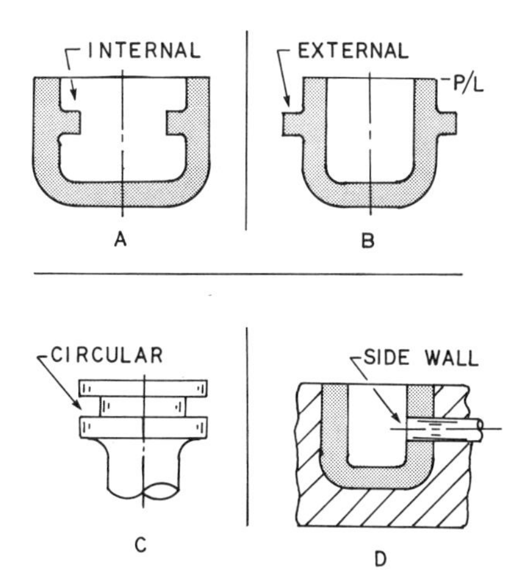

Figure 5–18. There are many different types of undercuts in molded plastic parts.

the part before the part can be extracted from the mold. Because wedges or pull pins must be used to mold undercuts, flash or parting lines will be evident where the movable sections meet or where they meet the fixed mold sections.

Internal undercuts (Fig. 5–18a) are impractical and expensive and should be avoided. Whenever undercuts are encountered, it is best to design the part in two halves and assemble the two parts after they have been molded. An internal undercut (Fig. 5–19a) can be produced by using a removable ejector wedge. This calls for elaborate tooling.

External undercuts (Fig. 5–18b) are located in the outside contours of the piece. It would be impossible to withdraw a piece of such a shape from a one-piece mold cavity. In order to mold such a part, it is necessary to build the cavities of two or more loose members. After the part has been molded, the loose members are parted, and the molded piece is removed. It must be remembered that a parting or flash line will be visible on the molded part.

Circular undercuts (Fig. 5–18c) frequently can be made less expensively by grooving the part in a lathe with a properly designed cutting tool.

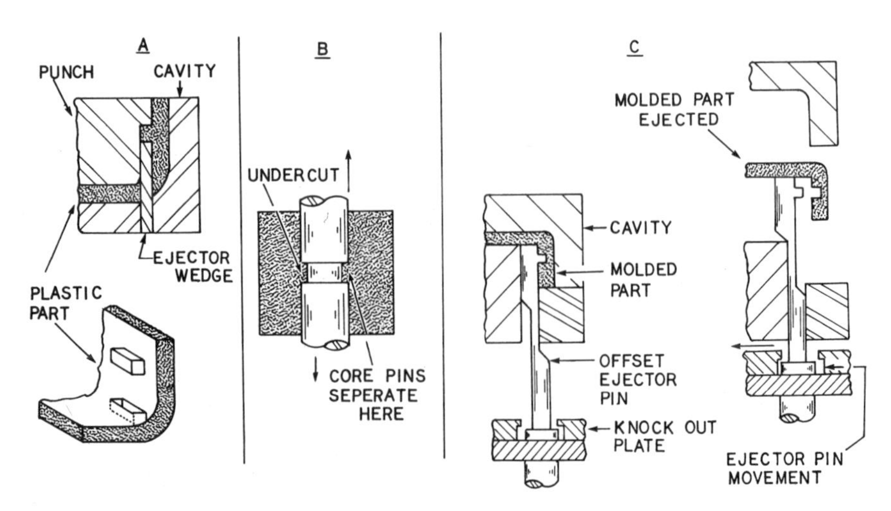

Figure 5–19. (A) An undercut in a molded plastic part produced by using a removable ejector wedge. (B) An internal undercut can be molded by using two separate core pins. (C) Offset or jiggler pins may be used for internal side wall undercuts or holes.

Undercuts in the side wall of a molded part (Fig. 5–18d) are produced by retractible core pins. Injection molds may be designed with cam-operated side cores for automatic molding. Side cores for transfer and compression molds are generally operated manually, although in some cases they are run by pressure cylinders on the side of the mold.

Internal undercuts can be molded by using two separate core pins, as shown in Fig. 5–19b. This is a very practical method to use, but flash will sometimes occur where the two core pins meet.

Offset or jiggler pins may be used for internal side wall undercuts or holes (Fig. 5–19c). This method of molding undercuts requires an increase in the tool costs and mold maintenance.

Some molded parts may be stripped from the mold without damage to the molded piece. (Fig. 5–20). All plastic materials, however, do not lend themselves to a stripping operation. It is advisable to contact the material companies for specific tolerances given for stripping. Most flexible thermoplastic materials can tolerate a 10% strain in the mold ejection and not encounter permanent deformation.

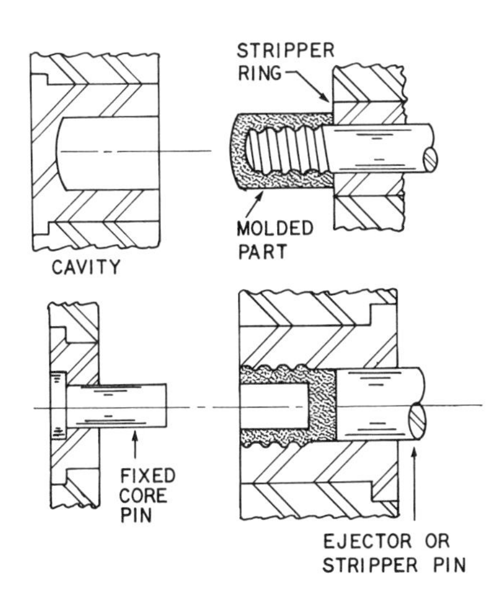

Figure 5–20. Many thermoplastic materials can be stripped from a mold.

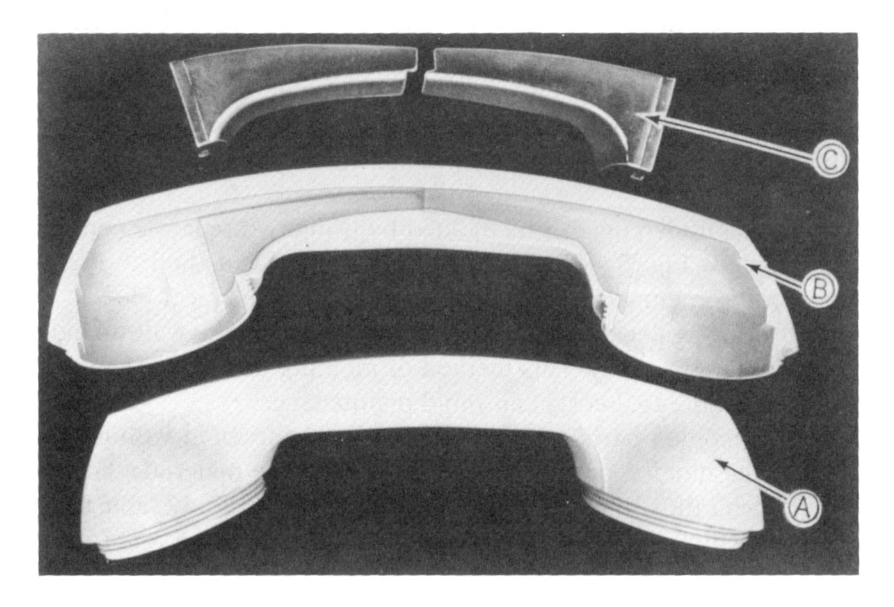

Figure 5–21. A cross-sectional view of an injection molded telephone handset handle showing the configuration of a complicated shape. (A) Shows the molded handset. (B) Shows the handset cut in half revealing the cored out sections. (C) Shows the two piece metal core detail. (Courtesy Western Electric.).

CORED-OUT SECTIONS IN MOLDED PARTS

Molded holes or cored-out sections in a plastic part can be of many geometric shapes, as shown in Fig. 5–21 of the telephone handset handle. The injection molded ABS telephone handle is hollow on the inside for the placement of electrical hardware. Note the two-piece core details (Fig. 5–21c). The two-piece core details fit into the ball-type details in Fig. 5–22a, which complete the metal component parts for the cored out section of the handle.

Fig. 5–23 illustrates six molded telephone handles ready to be removed from the injection mold. The two-piece core details will be removed from each of the molded handles by a special jig or fixture. Two sets of core details are used so that the die can be reloaded immediately.

At the base of the ball, detail in Fig. 5–22b, rotary thread rings are

Figure 5–22. A view of one half of a metal mold used to make injection molded telephone handsets. Note (A) the two piece core detail. Rotary thread rings (B) are used to permit quick removal of the part. The thread rings are not visible. *(Courtesy Western Electric)*

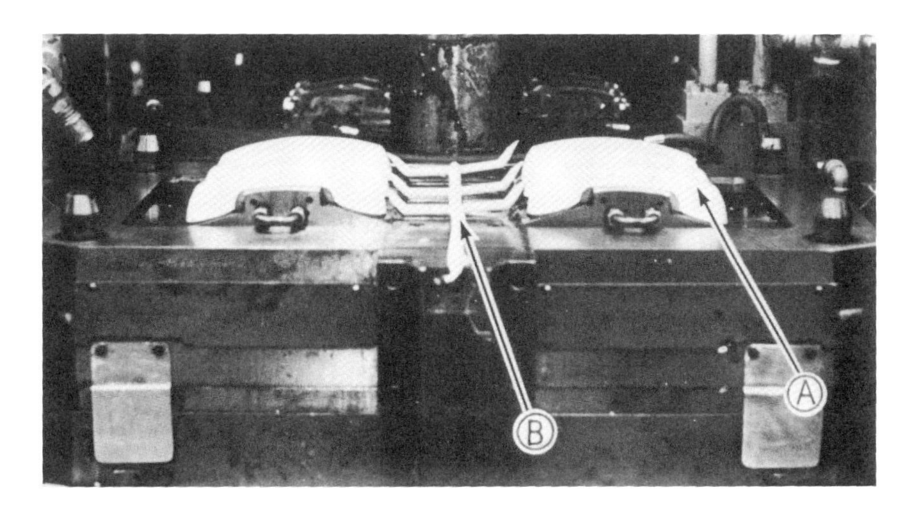

Figure 5–23. This picture illustrates six ABS injection molded telephone handset handles. (A) Shows the molded parts on top of the open mold. (B) Shows the runner system. Note the round runners and the pinpoint gating. *(Courtesy Western Electric)*

used to automatically unscrew the threads on the molded handset. The rotary thread rings are not visible in the picture.

A COLLAPSIBLE CORE

A new and uniquely designed collapsible core is available for molded parts that have internal threads, undercuts, or cutouts. The collapsible core is made up of a center core pin, a collapsible sleeve, and a collapsible bushing (Fig. 5–24a and b). The core pin is assembled inside the collapsible sleeve, and the positive collapsed bushing is assembled around the collapsible sleeve. These three parts together make the core side of the mold. This is the inside of the part such as, a threaded cap. The function of the positive collapsed bushing is to insure that the sleeve has collapsed. This is to prevent adhesion of any molded material that may flash into the collapsible core.

The collapsible sleeve consists of a series of 12 matching, vertical segments, extending completely around its diameter (Fig. 5–24c), for approximately one half its length, the remainder being solid or unsegmented. The segments are narrow and wide. These two types of vary-

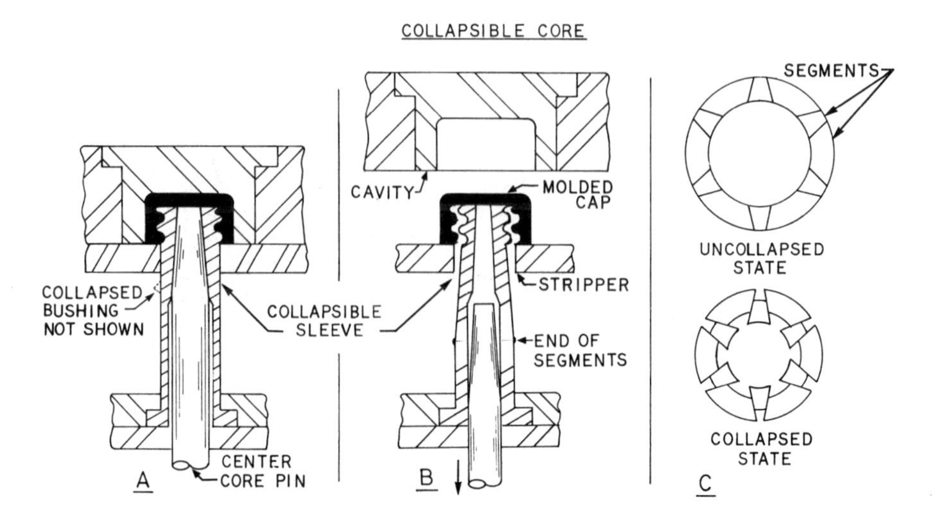

Figure 5–24. A partial schematic assembly of a collapsible core. *(Courtesy D-M-E Corp.)*

COLLAPSIBLE CORE

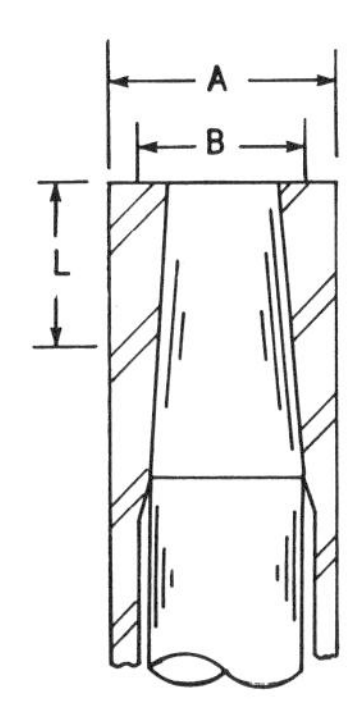

TYPE OF CORE	200	202	302	402	502	602
A-MAX. O. D. OF THREAD	1.270	1.390	1.740	2.182	2.800	3.535
B-MIN. I. D. OF THREAD	.910	1.010	1.270	1.593	2.060	2.610
L-MOLDED LENGTH-MAX.	.975	.975	1.225	1.535	1.750	2.125
COLLAPSE PER SIDE (NOT SHOWN)	.043	.055	.068	.090	.115	.140

Figure 5–25. This illustrates a range of sizes in inches for the collapsible core. *(Courtesy D-M-E Corp.)*

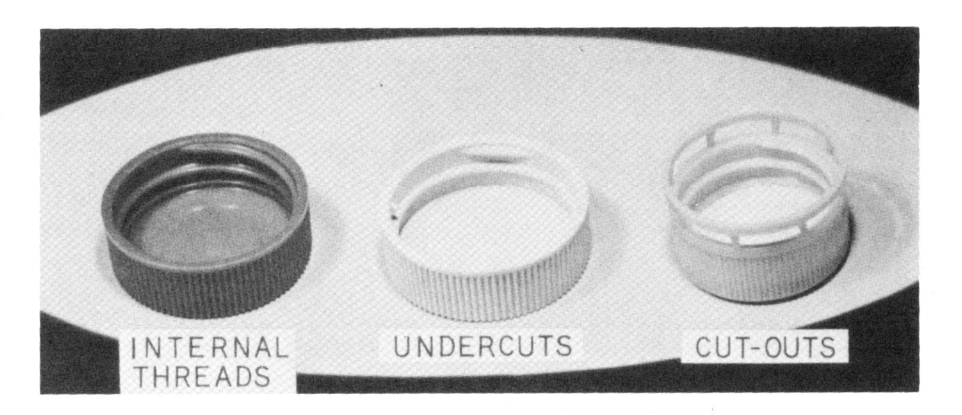

Figure 5–26. Three different plastic caps that have been molded by using the collapsible core.

ing width segments alternate around the outer surface of the sleeve. These narrow and wide segments are at only one end. The free end has the undercut of thread machined into it. These segments are perfectly matched so that no molding material can flash between them. When the center core pin is withdrawn at ejection, the segments collapse. The distance that the collapsible sleeve can collapse, per side, is shown in Fig. 5–25. Not all sizes of threads can be molded with one size of collapsible core. A number of standard sizes have been built and are available.

The new collapsible core can be used in place of conventional unscrewing molds, for molding internally threaded closures, and also for a new field of application of hitherto unmoldable undercuts. Fig. 5–26 shows three plastic caps that have been molded by using the collapsible core. These caps illustrate internal threads, undercuts, and cutouts.

6

Design of Threads

Threads are used in plastics for the purpose of providing a secure anchorage or locking device for a mating part. In all threads, the principle of the wedge is used to lock or anchor the two parts. Modern molding methods lend themselves to the molding of external or internal screw threads. An external thread is on the outside of the part (e.g., a threaded plug). An internal thread is on the inside of a part (e.g., a threaded hole). Threads in a plastic part are obtained by four methods: (1) they may be tapped; (2) they may be molded into the plastic; (3) threaded metal inserts may be molded in the part; and (4) threaded inserts may be pressed or cemented into place in the part after molding.

THREAD CLASSES AND FITS

The plastic industry uses essentially seven different types of threads. They are the American Standard, square, acme, buttress, bottle type, sharp "V" and a unified screw thread. The sharp "V" thread is not recommended, because of the stress concentration and notch sensitivity created in most plastic materials by molding this type of thread. Fig. 6–1 illustrates the six basic shapes of these threads, and Fig. 6–2 shows actual molded parts that have been molded with these threads. (Fig. 6–9 shows a unified screw thread).

The fitting of threads is classified as follows:
Class 1. A loose fit for quick and easy assembly.
Class 2. A moderate or free fit for interchangeable parts.

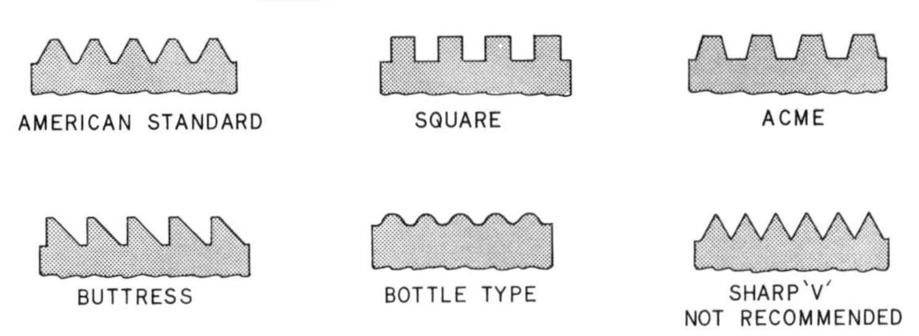

Figure 6–1. Profiles of the six different types of threads used in molded plastics.

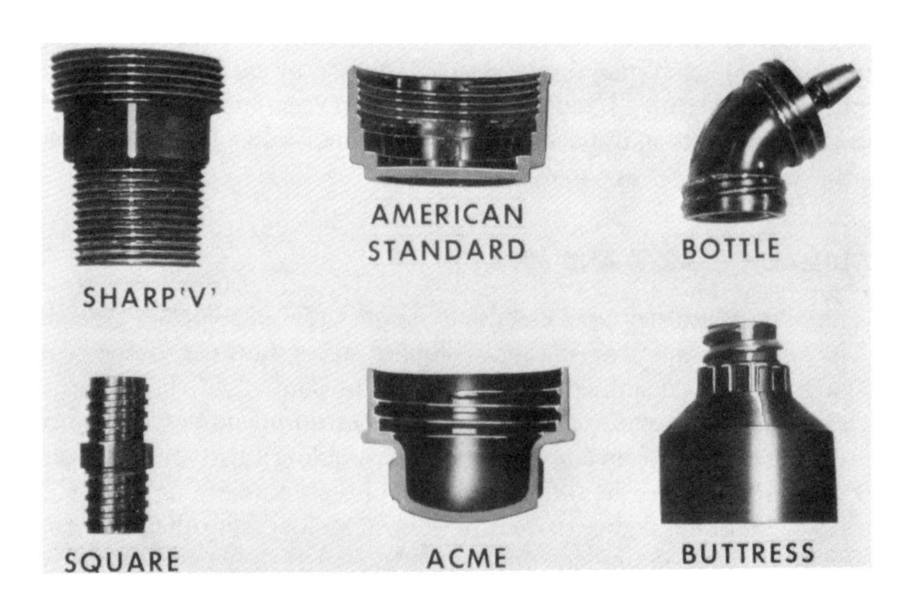

Figure 6–2. Different types of threads are used in molded plastic parts. The sharp "V" thread is an ABS pipe fitting. The American Standard thread is a cellulose acetate butyrate end cap for a flash light. The bottle thread is a phenolic pharmaceutical part. The square thread is a Saran pipe fitting. The acme thread is a polystyrene jar cap. The buttress thread is a high density polyethylene end for a caulking gun cartridge.

Class 3. A semiprecision or medium fit.
Class 4. A precision or snug fit for parts assembled with tools. The parts are not interchangeable. This type of thread fit is not recommended for plastics.

Threads of classes 1 and 2 are adequate for most applications in molded plastics.

The major diameter of a thread is the largest diameter of the thread of the screw or nut. The minor diameter of a thread is the smallest diameter of the thread of the screw or nut.

American Standard or American National Form Thread

The American Standard, sometimes called the American National Form thread, is used mostly in molded plastics. The American Standard thread is easily molded or tapped. This thread, in both thermosetting and thermoplastic molded materials is made 75% of full depth. Tapped threads average 70% of full depth (Fig. 6–3). This thread gives a better fit between the thread and its mating part. For example: turn the plastic closure on a medicine bottle (a bottle type thread) slightly and notice the play between the closure and the glass bottle; in contrast, unscrew a telephone receiver earpiece (American Standard) and observe the nice fit. There is very little jiggling or play after the first turn of the thread.

The American Standard thread is recommended for general use in plastics engineering work. It is used in threaded components where quick and easy assembly of the part is desired, and for all work where conditions do not require the use of fine-pitch threads.

Square Thread

A square thread is used where the highest strength is desired, e.g. in pipe fittings.

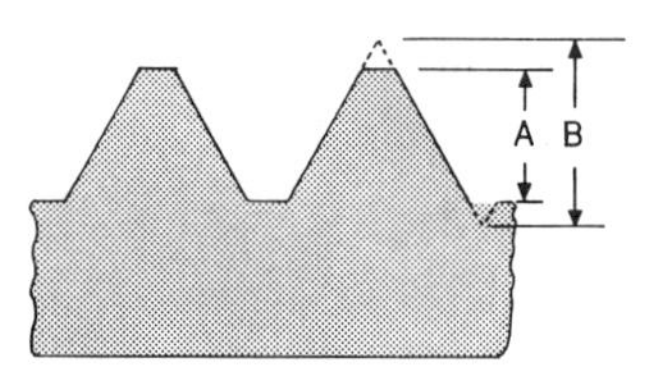

Figure 6–3. The American Standard Thread is easily molded or tapped. "A" is 75% of "B" when molded and 70% when tapped.

Acme Thread

The acme thread is similar to the square thread and is used in applications requiring strength. This type of thread is much easier to mold or cut than the square thread. Fig. 6–4 illustrates molded acme threads used in a pump housing.

Figure 6–4. An exploded view of a centrifugal pump. The precision injection molded parts are made of chlorinated polyether. An acme thread was used.

Buttress Thread

The buttress thread is used for transmitting power or strength in only one direction. It has the efficiency of the square thread and the strength of the "V" thread. It produces exceptionally high stresses in one direction only, along the threaded axis. Examples of actual applications are caps for tooth paste tubes and caulking gun cartridges.

Bottle Thread

The bottle thread is the type commonly used on glass containers. It is the accepted design standard set up by the Glass Container Association of America. All types of threads cause stress points in plastic materials, but the bottle thread results in the least. This type of thread has been developed to give the greatest ease in screwing or unscrewing mating parts. (Fig. 6–5). The round profile thread has been

Figure 6–5. Injection molded polypropylene bottle caps. Note the round type bottle threads used. *(Courtesy Dow Chemical Co.)*

Figure 6–6. An injection molded polypropylene container. Note the developed one quarter turn round type bottle thread. *(Courtesy Dow Chemical Co.)*

found to be very practical in plastic closures for glass containers, because of variations that occur in cast glass threads. A half thread, or a developed one-quarter-turn round-type bottle thread is also used quite frequently (Fig. 6–6).

Plastic parts incorporating round profile threads may be stripped from the mold if no undercuts other than threads exist. The plastic is actually stretched over the threads, and the method is limited to certain types of plastics (Fig. 6–7). Special grades of plastics suitable for stripping are offered by the material suppliers in the ureas and phenolics, both thermosetting rigid materials. Most thermoplastic materials can be stripped, due to their more deformable nature.

STRIPPING OF THREADED UNDERCUTS

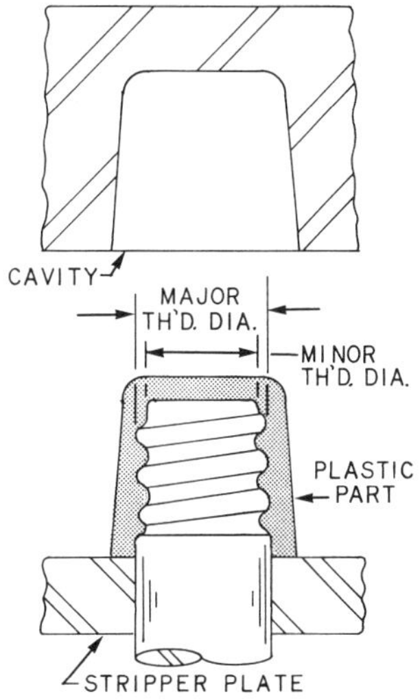

CAVITY

MAJOR TH'D. DIA.

MINOR TH'D. DIA.

PLASTIC PART

STRIPPER PLATE

$$\frac{\text{MAJOR TH'D. DIA.} - \text{MINOR TH'D. DIA.}}{\text{MINOR TH'D. DIA.}} \times 100 = \text{PERCENT STRAIN}$$

EXAMPLE

MAJOR DIA. 1.250″ MINOR DIA. 1.157″

$$\frac{1.250'' - 1.157''}{1.157} \times 100 = 8\% \text{ STRAIN}$$

MATERIAL	% STRAIN AT 150°F.
ABS	8
SAN	N.R.
POLYSTYRENE	N.R.
ACETAL	5
NYLON	9
ACRYLIC	4
POLYETHYLENE L.D.	21
POLYETHYLENE H.D.	6
POLYPROPYLENE	5
POLYALLOMER	15
POLYCARBONATE	N.R.
NORYL	N.R.
SURLYN	10

Figure 6–7. Stripping of threaded undercuts is permissible with certain types of plastic materials. A strain of 10% is generally the maximum that is allowed.

Sharp "V" Threads

Although the standard "V" type thread is sometimes used in molded plastic parts, it is not recommended. The sharp "V" points create stress points, making the plastic part notch sensitive and subject to breakage in these areas. Molded pipe fittings that must match metal pipe fittings use this type of thread. The conventional "V" type thread is generally used for mechanical assemblies.

Tapered or pipe threads should be avoided unless an ample wall section is provided around the threaded hole. The mating thread may be forced so tightly into the threaded hole that cracking will occur between the thread and the outer wall (Fig. 6–8).

Unified Screw Thread

The developed unified screw thread is used frequently (Fig. 6–9). It should be noted that the root of the thread has a radius and does not have a "feather edge." The tip or crest of the thread is flat and does

Figure 6–8. An injection molded glass-reinforced nylon housing called a "volute", used in a swimming pool cleaner pump. Note the large ribs and thick walls used to reinforce the female threaded pipe sections. *(Courtesy duPont)*

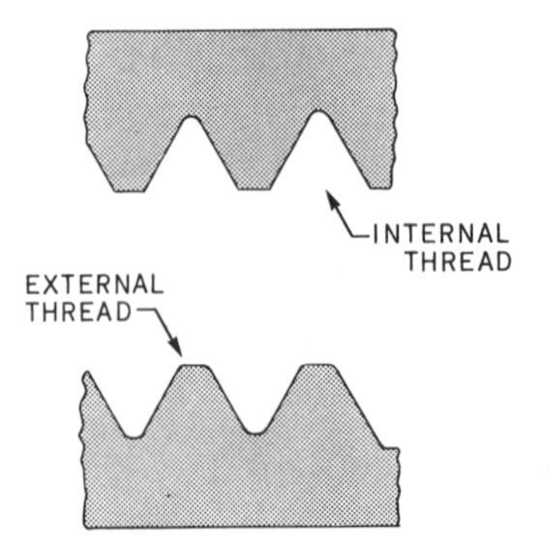

Figure 6–9. This type of thread does not have feathered edges. The root of the thread has a radius and the tip or crest is flat.

not have a "feather edge." The exact design of the thread can be found on pages 1103 to 1146 of the 17th edition of Machinery's Handbook. *

Molded Threads

A molded internal thread (Fig. 6–10) may be unscrewed from a threaded pin in the mold. A molded external thread (Fig. 6–11) may be unscrewed from a threaded recess in the mold, or a split mold section may be used (Fig. 6–12). If flash from the parting line develops between the threads, difficult assembly with the mating part is the result (Fig. 6–13). The thread may be chased, but this is an extra operation. To avoid the possibility of flash developing between the threads, it is best to use a split-mold section only when molding the thermoplastic materials by injection molding.

Molded threads, particularly internal ones, should be designed so that they can be removed quickly from the mold (Fig. 6–14). Most of the plastics compounds will shrink in the mold. Internal threads are made by a threaded pin in the mold, and the threads will shrink

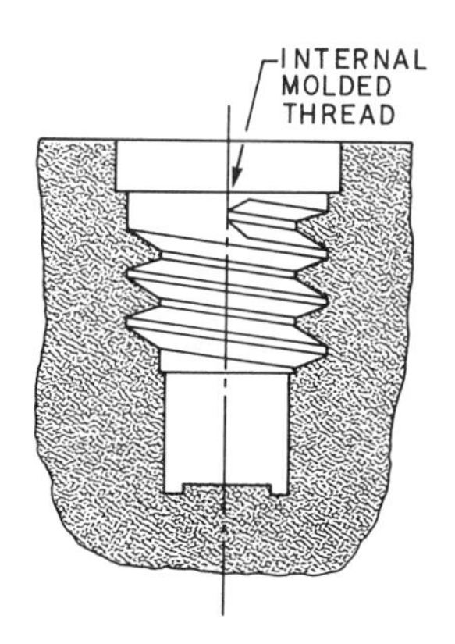

Figure 6–10. An American Standard molded internal thread. The thread starts and stops abruptly. The bottom unthreaded portion of the hole has a diameter equal to or less than the minor diameter of the thread.

*Machinery's Handbook. 17th ed., edited by Holbrook L. Horton and others. New York: Industrial Press, 1964.

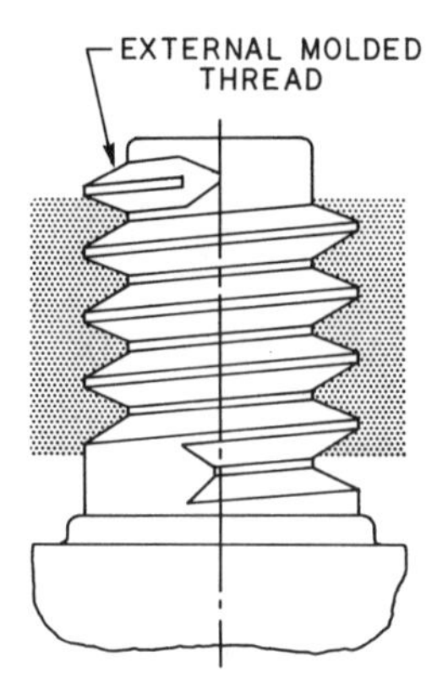

Figure 6–11. An American Standard molded external thread. The thread may be unscrewed from the mold or a split mold may be used. The thread starts and stops abruptly. The unthreaded bottom portion of the stud should have a diameter equal or greater than the major diameter of the thread.

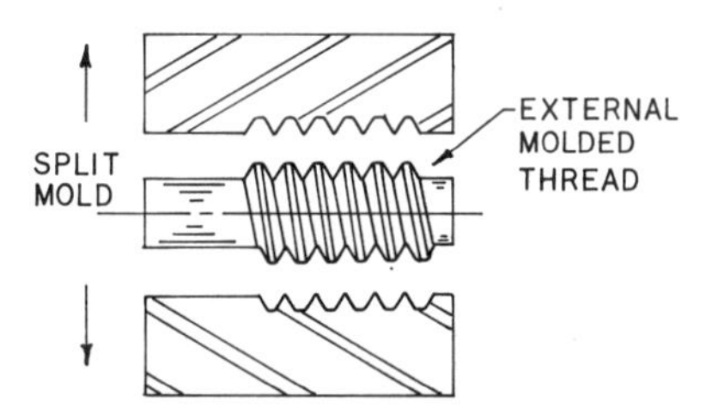

Figure 6–12. A split thread does not mean that the thread is split, but that the mold that made the thread is split or made in two parts. A split thread generally leaves a visible parting line.

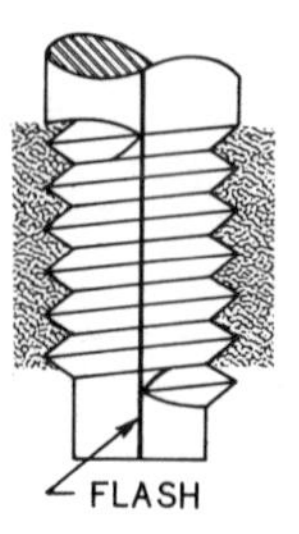

Figure 6–13. An external thread may be made by split cavity sections. Flash will develop where the cavity sections meet if the thermosetting materials are used.

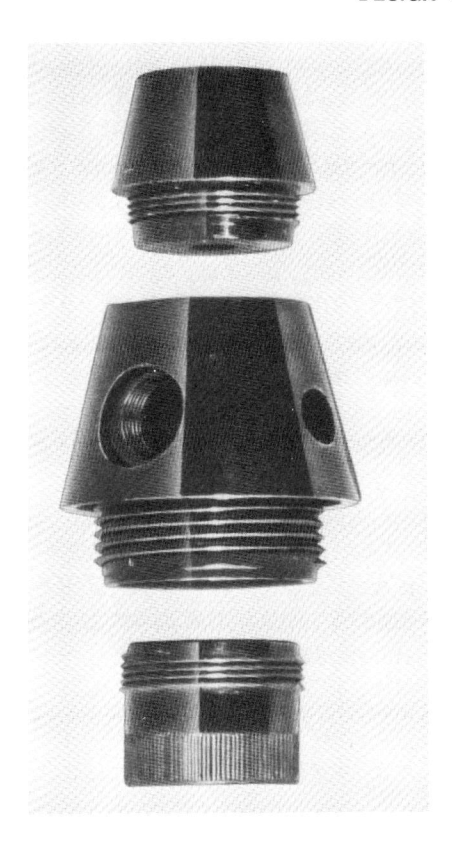

Figure 6–14. Three parts of a transfer molded phenolic M –52 fuse. The three parts screw together. The center part has three internal molded threads and one external molded thread. There are six molded threads in this molded fuse that must be held to close tolerance. The internal threads were designed so that they could be removed from the mold very quickly.

quickly around the pin. It is advisable to design molded threads no finer than 32 threads per inch and no longer than 1/2 in., or too much time will be taken by the operator when unscrewing the part from the mold. If more threads per inch are required, in order to increase thread strength, it may be wise to consider double or triple threads. On a single thread, the lead and pitch are equal. On a double thread, the lead is twice the pitch, and on a triple thread, it is three times the pitch. This will lessen the time of unscrewing from the mold and will provide greater thread strength.

If an internal thread is too fine and too long, the part may shrink so tightly around the threaded metal pin that it may crack before the operator removes the part from the mold. Shrinkage will also change the pitch of a long thread and cause difficult engagement. If a part having more than 32 threads to the inch is molded with cloth or glass reinforced thermosetting plastic, the filler or reinforcement may fail to enter the tip of the thread and may leave a weak tip filled only with the resin binder.

If threads are required that are finer than 32 threads to the inch and more than 1/2 in. in length, it is advisable that they be tapped, providing they are no larger than 1/2 in. in diameter.

When threads are tapped in a plastic part, the cutting of the thread is done across minute weld lines in the plastic. This often results in cracking or chipping of the thread, which can be seen only on very close examination. As a rule, closer tolerances can be held with a molded thread than with a tapped thread. This is due principally to the fact that in a molded thread, the exact impression of the thread will be made from the mold itself.

Molded Thread Design

A molded thread design is different from the design of a screw-machine thread or a tapped thread. The fundamental difference is that a molded thread starts and stops abruptly (Fig. 6–15). An internal and external thread should not feather out. A thread that feathers means a weak mold section that may break after repeated use. The thread itself will also be weak at this point. Fig. 6–16, A and B illustrates a thread that feathers out.

The bottom unthreaded portion of a molded male threaded stud should have a diameter equal to or greater than the major diameter of the thread (Fig. 6–11). If the unthreaded portion is less than the major diameter, the threads will be stripped in removing the part from the mold. Fig. 6–17a illustrates an incorrectly designed thread made by the ring section of the mold. The operator must unscrew the ring upward. If the thread is designed as illustrated, unscrewing

Figure 6–15. An injection molded two-inch diameter pipe fitting made from Saran. Note the American Standard type thread (1), which starts abruptly. Four metal inserts (2) are molded in place. *(Courtesy Dow Chemical Co.)*

the ring will result in stripping the threads to diameter "A" on the ring.

The bottom unthreaded portion of a female threaded hole should have a diameter equal to or less than the minor diameter of the thread (Fig. 6–17b). If the unthreaded portion is more than the minor diameter of the thread, the plug or pin will strip all of the threads as the screw moves upward.

If it is required that the mating part seat or meet flush with the companion plastic part, annular grooves should be provided on both male and female threads (Figs. 6–18a and b). Seating of a companion part is advisable if total tolerances on sub-assemblies are to be maintained and if maximum overall strength of the two parts is required.

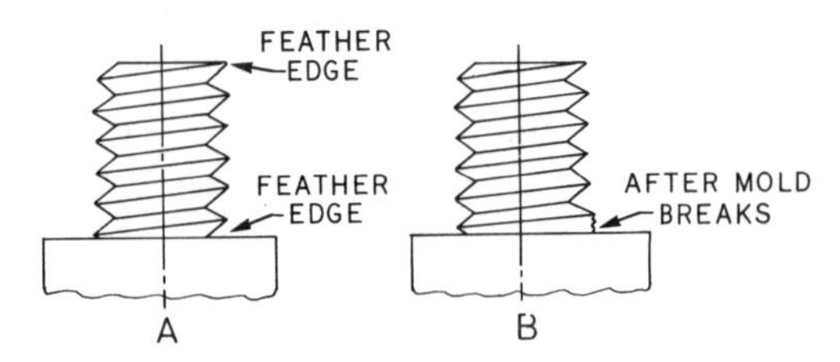

Figure 6–16. Avoid plastic threads that feather out, as shown at "A" at the bottom thread. This will produce a weak mold section which may break, causing a heavy molded thread as shown in "B".

Removing Threaded Parts From The Mold

Injection-molded parts having molded threads are often removed automatically, but when designing compression-molded parts having molded threads, it is advisable to consider how the part will be removed from the mold. Compression-molded parts are usually removed by hand after the mold opens. There are four commonly used methods for this operation.

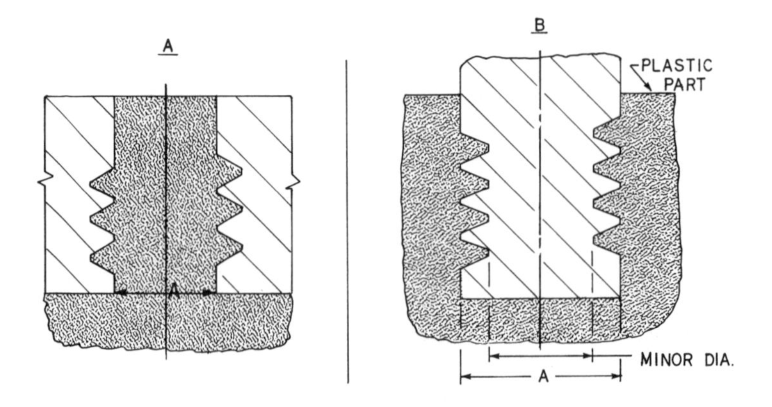

Figure 6–17. (A) An incorrectly designed thread molded by a loose ring in the mold. As the operator unscrews the ring upward, the threads will strip to the diameter marked "A". (B) If diameter "A" is larger than the minor diameter of the thread, the mold section cannot be unscrewed from the part without stripping the threads.

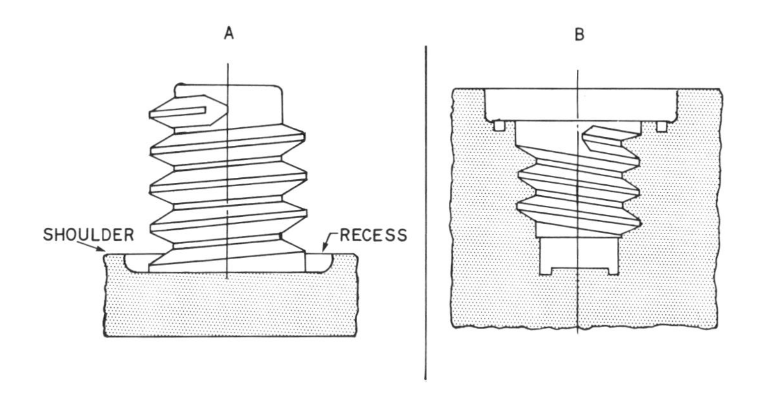

Figure 6–18. (A) Male thread. If the mating part is to seat properly with its companion plastic part, an annular groove should be provided at the bottom of an externally threaded stud. Also the recess at the base of the thread will allow the thread to end without having a "feather edge". (B) Female thread. If the mating part is to seat properly with its companion plastic part, an annular groove should be provided.

1. The operator may use a spanner wrench inserted in holes in opposite sides of the part.
2. He may use a screwdriver inserted in a slot provided for that purpose.
3. He may use a wrench, gripping against flat sides placed 180° apart.
4. He may unscrew by hand.

Internal and external threads are generally molded by means of a threaded metal ring. This ring is removed with the part when the mold is opened and an identical metal ring is placed in the mold so that the mold may be reloaded at once. During the molding cycle, the operator will unscrew the previously molded part from the metal ring.

TAPPED THREAD DESIGN

Tapped threads are not as strong, nor can they be held to as close a fit, as a molded thread. For reasons of overall economy, self-tapping screws should be used, if the screw is not to be removed and replaced

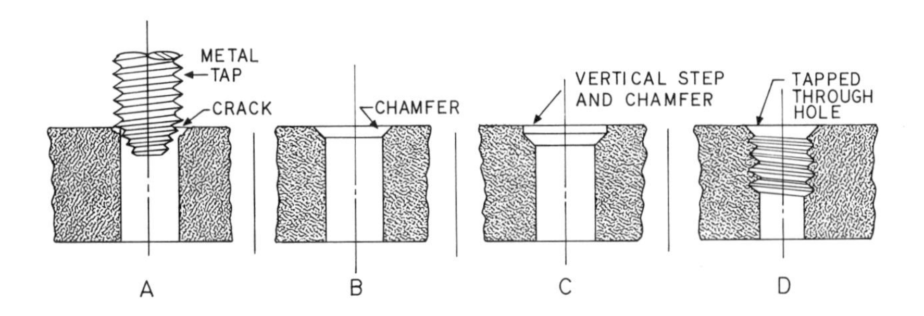

Figure 6–19. Molded or drilled holes should be countersunk or chamfered if they are to be tapped. This will avoid cracking at the hole entrance. (A) A plane hole will crack when tapped. (B) A countersink or chamfer will avoid cracking when tapped. (C) A vertical step and a chamfer will avoid cracking when tapped and also help in locating a mating part. (D) A through hole affords easier tapping, since chips resulting from tapping can fall through.

frequently. Repeated unions of the two threads eventually will damage the thread made in the plastic by the self-tapping screw.

Holes to be tapped should be slightly larger than those used in metal. The molded or drilled hole to be tapped should be slightly counter-sunk or chamfered (Fig. 6–19) to avoid chipping at the edge of the hole. A vertical step and chamfer (or counter sink) will be better if the location of the hole to a mating part is to be encountered. A through-hole affords easier tapping than a blind hole, since chips resulting from this operation will fall through the hole. A blind hole may require frequent extraction of a tap in order to remove the chips.

7
Inserts

Inserts may be used in plastic parts to take wear and tear, to carry mechanical stresses that are beyond the limits of the plastic material, to decorate the part, to transmit electric current, and to aid in subassembly or assembly work. Inserts generally serve an important functional purpose, but should be used sparingly because of the costs involved. Although inserts may be made of brass, aluminum, or steel, other materials, including ceramics and plastics are used. Brass is used most frequently because it does not rust or corrode and is inexpensive and easy to machine. Inserts must be designed to insure a secure anchorage to the plastic part, to prevent rotation as well as pulling out. Usually a medium or coarse diamond knurl provides adequate holding strength. Fig. 7–1 shows a selection of metal inserts used in molded plastic products.

FACTORS TO BE CONSIDERED WITH MOLDED-IN INSERTS

1. The insert must provide the required mechanical strength. The insert should be of sufficient size to resist forces likely to be met by the part in service. Sufficient anchorage must be provided to prevent the insert from pulling out of the plastic.
2. It is not feasible to mold inserts in all plastic materials. Some plastic materials will crack around the insert after they have aged. Other materials will creep in aging, or if two inserts are rigidly located together in a mating part, one or the other of the inserts will pull out as the plastic ages and shrinks.
3. The flow of the plastic material in the molding process should not dislodge the insert. It is not advisible to place fragile inserts

Figure 7-1. A selection of metal inserts used in molded plastic products.

in the path of the flow of material from the gates of transfer or injection molds. Too heavy an impact from the flow of compound against a fragile or delicate insert will dislodge or break the insert.

4. Sufficient wall section should be allowed around the insert to prevent cracking of the plastic as it cools. Plastic materials have a higher coefficient of thermal expansion and contraction than the metals commonly used for inserts.
5. Inserts may require retapping, facing, or other expensive cleaning operations after they have been molded in the part. If there is poor design or improper location of the insert in the part, cleaning operations to remove objectionable flash will be required. This expensive operation can be eliminated in many cases by a proper design of the insert.

SHAPES OF MALE AND FEMALE INSERTS

The great majority of inserts used in plastic parts are made by either automatic screw machines or metal stamping machines. Fig. 7-2 il-

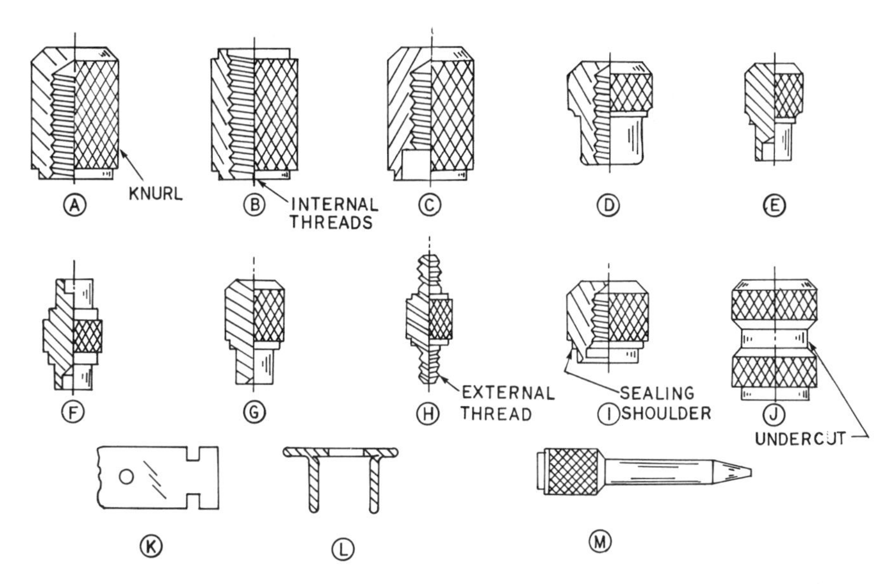

Figure 7–2. Many types of inserts are used in molded plastics. (A) Blind hole female insert with internal threads. (B) Open ends female insert with internal threads. (C) Blind hole female insert with internal threads and counter bore. (D) Male stud with internal threads. (E) Eyelet projecting. (F) Eyelet both ends projecting. (G) Projecting rivet. (H) Double projecting insert with projecting threads. (I) Blind hole female insert with internal threads and double sealing shoulder. (J) Female insert with undercut. (K) Metal stamped insert. (L) Drawn eyelet. (M) Rod or pin type insert.

lustrates and gives the proper name to most of the metal inserts used in the plastic industry. Typical male and female inserts are shown in Fig. 7–3. Note that the end of the insert to be imbedded in the plastics is chamfered or rounded. A chamfered or rounded end is desirable so that the plastics will flow easily around the insert. Embedded sharp corners on inserts may cause the plastics to crack at the corners. Tolerances on the minor thread diameters of female inserts should be held to plus 0.0025 in. and for precision work minus 0.0005 in. Tolerances on the major thread diameters of male inserts should be held to plus 0.0025 in. and for precision work minus 0.0005 in. Close tolerances on this dimension give the insert a positive location in the molded part and help to prevent compound from flowing into the insert during molding.

Inserts should be located so as to be parallel to the movement of

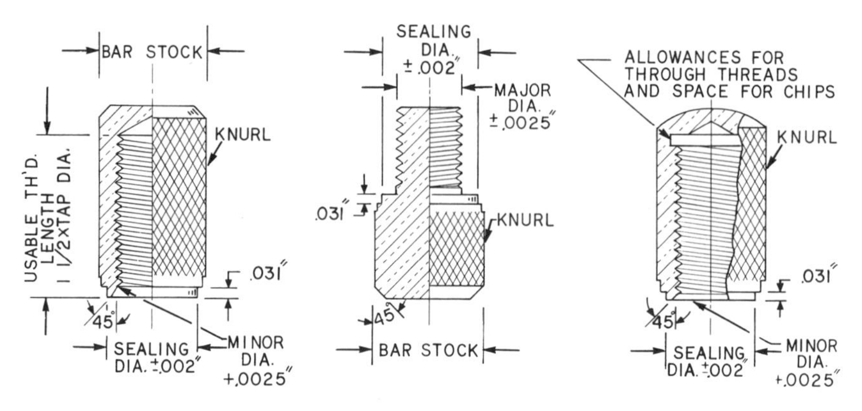

Figure 7–3. Standard designs for male and female inserts.

the mold as it opens and closes. Inserts located at right angles and oblique angles are difficult and expensive to mold (Fig. 7–4). This also holds true with molded holes.

Some female inserts are open at both ends and are molded through the part (Fig. 7–5a). The length of these inserts should be 0.001 to 0.002 in. oversize, if their axes are molded parallel to the draw. The extra length on the end of the insert aids greatly in preventing plastic compound from covering the ends of the insert and getting inside. More extra length may cause the insert to break as the mold comes together and pinches the insert. If the outside diameters of the insert are used to hold the insert in place during the molding operation, tight tolerances are required (Fig. 7–5b.).

Female inserts molded through the part frequently are not

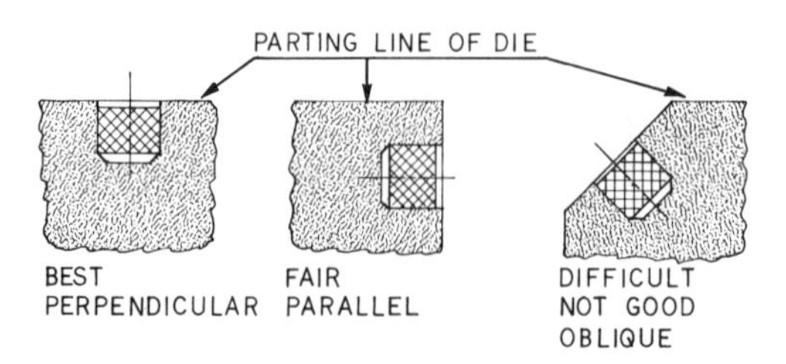

Figure 7–4. Inserts should be located so as to be parallel to the movement of the mold as it opens and closes. Inserts located at right angles and oblique angles are difficult and expensive to mold.

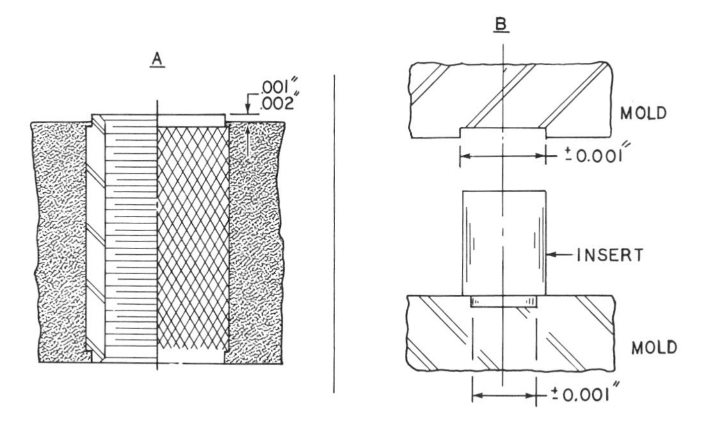

Figure 7–5. (A) Female inserts which are open at both ends should be designed to have an axial tolerance of plus 0.001 to 0.002 inch, compared with the part. This precaution will help to keep the plastic compound from entering the thread. (B) If the outside diameters of an insert are used to hold the insert in place during the molding operation, tight tolerances are required.

threaded before molding when a thermosetting material and compression molding are used. Plastics material may flow into the threads, necessitating a retapping operation to remove the compound. Through female inserts molded by the injection or transfer method are frequently threaded before molding. In these cases, the mold is closed on the insert before compound is forced into the mold. A well-constructed mold and good inserts allow little or no flash to get into the threads. A female insert located on a mold pin or insert pin requires tight tolerances in order to prevent misalignment (Fig. 7–6).

Male and female inserts should be provided with a shoulder in order to prevent plastic compound from flowing into the threads (Fig. 7–7). An adequate sealing shoulder should be allowed.

Male inserts as bolts that are not provided with a shoulder (Fig. 7–8) should be avoided, because compound will flow up into the threads during molding. A single or double-sealing shoulder is better. The necessity of chasing these threads after molding may be eliminated on male inserts not having shoulders, if threaded lugs or bullets are used (Fig. 7–9a). This procedure increases the parts cost, however, because the threaded mold section must be unscrewed from the insert after each part is molded.

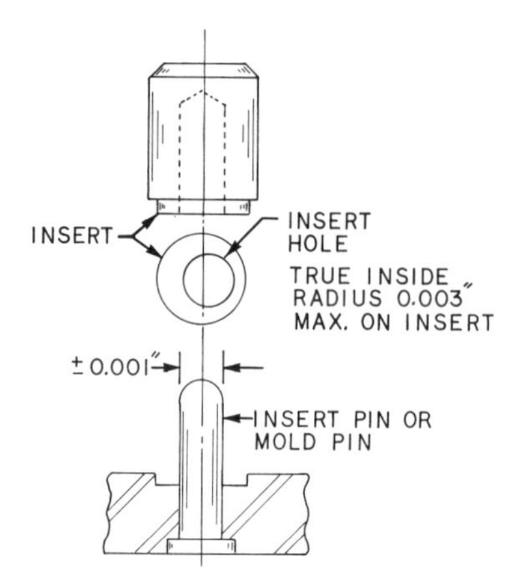

Figure 7–6. A female insert located on a mold pin will require close tolerances in order to prevent misalignment. It will be noted in the drawing that the true inside radius of the insert must be held to 0.003 inch maximum in relation to the outside radius of the insert.

Some designs require that a portion of the insert extend above the surface of the part (Fig. 7–9b). The extended portion should be round, since that portion of the insert must fit into a recess in the mold while the part is being molded. Anything but a round recess is expensive to machine and should be avoided. Also, a round shoulder on the top of a hexagonal insert eliminates the necessity to machine a hex hole in the mold (Fig. 7–9c).

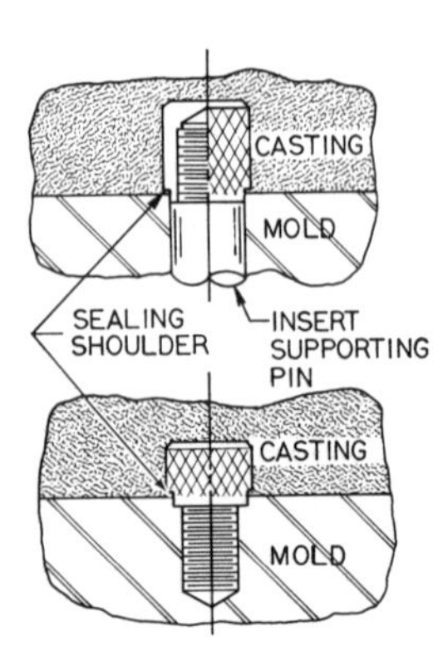

Figure 7–7. Male and female inserts should be provided with shoulders to help prevent the plastic compound from flowing into the threads.

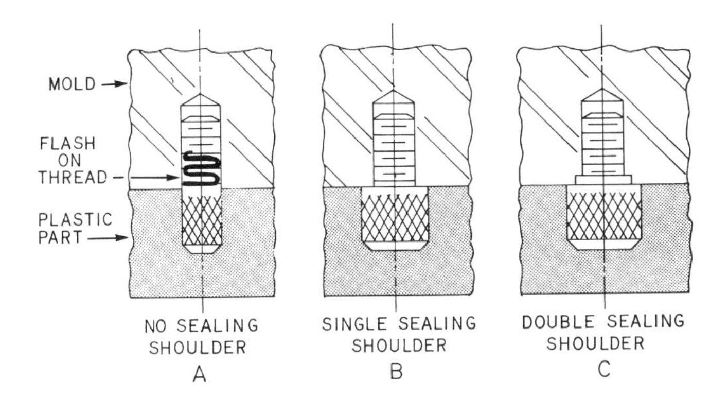

Figure 7–8. Avoid using an insert that has not been provided with a shoulder as shown in (A). A single sealing shoulder (B) is better. A double sealing shoulder is the best (C) but is more expensive.

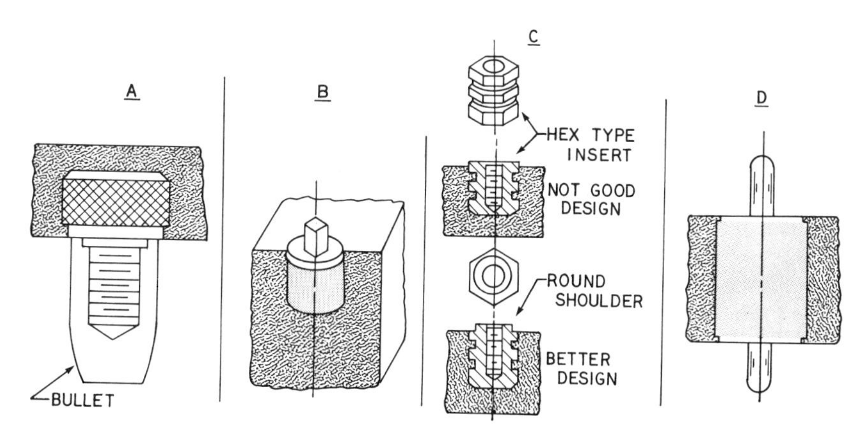

Figure 7–9. (A) Plastic flash may be kept out of threads of an insert by means of a threaded lug or bullet. (B) The portion of a molded-in insert that extends above the surface of the part should be round. Square holes are difficult and expensive to machine in the mold. (C) A round shoulder on the top part of a hexagonal insert eliminates the necessity to machine a hexagonal hole in the metal mold. (D) Avoid male inserts above the top and below the bottom of the part. Mold misalignment may cause damage to the insert as well as the mold, when the mold closes.

Male inserts extending through the top and bottom of the part and into recesses in the mold (Fig. 7–9d) are considered poor design. In cases of mold misalignment of a maximum of 0.006 in., the mold in closing may scratch one of the extended portions of the inserts.

Female spun-over inserts are used to provide a permanent assembly of contact strips and washers to the molded part. (Fig. 7–10a). The design features are essentially the same as those used in threaded inserts, except for the tubular projection.

No insert is so well sealed in the plastic that gas, under sufficiently high pressure, or a liquid, will not pass between the insert and the compound. To aid in preventing seepage around the insert, a rubber "O" ring may be molded with the insert in the plastic part (Fig. 7–10b). Sometimes the rubber "O" ring is placed around the insert and assembled after molding (Fig. 7–10c).

If an insert is subjected to excessive axial strain, a firmer anchorage of the insert to the compound may be obtained by grooving (Fig. 7–11).

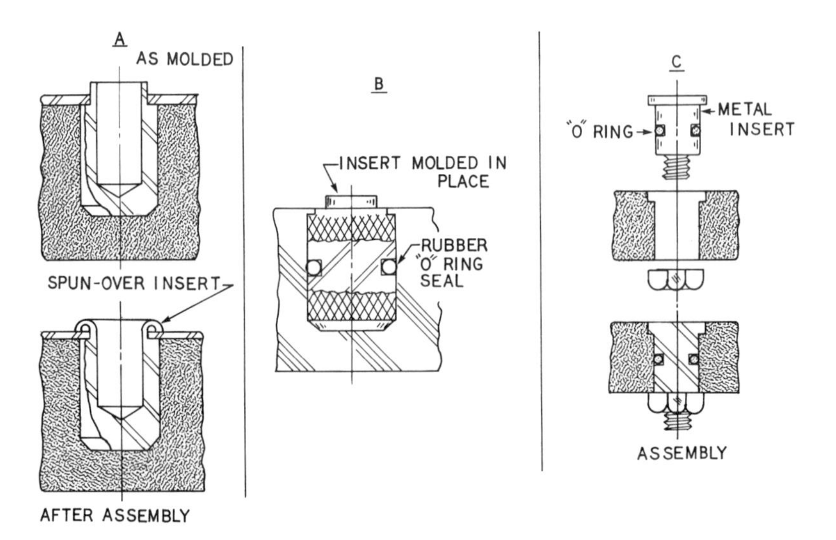

Figure 7–10. (A) Female spun-over inserts are used to provide a permanent assembly of contact strips and washers to the molded part. (B) Rubber "O" rings are sometimes molded with the metal insert to prevent gas or liquid seepage around the insert. (C) Rubber "O" rings are placed around inserts and assembled after molding when it is impractical to mold-in the insert.

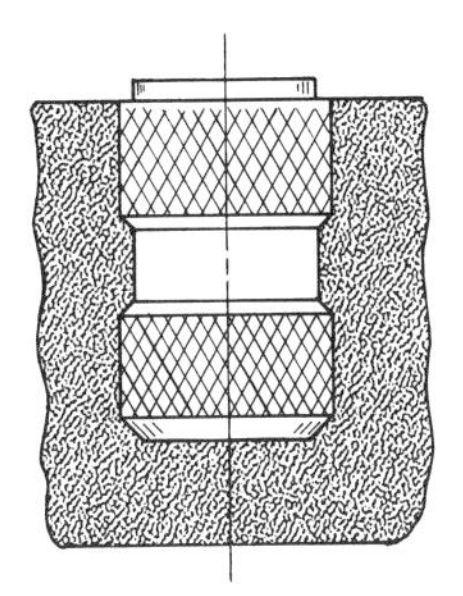

Figure 7-11. An insert subject to excessive axial pull should be grooved in addition to being knurled to aid in providing good anchorage.

EFFECT ON MOLD STRENGTH

The problems involved with the shape of inserts and their effect on mold strength are similar to these encountered with holes (discussed in Chapter 5).

If compression molding is used, it is advisable to have the length of the embedded portion of a closed-end insert no longer than twice its diameter, when the insert is molded parallel to the draw (Fig. 7-12a). It is best to have a through insert that is no longer than twice its diameter, when molded parallel to the draw (Fig. 7-12b).

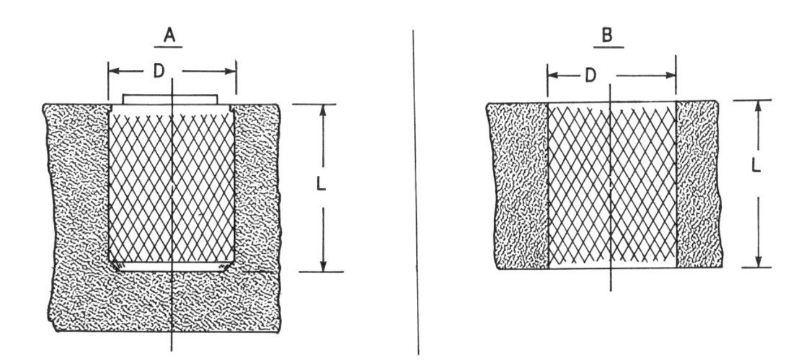

Figure 7-12. (A) When compression molding is used, the length (L) of the embedded portion of an insert should be no longer than twice its diameter (D) when molded parallel to the draw. When molded perpendicular to the draw, the length should be no longer than the diameter. (B) When compression molding is used, a through insert when molded parallel to the draw should be no longer than twice its diameter.

LOCATION OF INSERTS IN THE PART

Inserts that are improperly located in a part, from a molding standpoint, may result in objectionable decorative effects, poor electrical properties, a weak part, a weak mold, or excessive finishing costs.

Because of the differences in the coefficients of expansion and contraction between metals and plastics, sink marks or concaved depressions on a part may result if the end of the insert is too close to the opposite wall (Fig. 7–13). A sink mark may be objectionable from a decorative standpoint. If letters or designs are to be hot-stamped on the surface having the sink mark, they will not be as deep on the sink mark area as on the rest of the surface. If sink marks are to be avoided, the thickness of the plastic compound at the end of the insert should be at least one-sixth the diameter of the insert.

Inserts that are used in making and breaking an electrical circuit are usually molded so that they extend above the part (Fig. 7–14). The inserts should extend sufficiently high so that the length of the arc or spark (at the rated voltage) will not have a tendency to ground itself on the plastic. An automobile distributor cap is a typical example of the application of this rule (Fig. 7–15). If the insert is set flush with the part, the heat of the arc that forms as the circuit breaks may carbonize the plastic. Carbonization of the plastic results in eventual grounding of the insert. The insert then becomes loose.

The location of male and female inserts in a molded part may af-

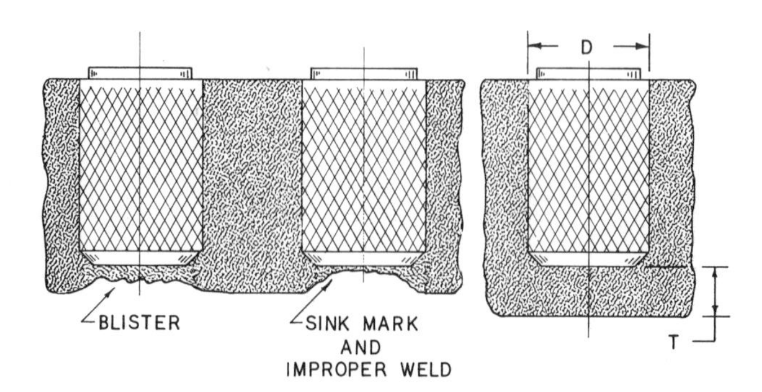

Figure 7–13. If sink marks and blisters are to be avoided at the end of inserts, the thickness (T) of the plastic at the embedded end of the insert should be at least 1/6 the diameter (D) of the insert.

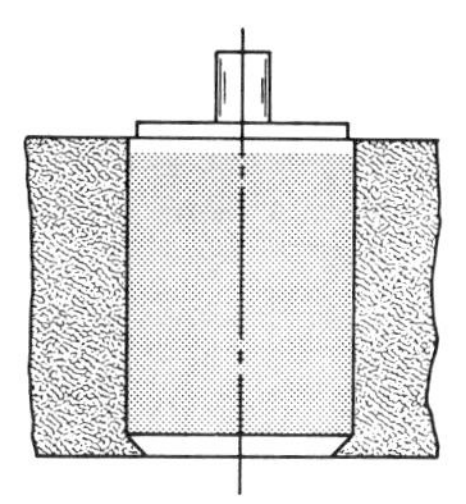

Figure 7–14. Inserts used making and breaking electrical circuits should extend above the plastic part.

fect the mold strength. Female inserts, for example, are held in place by a steel rod or pin that has been inserted in a hole in the mold. If the hole through which the pin is inserted is less than 0.020 in. from the face of a mold section, the mold may crack as this point (Fig. 7–16a).

Inserts used in bosses should extend to within one material thickness of the opposite wall, and ribs should be used to support the boss (Fig. 7–16b). The end of the insert should be rounded or chamfered to avoid concentration of stresses at the sharp edges.

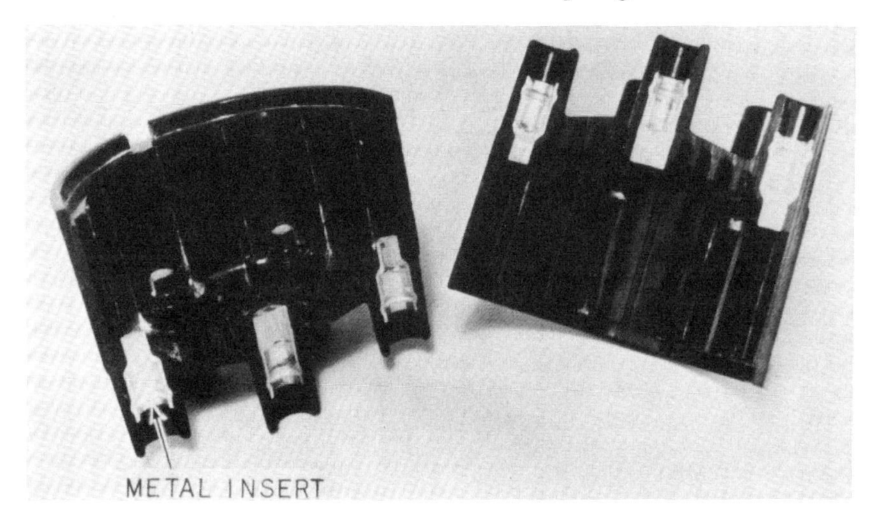

Figure 7–15. A distributor cap that has been cut into two halves. The cap was compression molded out of phenolic arc and electrical insulating material. Note the molded in place aluminum insert.

Molded-in inserts are used to carry mechanical stresses that are beyond the limits of the plastic material. When heavy loads are encountered, the insert and not the plastic should carry the load (Fig. 7–16c).

Long male inserts generally present no problem as far as mold strength is concerned. This is true because the larger portion of the length is firmly held by a recess in the mold and cannot be subjected to the flow of the compound. As previously discussed, however, the distance between the mold insert recess and the side of the mold should be greater than 0.020 in. (Fig. 7–16d).

It has been stated before that both male and femal inserts should be provided with shoulders to aid in preventing plastic compound from working into the threads or over the insert during the molding process. Shoulders may be located to afford a vertical as well as a horizontal seal. Employing seals in both directions is the best preventive measure against the flow of compound over the threads or over the insert.

Figs. 7–17 and 7–18 illustrate various methods of designing inserts

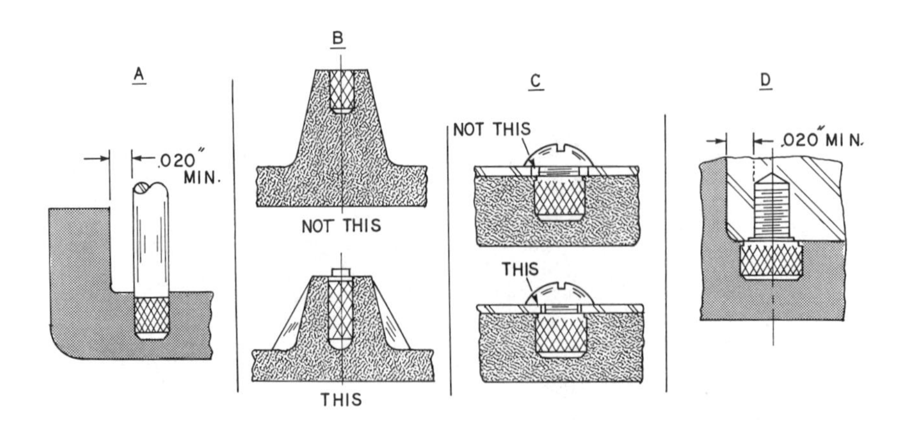

Figure 7–16. (A) Avoid locating inserts too near the edge of the part. If insert supporting pins are used, the thin adjoining mold section may crack. Minimum allowable thickness for the mold section is 0.020 inch. (B) Metal inserts in bosses should extend to within one material thickness of the opposite wall and ribs can be added for additional support to the base. (C) When stresses are encountered, the insert and not the plastic should carry the load. (D) Mold recesses for male inserts should be at least 0.020 inch from the edge of the mold if cracking of the mold is to be avoided.

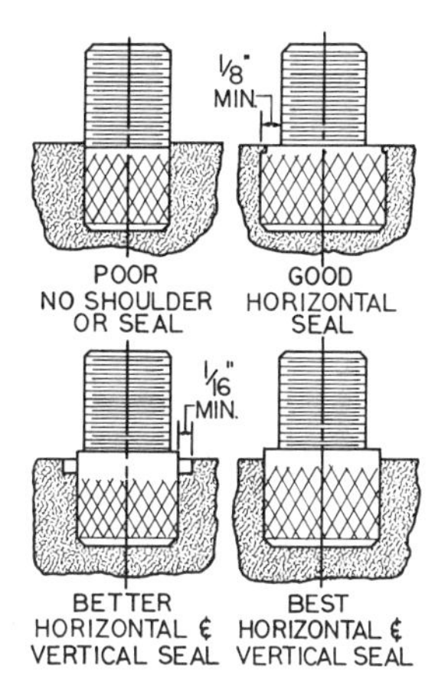

Figure 7–17. Male inserts should be designed so that they have a shoulder in order to seal out any plastic material that might be forced around the insert during the molding operation.

for effective seals. Note that shoulders or seals are always flush with or in contact with a portion of the mold (Fig. 7–19). Seals are located to prevent material from flowing into the threads. Female inserts generally are supported by insert pins, and male inserts are held in place by recesses in the mold.

If a female insert is used without any sealing shoulder, the minor thread diameter of the insert must be held to plus or minus 0.001 in., in order to keep plastic material from flashing into the threads. This

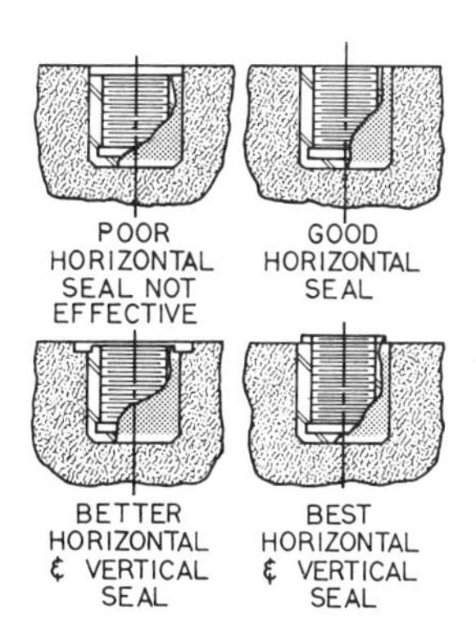

Figure 7–18. Female inserts should be designed so that they have a shoulder in order to seal out any plastic material that might be forced into the insert during the molding operation.

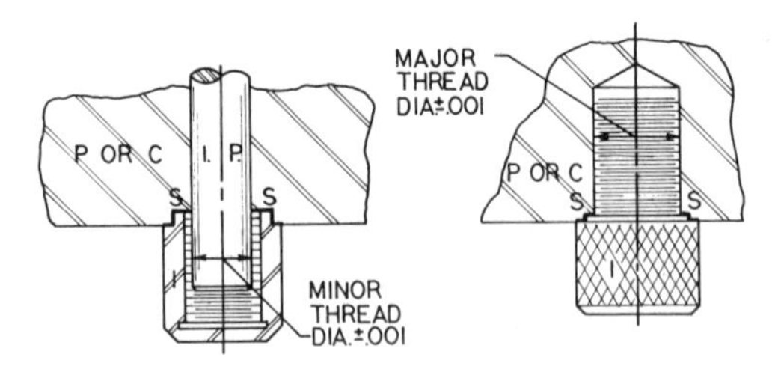

Figure 7–19. Shoulders or seals are always flush or contact a portion of the mold. P or C, plunger or cavity; IP, insert pin; I, insert; S, sealing surface, shown heavy.

close tolerance makes a tight fit or seal between the threads on the insert and the locating pin in the mold.

Male inserts molded without any sealing shoulders should have a tolerance of plus or minus 0.001 in. on the major diameter, in order to help prevent material from flashing into the threads on the insert and locating pin in the mold.

CRACKING AT THE INSERTS

Metal inserts molded in either thermosetting or thermoplastic material require a wall of compound around them of sufficient thickness to prevent cracking upon alternate heating and cooling and aging of the part. Table 7–1 gives what is considered the minimum wall thickness requirements for both the thermoplastics and thermosetting com-

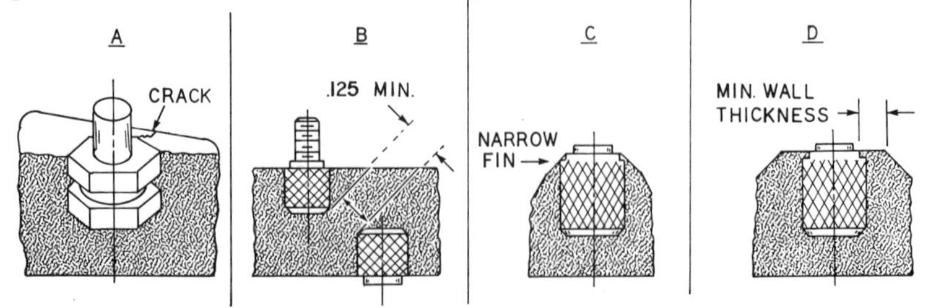

Figure 7–20. (A) Inserts with sharp corners should be avoided. Such inserts may cause cracking of the part. (B) Inserts molded in opposite sides of a thermosetting part should be no closer than 0.125 inch. (C) Avoid bringing a boss down to a narrow fin around an insert. This may cause cracking of the plastic material. (D) Inserts molded in bosses should have enough compound around them to prevent cracking. Minimum wall thickness can be obtained from Table 7–1.

pounds. In some cases, an insert with sharp embedded corners is molded in the plastic. Sharp corners increase the danger of cracking through the wall (Fig. 7–20a).

Inserts that are molded in opposite sides of a thermosetting part should be no closer than 0.125 in. (Fig. 7–20b). Cracking of the compound between the inserts may occur if they are closer. If inserts molded in this position carry electrical current, they may short circuit through the crack.

If a boss supporting insert is required, it may be necessary to bring the boss down to a narrow fin around the insert (Fig. 7–20c). This is to be avoided. The same results may be accomplished by cutting the

TABLE 7–1. SUGGESTED MINIMUM PLASTICS WALL THICKNESS FOR INSERTS OF VARIOUS DIAMETERS.

DIAMETER OF INSERTS						
INCHES	.125	.250	.375	.500	.750	1.00
PLASTIC MATERIALS						
ABS	.125	.250	.375	.500	.750	1.00
Acetal	.062	.125	.187	.250	.375	.500
Acrylics	.093	.125	.187	.250	.375	.500
Cellulosics	.125	.250	.375	.500	.750	1.00
Ethylene vinyl acetate	.040	.085	N.R.	N.R.	N.R.	N.R.
F.E.P. (fluorocarbon)	.025	.060	N.R.	N.R.	N.R.	N.R.
Nylon	.125	.250	.375	.500	.750	1.00
Noryl (modified PPO)	.062	.125	.187	.250	.375	.500
Polyallomers	.125	.250	.375	.500	.750	1.00
Polycarbonate	.062	.125	.187	.250	.375	.500
Polyethylene (H.D.)	.125	.250	.375	.500	.750	1.00
Polypropylene	.125	.250	.375	.500	.750	1.00
Polystyrene	Not Recommended					
Polysulfone	Not Recommended					
Surlyn (ionomer)	.062	.093	.125	.187	.250	.312
Phenolic G.P.	.093	.156	.187	.218	.312	.343
Phenolic (medium impact)	.078	.140	.156	.203	.281	.312
Phenolic (high impact)	.062	.125	.140	.187	.250	.281
Urea	.093	.156	.187	.218	.312	.343
Melamine	.125	.187	.218	.312	.343	.375
Epoxy	.020	.030	.040	.050	.060	.070
Alkyd	.125	.187	.187	.312	.343	.375
Diallyl phthalate	.125	.187	.250	.312	.343	.375
Polyester (premix)	.093	.125	.140	.187	.250	.281
Polyester T.P.	.062	.125	.187	.250	.375	.375

fin back, as illustrated in Fig. 7–20d. Minimum compound wall thickness, as covered in Table 7–1, should be allowed.

A few applications may call for a plastic material to be molded in or around the inside of a metal insert. Since the plastic material shrinks on cooling, the metal insert should be undercut to prevent it from falling off (Fig. 7–21a).

Sometimes, inserts will be covered with a certain amount of flash after they have been molded in the part. This is true even though the inserts have been provided with shoulders. All inserts that are to be faced after molding should project above the surface of the part at least 0.015 in. (Fig. 7–21b). It may be necessary to face off an insert after molding if a plastic material with high shrinkage values has been used (Fig. 7–21c).

PRESSED-IN INSERTS

Inserts may be pressed into holes provided for them immediately after the part has been molded. As the part cools, it shrinks around the insert, holding it securely in place. Pressed-in inserts are not generally recommended for use with thermoplastic molded parts. These parts have been cooled slightly before they are removed from the mold.

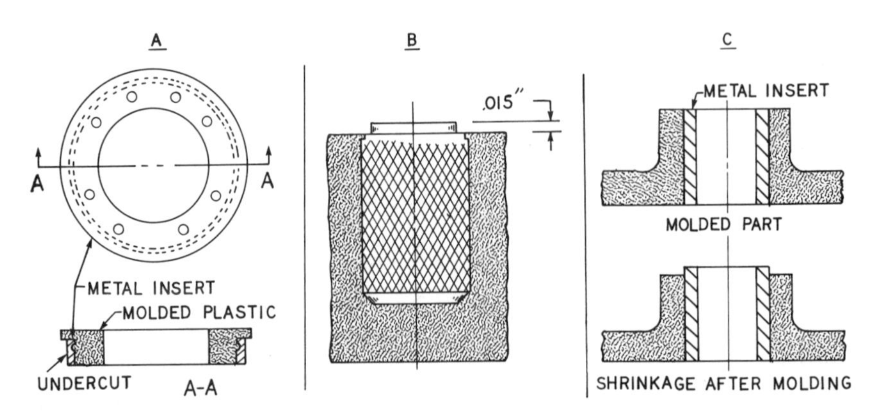

Figure 7–21. (A) A metal insert molded around a plastic part should be provided with proper undercuts in order to hold the plastic and metal together. (B) Inserts to be faced after molding should project above the part at least 0.015 inch. (C) It may be necessary to face off an insert after molding if a plastic material with high shrinkage values has been used.

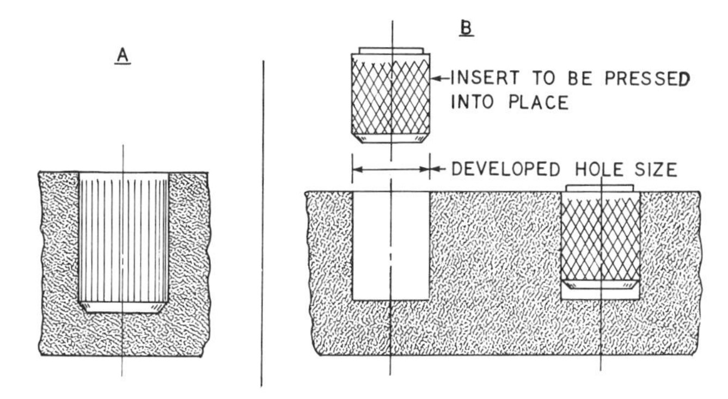

Figure 7–22. (A) Inserts that are pressed in after molding should be provided with a straight knurl. (B) It may be necessary to develop the hole size for an insert that is to be pressed in place. The mold pin can be made oversize and turned down until the desired press fit is obtained.

Care must be taken to prevent exceeding the elastic limit of the plastic material. Pressed-in inserts are not used in the more brittle plastics. Inserts that are to be pressed in after molding require a straight knurl (Fig. 7–22a). The holding qualities of these inserts are not as good as those of molded-in inserts, because of the straight knurl feature. Sometimes it is necessary to develop the hole size for an insert that is to be pressed in place (Fig. 7–22b). This means that the metal mold pin is made over-size and then gradually reduced in size by removing metal from the pin until the exact dimensions are obtained in the molded part. If a diamond knurl is used, it may be necessary to use a suitable adhesive to help hold the insert in the plastic.

METAL-STAMPING AND ROD-TYPE INSERTS

Metal-stamping inserts (Fig. 7–23a) are not advisable, as it is difficult to hold them to proper size for a close mold fit. The metal used in stamping the insert may be too soft to withstand flow pressures. Flash will flow over the insert and must be cleaned by a finishing operation. Metal-stamped inserts should be placed at the parting line or below, if scouring or pinching is to be prevented. Also, some means should be provided for anchoring it solidly in the mold and the part (Fig. 7–23b).

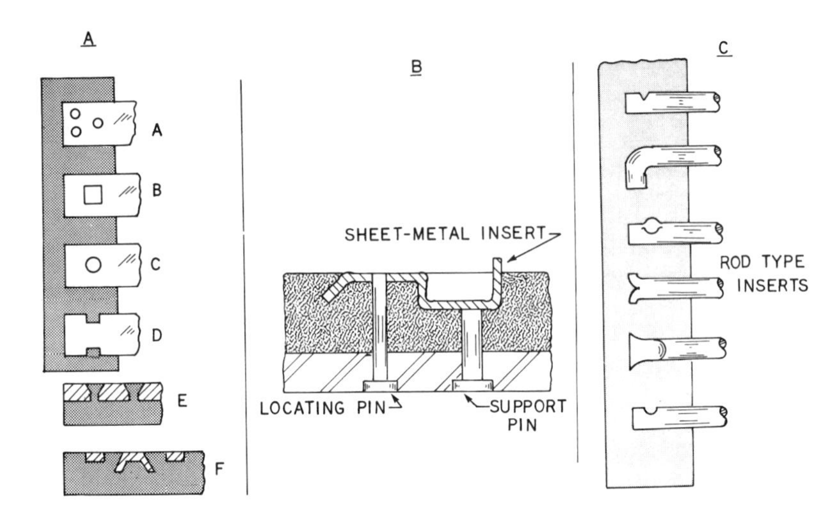

Figure 7–23. (A) Metal stampings or sheet metal inserts may be secured to the molded plastic material through the use of punched holes, (A,B,C), notches (D,E), and bent tabs or flanges (F). (B) Sheet metal inserts and metal stampings should be anchored securely by locating pins and supported with die pins. (C) Rod type inserts are generally restricted to injection or transfer molding. The rod insert may be anchored by notching, bending, swaging, grooving, etc.

Rod-type inserts (Fig. 7–23c) should be used with fairly soft or free-flowing plastic molding material. Injection or transfer molding with low molding pressures are required, or the rod insert will tend to bend readily or, in some cases, shear off. Rod-type inserts are also difficult to hold in position and will lift out of location, due to the flow of the plastic material. Fig. 7–24 illustrates an automobile steering wheel that is a successful rod-type insert and that has been used for many years. Injection molded steering wheels have been made out of cellulosics, PVC, and polypropylene. Most steering wheels are compression molded from hard rubber and then painted. The same rod-type insert is used for both injection and compression molding.

Carriage bolts, stove bolts, machine screws, rivets, etc., are very similar to rod-type inserts and are not recommended, as they are difficult to anchor in the plastic and generally require a clean operation after molding (Fig. 7–25).

Modern automobiles and many electrical appliances are dependent on plastic-molded parts. Fig. 7–26 shows three molded collector

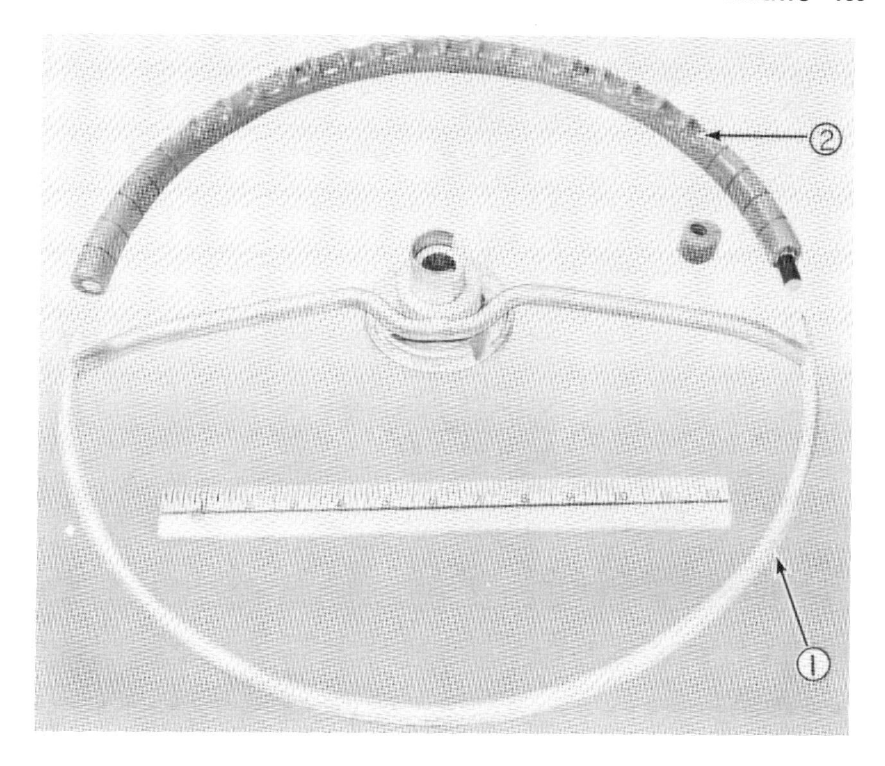

Figure 7–24. A cross section of an automobile injection molded steering wheel using a rod type insert. (1) The rod insert. (2) A cellulosic type plastic material. Many steering wheels are compression molded from hard rubber and then painted. The same type rod insert is used.

rings for alternators and generators. The collector rings are composed of metal contacts integrally molded with plastic as the supporting material. Alkyd material used in these parts must have unlimited insulation qualities and dimensional stability over a wide range of temperatures. Metal inserts used in electrical parts are made of copper, and the transfer molding process is used to mold the alkyd around the insert.

PLASTIC INSERTS INSTEAD OF METAL INSERTS (Outsert Molding)

Instead of molding plastic materials around metal inserts, a new concept of molding small component parts into a metal stamping has

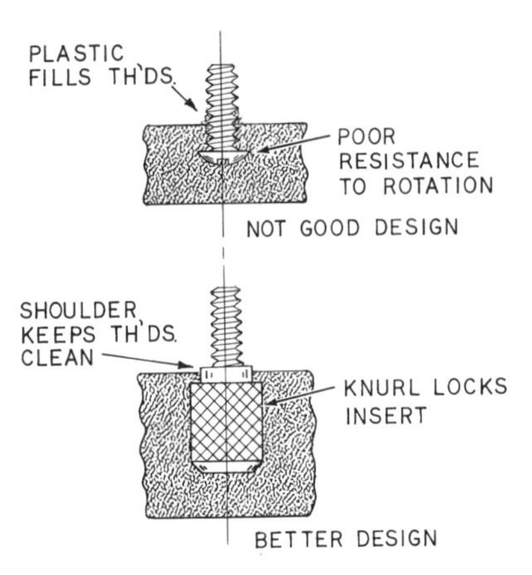

Figure 7–25. Avoid using standard threaded bolts as they will generally require a cleaning operation after molding.

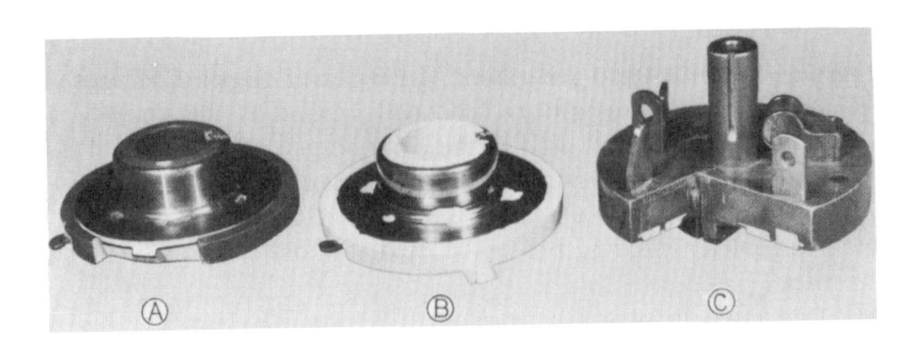

Figure 7–26. Slip rings for generators, alternators, and the like made from alkyd material. Parts (A) and (B) are the same except for different color material. Slip ring (B) is shown as it came from the transfer mold. Note the flash around the copper insert. Slip ring (A) has the flash removed and shows a cut away section to illustrate the anchorage of the insert in the plastic. Slip ring (C) shows a cut away section to illustrate the complicated molded inserts.

been developed. The new molding process is called "outsert molding". Outsert molding gets its name from the fact that small plastic parts (bosses, hubs, pins, bearings, sockets, posts, bushings, gears, and cams) are molded into a metal plate. Figure 7–27 shows a clock mechanism that has been fabricated by the outsert molding process. The metal stamping with the molded-in posts and bearings for the clock is shown in Figure 7–28. Closer tolerances can be held by outsert molding than can be obtained in a one-piece plastic molding.

In outsert molding, the metal plate forms the major component of the part, and the plastic is used for other types of support. The metal plate thickness is between .040 to .080 in. and is steel or aluminum. The metal plate is prepunched and held in the mold by locator pins. The molds for outsert moldings are either three-plate or hot runner with multiple pin gates feeding various component parts, as shown in

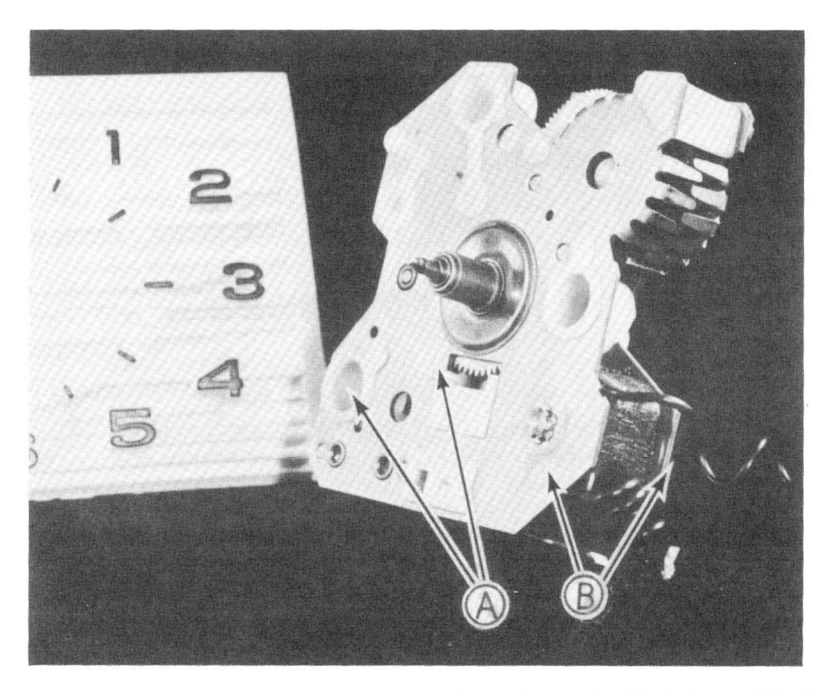

Figure 7–27. An electric clock motor. The bearings and mounting blocks (A) are injection molded of acetal resin in one shot into the metal front and back plates (B). The two sides of the metal front and back plates are folded together to make the clock motor. The whole frame is the field stator. A small flywheel in the squirrel cage makes up the motor. See Figure 7–28.

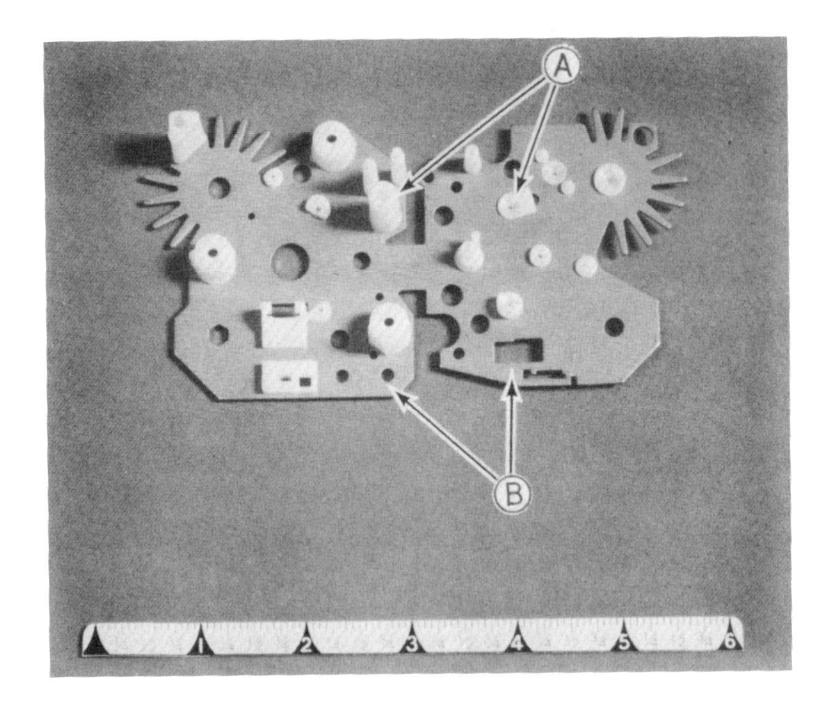

Figure 7–28. An electric clock frame plate. The bearings and mounting blocks (A) are injection molded in one shot from acetal resin into the metal front and back plate (B). The back plate is used as an insert in the mold (see Figure 7–27).

Figure 7–29a. The basic runners and gate system for three-plate molds can be applied in outsert molding. By using sub-runners between the various plastic parts it is possible to fill a number of complex cavities across the face of the metal section. Sub-runners may cause some problems such as warpage or bending of the metal plate as shown in Figure 7–29b. To avoid this and to absorb the shrinkage, it is advisable to use a curved runner instead of a straight runner. The part shrinkage allowance is figured for each individual molding on the metal plate.

Another application of molding plastic inserts into a metal part is shown in Figure 7–30. This is a telephone-dial frameplate. It is used in a telephone-dial handset. The plastic inserts were injection molded from molybdenum disulfide filled nylon. Outsert molding is used in oven and washer timers, telephone dial base plates, clock bases, record players, and tape cassettes.

Another application of a plastic insert instead of metal is shown in Fig. 7–31. Here, a dishwasher pump impeller is made with a phenolic hub and a polypropylene impeller. The hub is transfer molded and then placed in an injection mold as an insert. Polypropylene is then injection molded around the hub and held securely by the small through-holes provided in the hub.

A ball and socket fastener can be molded from two plastics instead of one plastic and a conventional metal insert, as shown in Fig. 7–32. The nylon ball is molded first and serves as an insert. Over the nylon ball insert, an acetal socket is injection molded. The ball and socket are completely interlocked in the molding process. There is no welding of the two plastic materials, because the melting point of the nylon is higher than that of acetal.

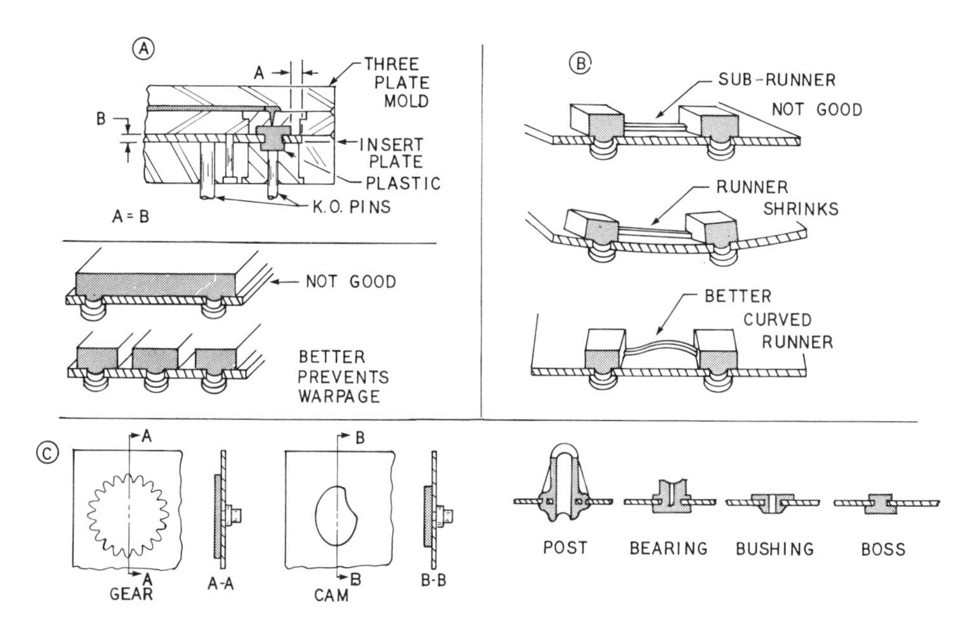

Figure 7–29. This drawing illustrates some of the design aspects of outsert molding. (A) Die section. (B) Sub-runner section. (C) Plastic parts that are molded.

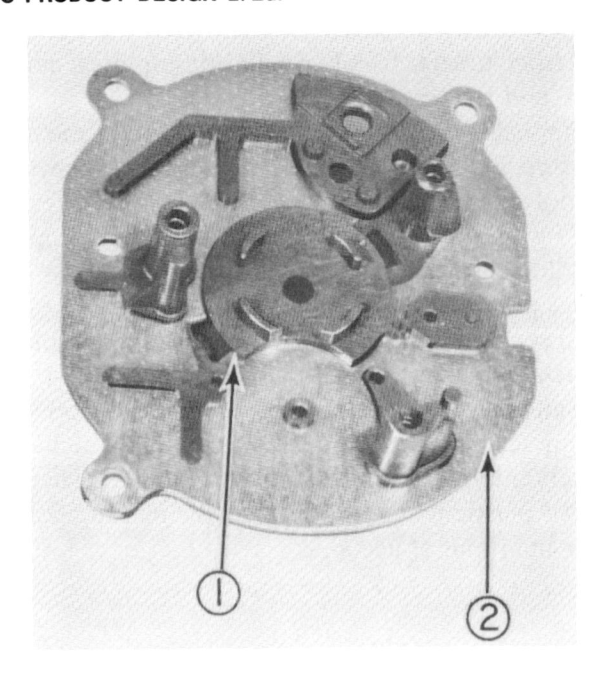

Figure 7–30. A telephone-dial frameplate used in a telephone-dial handset. This is a one-shot nylon injection molded part. The nylon (1) is injected into the metal insert (2).

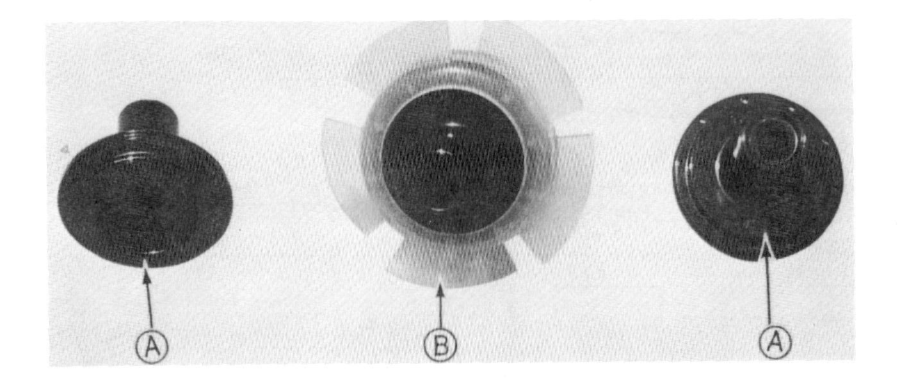

Figure 7–31. A dishwasher pump impeller made in two shots. (A) Shows a front and back view of the transfer molded phenolic hub. (B) The injection molded polypropylene impeller molded around the phenolic hub insert. *(Courtesy General Industries Co.)*

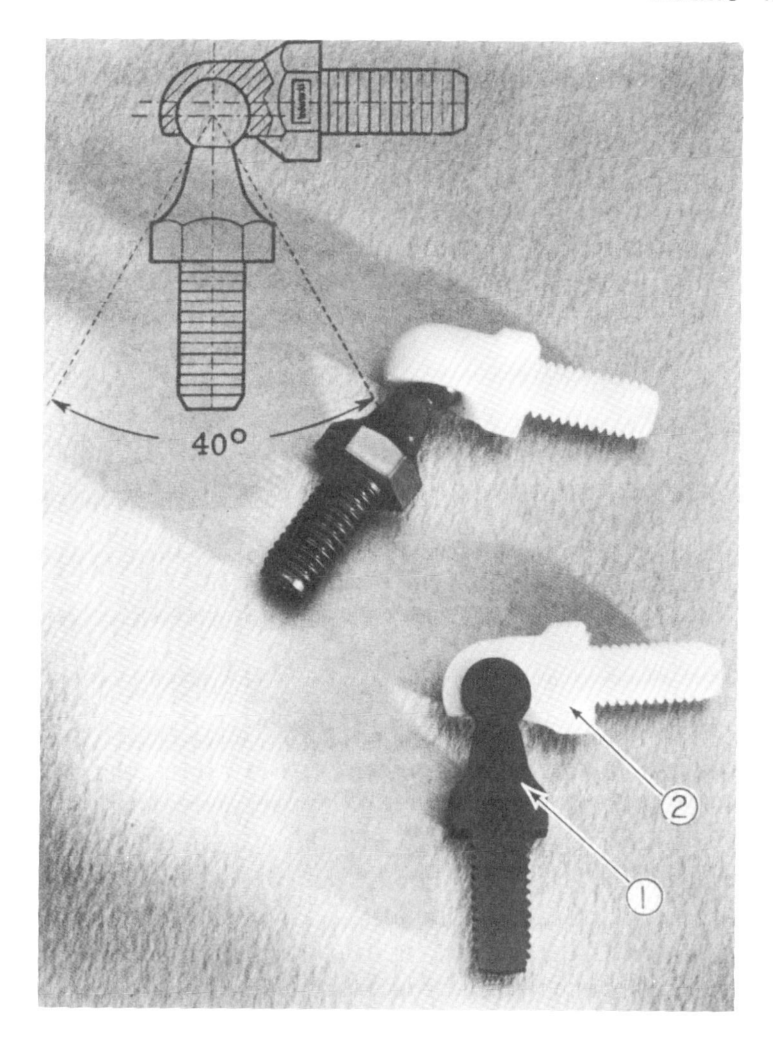

Figure 7–32. Ball and socket fastener. The nylon ball (1) is molded first. The acetal socket (2) is molded over the nylon insert. *(Courtesy Gries Reproducter Co.)*

ENCAPSULATION

Many intricate electronic devices are encapsulated or molded in diallyl phthalate molding compounds. Fig. 7–33 shows many types of metal protruding inserts that have been molded in plastic by the transfer process. Very low transfer molding pressures are used to pre-

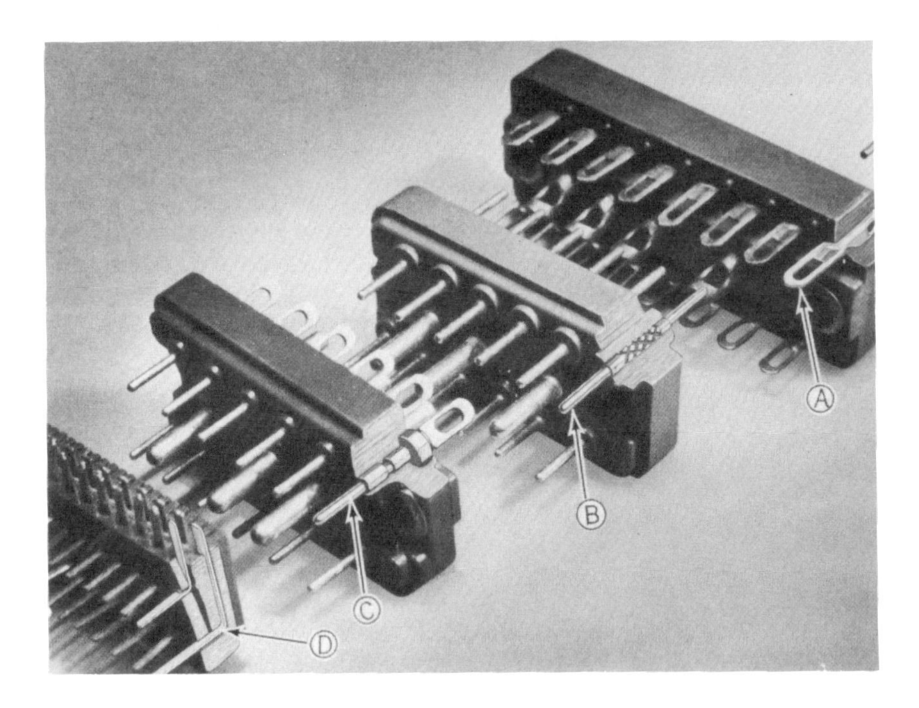

Figure 7–33. Intricate electronic parts showing molded in metal inserts. A section of each part has been cut away to show the insert. (A) An eyelet insert assembled after molding. (B) A knurled wire or pin type molded through insert. (C) An eyelet pin type molded through insert. (D) A flat ribbon wire molded through insert.

vent damage to the delicate inserts. Extreme accuracy is required in making the inserts and the molds. It has been the development of plastic materials that can be transferred at low pressure that has made possible the embedment of delicate electronic parts. The most widely used materials are diallyl phthalates, epoxies, and silicones. Fig. 7–34 shows a group of electronic molded parts with wire-type inserts that have been molded with a fast-cure phenolic resin. The transfer molding process was used.

Fig. 7–35 shows a linear motor that has been encapsulated in modified polyphenylene oxide by injection molding. Note the intricate wiring and metal inserts that have been encapsulated. The mold that made the part was two-cavity, employing kiss-off areas at the top and bottom of the part. The molded part is 9.5 in. long and 1.5 in. wide.

The linear motor is used to move a special drapery rod, opening and closing the drape attached to the rod without the need of pulleys.

The principle of the linear motor is one of electromagnetic fields. Two sets of copper wire coils, four to a set, are prime components of the motor. The copper wire in one set is wound to the left and in the other set, to the right. These are housed on a "stack" made of steel called motor stator. When current is applied through the coils, magnetic energy is created. The capacitor phases the magnetic flux on the set of coils to permit motion in either direction.

COMPOSITE PARTS

Composite molding is a molding process whereby two or more plastic materials are molded in one operation to make one molded part. The advantage of this type of molding is that it utilizes the different properties of the two plastic materials. Fig. 7–36 illustrates a gear

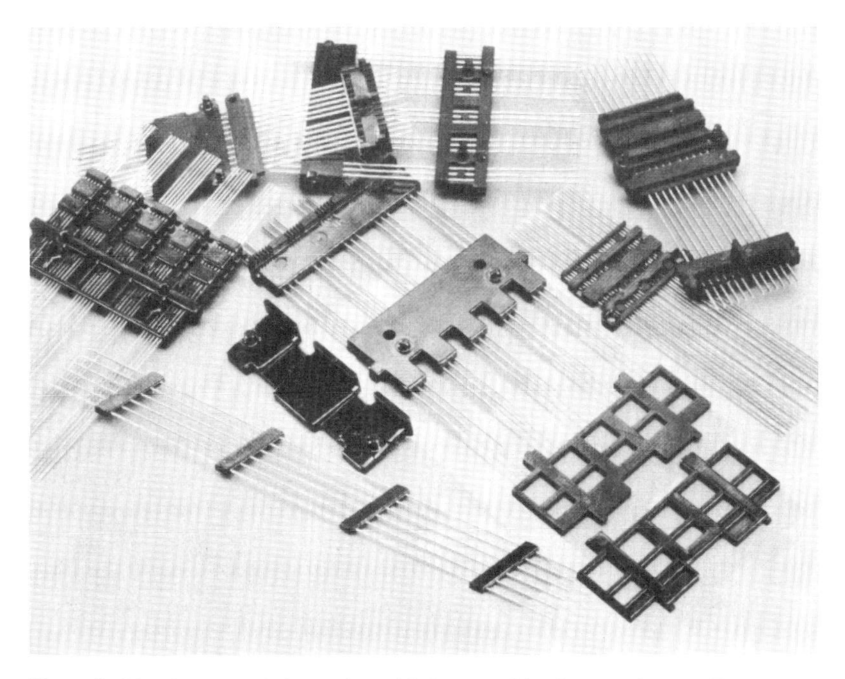

Figure 7–34. A group of electronic molded parts with wire type inserts. The parts have been transfer molded from a general purpose phenolic fast cure resin.

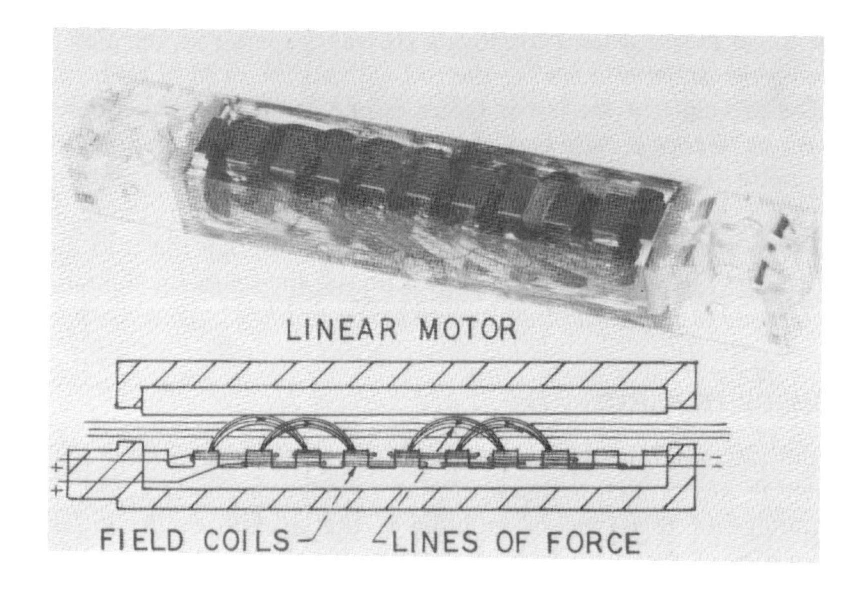

Figure 7–35. This illustrates a linear motor that has been encapsulated in modified polyphenylene oxide by injection molding. Note the intricate make-up of the motor. *(Courtesy Sturgis Molded Products).*

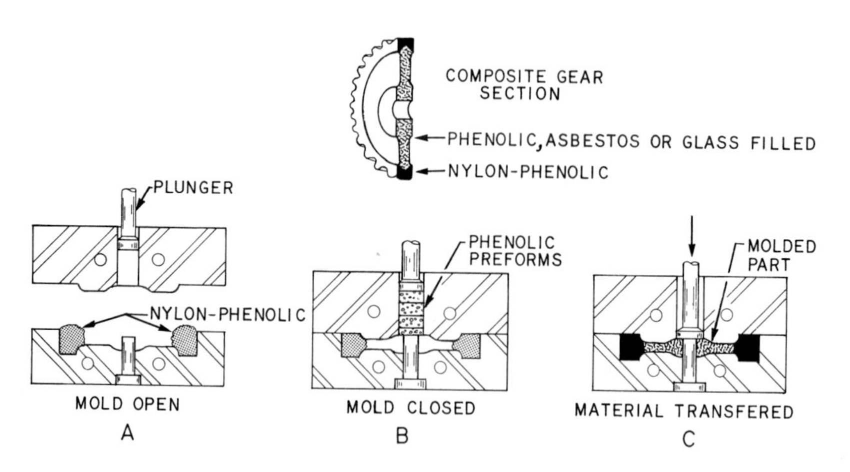

Figure 7–36. The composite molding of a gear. Nylon-filled phenolic is used for the teeth of the gear and glass-filled phenolic is used for the hub.

molded with two materials. The hub of the gear is made of glass-filled phenolic material, and the teeth of the gear are made of nylon-filled phenolic. Fig. 7–36a illustrates the first step in the molding process. The loose nylon-phenolic material is placed in the outer periphery of the open mold at the teeth area. Fig. 7–36b shows the transfer die closed with the glass-filled phenolic material loaded in the transfer chamber. Fig. 7–36c shows the material transferred into the mold and the plastic cured, thus making the composite gear. This process makes gears that are very stable to close dimensional tolerances.

Fig. 7–37 shows a nylon coupling gear that is used in conjunction with an automobile window lift motor. The whole assembly represents a composite gear made of nylon, rubber, and metal. The nylon gear is injection molded around a straight knurled metal shaft. The neoprene coupling shield is bonded to the nylon after molding, and a metal back plate is bonded to the neoprene rubber coupling shield. The neoprene rubber seals out any water that might get into the electric motor.

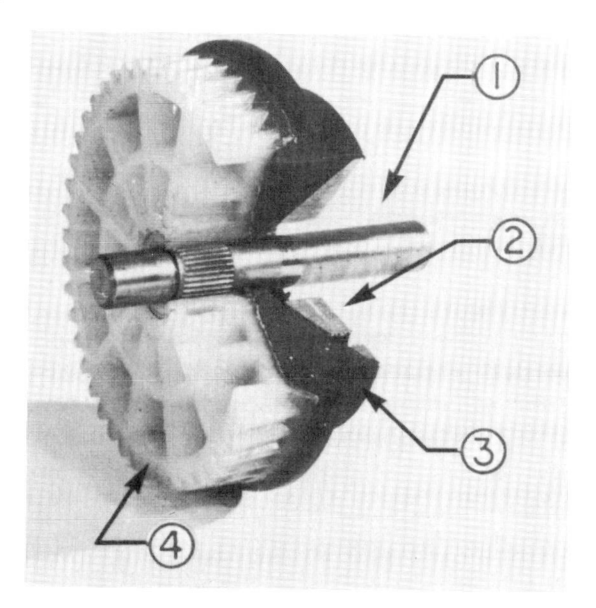

Figure 7–37. This picture illustrates a cross section of a nylon coupling gear used in conjunction with an electric motor for an automobile window lift. Note the straight knurled metal shaft. (1) Knurled metal shaft. (2) Metal face plate. (3) Neoprene coupling shield. (4) Nylon molded gear.

8

Fastening and Joining Plastics

Today, more than ever, the increased use of plastic components in all types of assemblies has initiated the use of new and varied methods of fastening and joining plastic parts. There are four broad techniques used for joining plastics to each other and to other materials: (1) mechanical fasteners, such as rivets, pressed in inserts, self thread-cutting screws, etc.; (2) mechanical means, snap-in fits, and press-in fits; (3) welding, such as spin welding, heat welding, ultrasonic welding, and electronic heat sealing; and (4) adhesives, including solvents, elastomers, monomers, and epoxies.

MECHANICAL FASTENERS

Screws

The self-tapping screw is perhaps the oldest type or method of fastening plastic parts. There are two types of self tapping screws: thread-forming and thread-cutting. The self-tapping screw tends to make its own threads. This is done either by compressing and extruding, as in thread-forming screws, or displacement and cutting away of the plastic material in thread-cutting screws.

In order to select the proper type of self-tapping screw, the designer should first select the type of plastic that is to be used. If the plastic material selected is a thermoplastic, a thread-forming type of screw should be used. If the plastic material selected is a thermoset, a thread-cutting screw should be used.

Thread-forming screws. The type of thread-forming screws recommended for most thermoplastic materials is shown in Fig. 8–1. The USA Standard "B" type screw is a blunt-point, spaced-thread screw. It is a fast-driving screw with tapered threads.

The "BP" screw is essentially the same as the type "B," except that it has a 45° included angle, unthreaded cone point. The cone point helps in aligning holes in assembly. The "U" type screw is a multiple-thread drive screw with a blunt point. This type screw is intended for making permanent fastenings and is not recommended where removal is anticipated. The side walls should be at least as thick as the diameter of the screw. Metal threaded inserts should be considered when frequent removal of small diameter screws with high pull-out strengths are required.

A special type of screw is used with nylon. It is called a type "L" screw and is a combination thread-forming and thread-cutting screw. The end of the screw has a tapered flat edge to start the thread, and then the remaining threads on the screw form the full diameter thread in the nylon.

Thermoplastic materials all have plastic memory and will attempt to return to their original shape if distorted. This is the key factor in using a thread-forming screw. As the screw is tightened, the thermoplastic material is forced out of the way by the thread engagement, but it continually tries to return to its original shape. This principle produces a secure locking and tight fit against the screw. On the other hand, if this same type screw were driven into a thermosetting plastic material, it would set up stresses and eventually cause the part to crack.

Fig. 8–1 illustrates an assembly of a thread-forming screw. The molded or drilled hole should have a chamfer in order to guide the screw into the hole and to prevent it from any misalignment. The chamfer also helps to prevent burring or swelling of the plastic part. If possible, the taper on the molded hole should carry the same taper as the thread-forming screw. If the thread-forming screw is to be located in a boss, the wall thickness of the boss should be equal to one screw diameter.

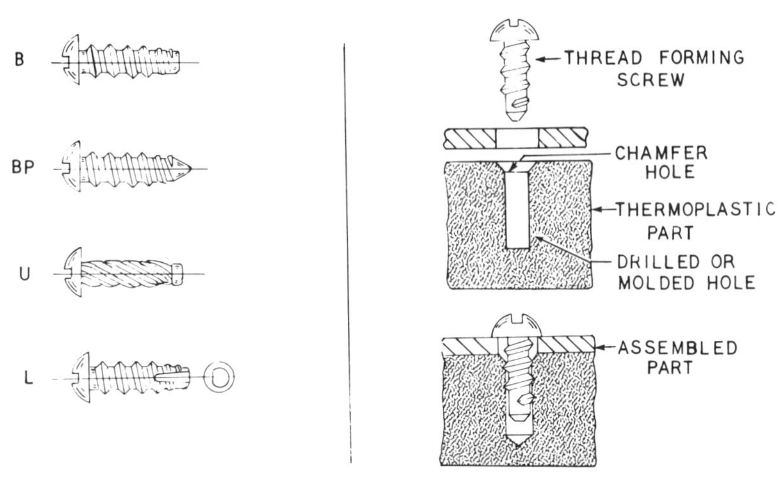

Figure 8–1. The four most often recommended USA standard thread-forming screws used in thermoplastic materials, and a metal thread-forming screw assembly in a thermoplastic material.

Thread-cutting screws. Thread-cutting screws have cutting edges and chip cavities that actually cut a thread in the rigid thermosetting plastic material. The type of thread-cutting screws recommended for most thermosetting plastic materials is shown in Fig. 8–2. The USA Standard "D" type screw has a blunt-point with threads of the same pitch as a standard machine screw. The flute on the end of the screw is designed to produce a cutting edge. This type of screw is very easy for a person to start to thread in a hole. The "F" screw is similar to a machine screw thread and has a blunt point. It has five evenly spaced cutting grooves and large chip cavities. It can be used in most thermosetting plastic materials. The "G" screw has a machine screw thread with a single slot that forms two cutting edges. The "T" screw has a blunt point with a wide flute that gives more chip clearance. The "T" type screw cuts easier than the "D" type. The "T" type screw is easy to start in a hole, and the threads are reusable. The "BF" screw is like type "B" (Fig. 8–1), with a blunt point, but has five evenly spaced cutting grooves and chip cavities. The recommended wall thickness for the "BF" screw is one and one half times the major thread diameter of the screw. This type screw drives faster and provides good pull out strength. The "BT" screw is the same as the "BF" screw, except for a single wide flute that provides room for large chips. The "BF" and "BT" screws require lower driving torque and develop lower boss stresses than thread-forming screws.

Thread-cutting screws are recommended and used in most thermosetting plastic materials. Fig. 8–2 illustrates an assembly with a thread cutting screw. It should be noted that the molded or drilled hole should have a chamfer in order to guide the screw into the hole and prevent it from any misalignment. If possible, the taper on the molded hole should carry the same taper as the thread-cutting screw. If the thread-cutting screw is to be located in a boss, the wall thickness of the boss should be equal to one screw diameter. Note the extra area located at the bottom of the hole. This provides a reservoir for thread-cutting chips.

It is advisable to contact the plastic material company and the

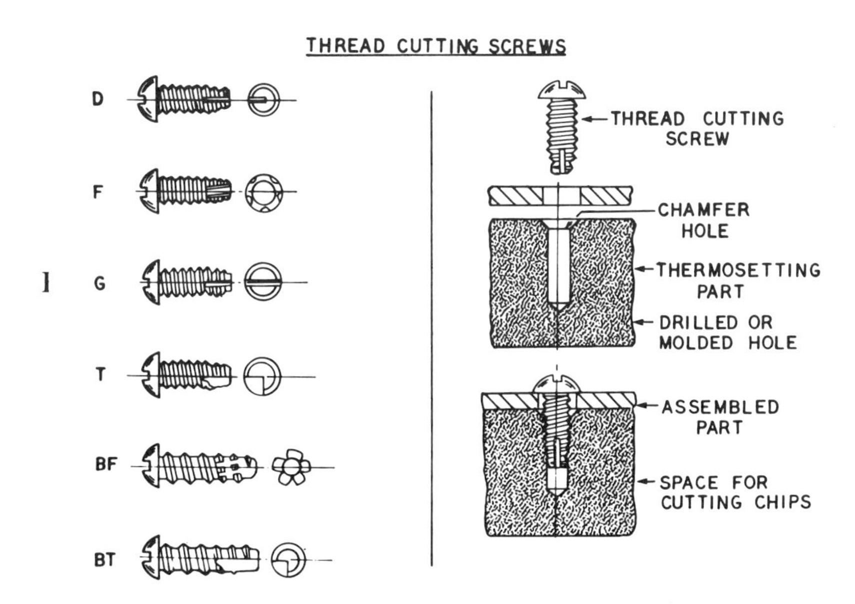

Figure 8–2. Six types of USA standard thread-cutting screws are recommended for use in thermosetting plastic materials, and a metal thread-cutting screw assembly in a thermosetting plastic material.

metal screw manufacturer to determine the exact type of screw sizes that will be needed in any fastening application.

Speed Nuts and Clips

A very rapid assembly method of fastening two component parts is the use of the speed nut and the speed clip (Fig. 8–3 a and b). The speed nut automatically locks itself to the thread of the screw or bolt and is vibration proof when it is fully torqued or tightened down. The speed clip is snapped over the boss or stud provided in the molded plastic part. This locks the parts together securely.

The designer should be cautioned and on the alert to the fact that plastic materials have a coefficient of thermal expansion five to six times greater than metals. Therefore, extreme care should be taken in attaching large plastic parts to metal if warping and buckling of the plastic part is to be eliminated. The placing of elongated slots in the plastic part is one method of overcoming this design problem. Fig. 8–4 illustrates several methods of attaching plastic to metal. Fig. 8–5 shows one method of fastening a plastic part to metal. It is the "pillar" effect. Note that a metal cap fastener is placed over a round molded boss or pillar. The metal cap contains a screw or threaded bolt which in turn attaches to the metal panel. This will allow the pillar to move in relation to expansion and contraction of the part, instead of permitting the whole plastic part to buckle or warp. In order to overcome the expansion of a plastic tail light lens on an automobile, the lens is mounted in a rubber gasket. The rubber gasket will also help to seal out any dirt that may get into the lens.

Wire-type Screw Thread Insert

Fig. 8–6a illustrates a wire-type screw thread insert. The coil wire-type insert is made from a diamond shape wire. The diamond shape wire will act as an internal and external thread when it is made into a coil form. It is installed by pushing the insert into a drilled or molded hole. This acts as a thread for the assembly screw. The thread of the insert is usually of standard size and thread form. This same type of

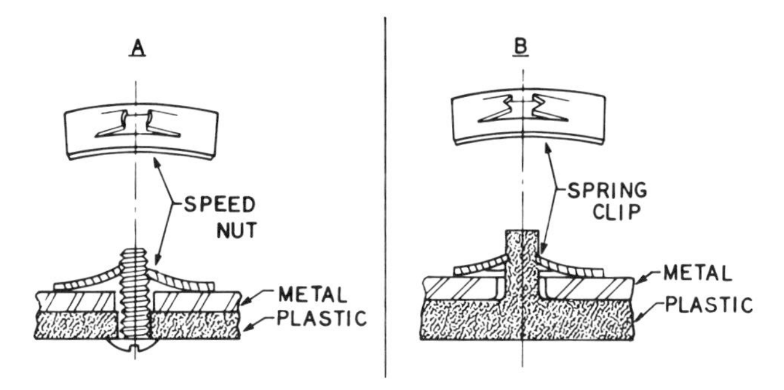

Figure 8–3. (A) A speed nut is a one piece self-locking spring steel fastener used with a threaded bolt to hold assembled parts together. (B) A spring clip or sometimes called a speed clip can be snapped over a plastic stud to lock parts together securely.

wire insert principle can be used in a blind hole by first taping the molded hole and then threading in the wire coil insert. A wire-type screw thread insert assures freedom from thread wear if an assembled part must be taken apart frequently. Fig. 8–6b shows a double threaded insert for attaching to a plastic part and to any other type substrate. Fig. 8–6c illustrates a double thread fastener. This unique thread design provides excellent holding power in plastics. The

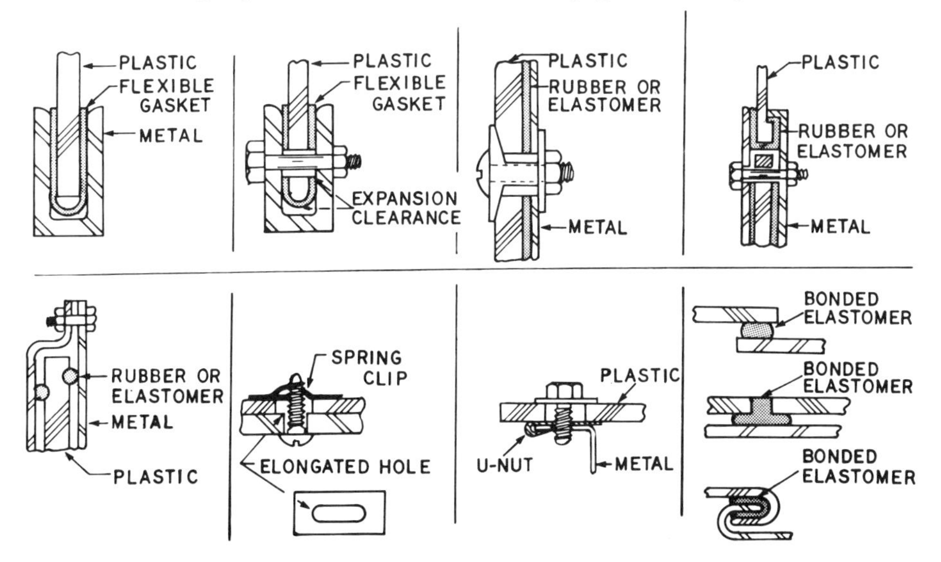

Figure 8–4. This illustrates some of the many ways that plastic materials and metal can be attached in order to overcome the differences in the coefficient of thermal expansion.

Figure 8–5. This illustrates one method of fastening a large plastic part to metal in order to overcome the thermal expansion between plastic and metal.

thread form is designed with a double lead consisting of a high and low thread. More plastic material can be packed between the threads to assure greater holding power and strong resistance to pull-out.

Expansion-type Metal Inserts

A standard type expansion insert is shown in Fig. 8–7a. The insert is placed in a molded or drilled hole and the tapered knurled bottom sections of the insert are spread apart by the metal spreader as it is forced down the four slots in the insert. The molded or drilled hole diameter is generally 0.002 in. greater than the insert. The molded

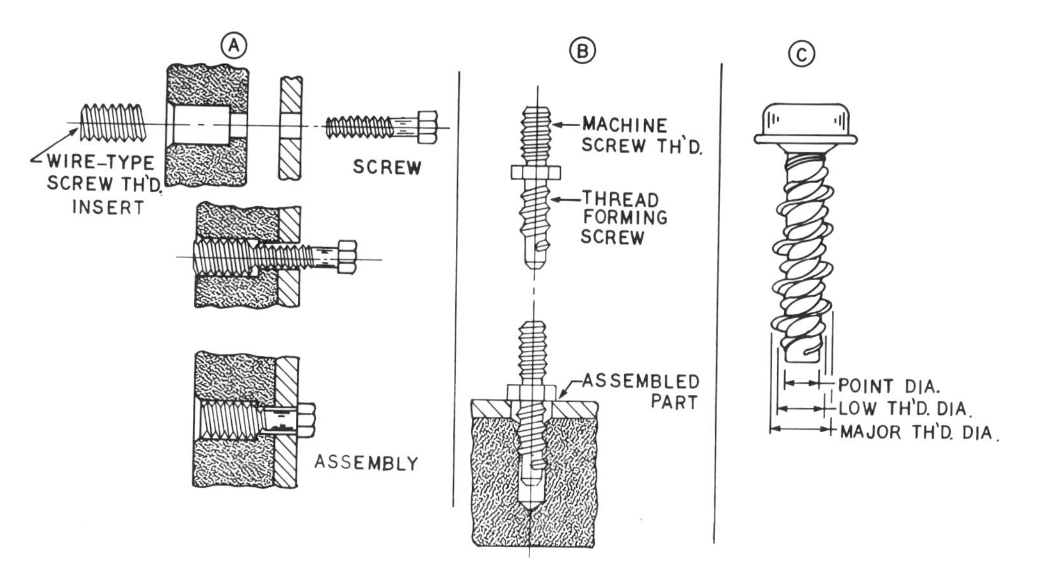

Figure 8–6. (A) A cross-sectional drawing of a wire-type screw thread installed in a molded through hole. (B) A double threaded insert. (C) A double threaded fastener.

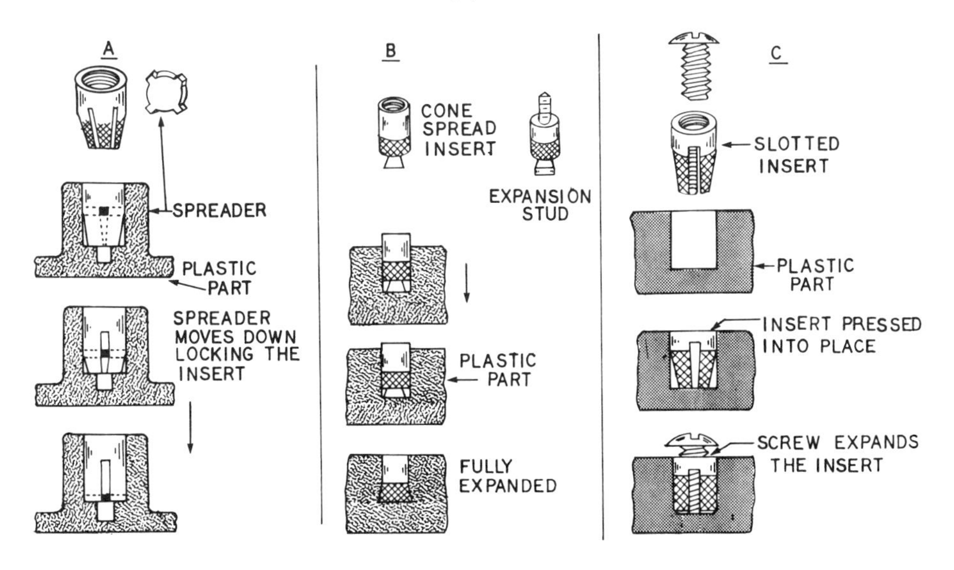

Figure 8–7. One type of an expansion metal insert that is used after a plastic part has been molded. (B) A cone-spread metal insert. This type of metal insert is used in plastic parts that have molded or drilled holes. (C) A metal expansion type insert. The double slotted insert is expanded when the screw is installed.

hole should be flat at the bottom in order to support and retain the spreader.

Fig. 8–7b illustrates a cone-spread metal insert. The insert is made in one piece. It has a knurled outside surface and threads on the inside. The insert has a spreader cone attached to the closed end. As the insert is pressed down into the molded or drilled hole, the spreader cone breaks and forces the external knurls on the insert to expand against the hole wall and lock the metal insert in place. The molded or drilled hole diameter should be equal to the insert body diameter.

Fig. 8–7c illustrates a double slotted metal insert. As the insert is pressed into the hole, it is compressed until the slot is closed. The spring tension holds the insert in place. When the screw is installed, it expands the slotted portion of the insert. The insert is then locked in place. This type of insert is suitable for use in soft plastic materials.

Rivets

Perhaps the oldest method of fastening plastic parts together is the metal rivet. Very few fasteners can match the advantages of the tubular and split rivets. Rivets can be used manually or on automatic bench riveting equipment, but they do not carry the precision tolerances of metal screws and inserts. However, rivets are not considered to be the best type of fastener to withstand tension. Their great advantage is that they are inexpensive and easy to install. Fig. 8–8a illustrates a semitubular rivet. The proportion for the distance of the rivet from the edge should be three times the shank diameter. The proportion for the clinch allowance should be six tenths (.6) times the shank diameter.

Blind Rivets

A blind rivet is used when it is impossible to have access to the reverse side of the joint. Blind rivets are available both in metal and plastic and are designed for installation from one side only. Essentially, the blind rivet consists of a hollow body and a solid pin (Fig. 8–9). The setting of the rivet is done by driving or pulling the solid

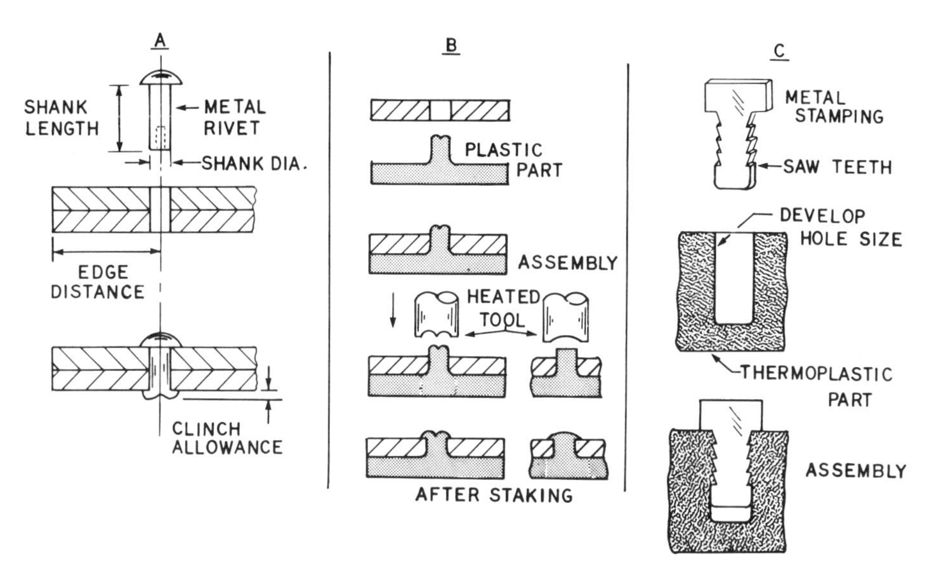

Figure 8–8. (A) Metal rivets are used to fasten plastic parts together. (B) The process of heat staking a thermoplastic assembly part. (C) A broaching assembly process. The metal insert is held in place by broaching.

pin through the hollow shank and flaring the shank on the blind side of the rivet and joint. This provides a positive locking action. There are a variety of proprietary designs for plastic rivets.

Hot Heading or Heat Staking

In this process, a heated metal staking tool is pressed down and over a thermoplastic stud (Fig. 8–8b). The thermoplastic stud melts, taking the form of the staking tool. This type of staking is not as fast as ultrasonic staking. The amorphous type of thermoplastic materials are much easier to heat stake than the crystalline materials.

Broaching

In this process, a metal stamping with saw teeth on both sides of the metal inserts broaches through the plastic as the part is assembled (Fig. 8–8c). It will be necessary to develop the hole size to determine the exact amount of undercutting that will be necessary for the flow

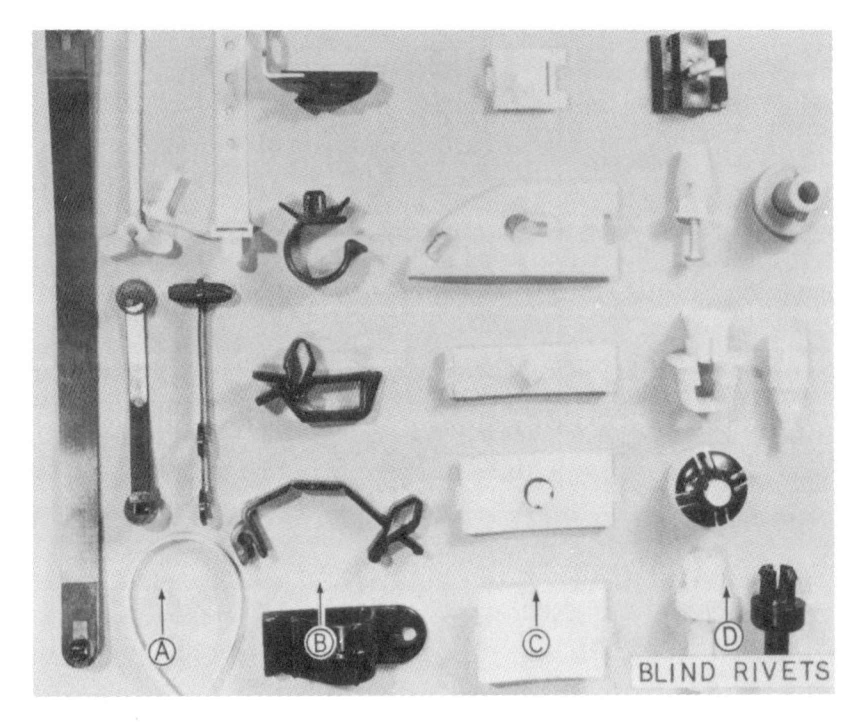

Figure 8–9. This picture illustrates an assortment of plastic fasteners. (A) A group of strapping harness fasteners for wires. (B) A group of plastic cable clamps for wires. (C) A group of plastic fasteners used for holding chrome strips to automobile bodies. (D) This illustrates a group of plastic blind rivets.

of the plastic into the teeth and the pull out strength. This means that the metal mold pin is made over size and then gradually reduced in size by removing metal from the pin, until the exact dimensions are obtained in the molded part. An interference of between 0.015 to 0.020 in. per side has been found to be adequate for most applications.

Hinges

No discussions on the subject of fastening would be complete without covering hinges, which act in effect as inserts to hold two plastic parts together. The two parts, top and bottom, of a plastic closure may be held together by a hinge. Three types of hinges that are standard in the plastic industry and have been used in the past are the Rathbun hinge, the piano hinge, and the lug and pin hinge (Fig. 8–10).

Rathbun hinge. The Rathbun hinge uses elliptical shaped steel spring clips to hold a box lid tightly closed or wide open. The steel spring clips are placed in small undercuts or slots provided in both top and bottom of the lid and box. Special tooling in the mold is required in order to mold the undercuts. This type of hinge provides a sturdy spring action and holds the cover of a box tightly in place.

Piano hinge. The piano hinge is made of metal and is the same type design and construction that is used to fasten the hinged cover of a piano. This type of hinge is used in the plastic industry on plastic containers, boxes, etc. The hinge is fastened to the plastic parts by

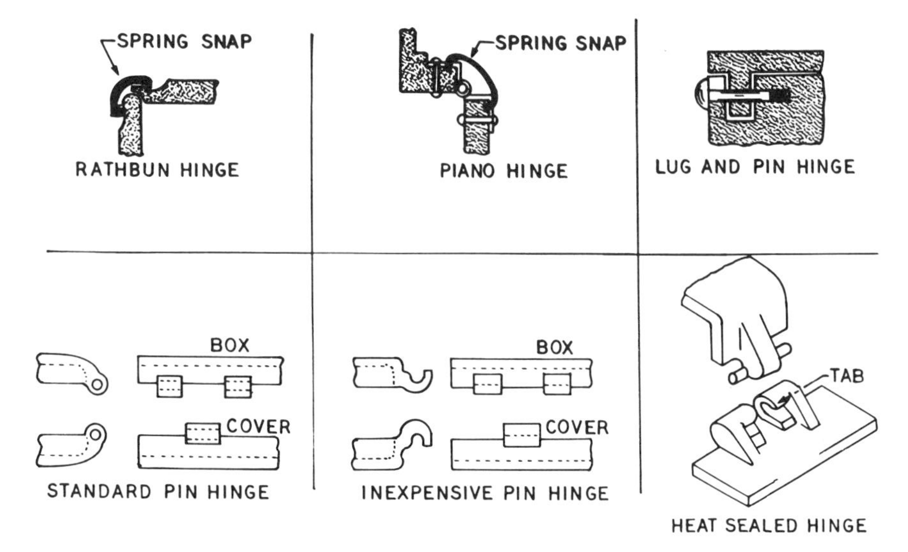

Figure 8–10. Hinges used in plastics. The Rathbun hinge with molded mating curved sections and molded special undercut grooves. The piano type of spring snap hinge assembled by rivets. The lug and pin hinge with molded mating lug and slot. The standard pin hinge will require molded or drilled holes. The inexpensive pin hinge design will eliminate molding or drilling of holes. The heat sealed hinge is assembled and the two tabs are heated and bent permanently around the two pins.

rivets or self tapping screws. The holes for the rivets and self tapping screws are generally drilled after the part has been molded.

Lug and pin hinge. The lug and pin hinge is the least expensive from a molding and assembly standpoint. This hinge requires a recess in the side of the box and a molded lug or prong on the cover. A hole is drilled through the side of the box and through the center of the lug, and then a metal pin is driven into the hole.

Standard pin hinge. The standard pin hinge will require molded holes or drilled holes. Molded holes are expensive and will require cams in the mold.

Inexpensive pin hinge. The inexpensive pin hinge design will eliminate the drilling or cam operation that is required in the standard pin hinge.

Heat sealed hinge. Heat sealed hinges are very strong and durable. One half of this hinge is molded with a pin and the other half is molded with two tabs. After the hinge is assembled, the lower part of the two tabs are heated and bent permanently around the two pins.

Integral hinges. Fig. 8–11 illustrates three rather modern type hinges that are classified as integral hinges. They are integral molded strap hinge, the integral molded hinge, and the integral coined hinge.

Integral molded strap hinge. The flexible strap hinge is generally molded out of polyolefins and is approximately 0.250 in. wide and 0.035 in. thick. The straps can be spaced any distance apart.

Integral molded hinge. The integral molded hinge is the most popular and practical hinge used today. It is generally molded out of polypropylene and can be flexed many hundreds of thousands of times without failure. In order to orient the molecular chains of the material across the hinge for increased strength and life, the part should be opened and closed a few times right after molding. The gate in the mold must be placed so that the flow of plastic is straight

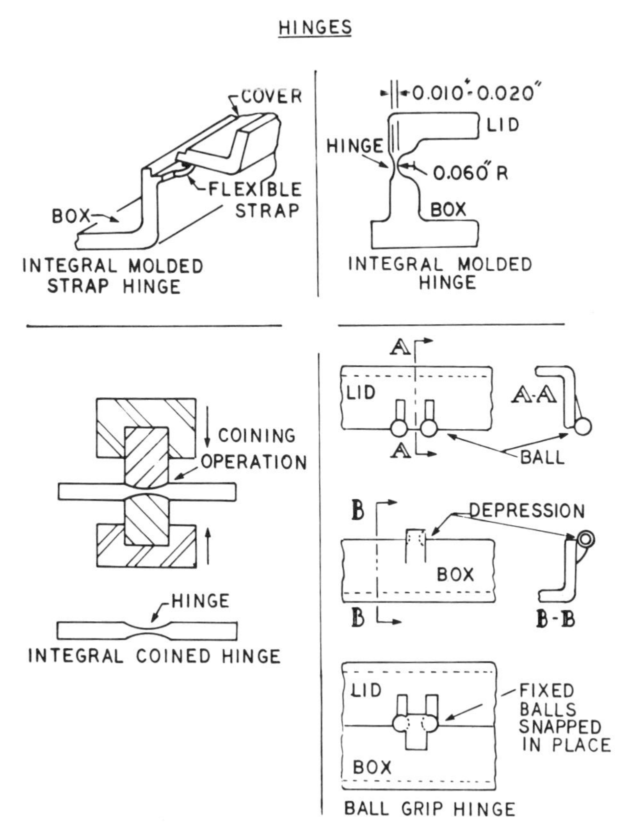

Figure 8–11. Four types of plastic hinges. The integral molded strap hinge and the integral molded hinge are made from the polyolefins. The integral coined hinge has been made from nylon, acetal and the polyolefins. The ball grip hinge is generally made out of polystyrene.

across the hinge, and not lengthwise. The gate should also be located in the heavier half of the molded part so that the flow is across the hinge to the lighter half of the piece. Hinge thickness is usually 0.010 to 0.020 in.

The integral hinge can be extruded by the extrusion process, however, the hinge has poor flex life as compared to the standard injection molded hinge. This is because the hinge is formed in the direction of plastic polymer flow, and as a result, it cannot be sufficiently oriented when flexed. Fig. 8–12 shows an integral molded hinge.

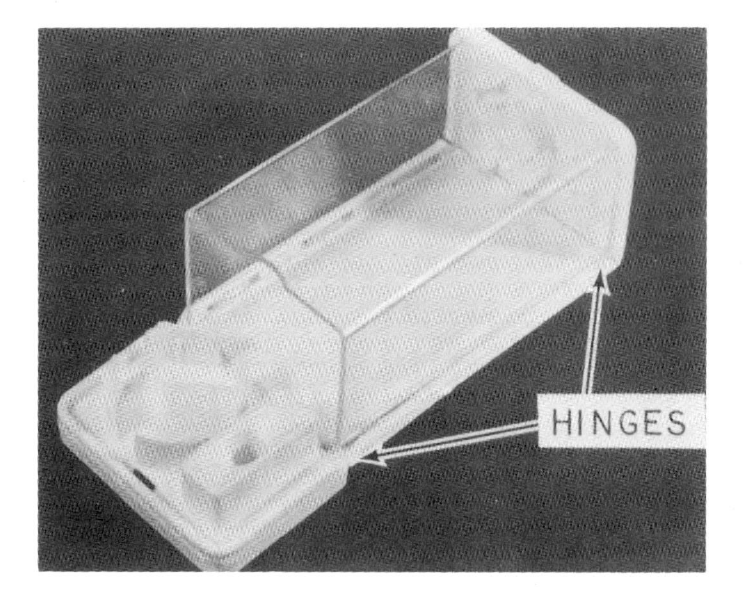

Figure 8–12. Two integral molded hinges. The display case is injection molded from polyethylene.

Integral coined hinge. A coined hinge is made by placing the part to be coined between two coining bars and applying enough pressure until the desired hinge thickness is reached. Heat is sometimes applied to the coining bars. The pressure is released and the hinge is removed. The hinge thickness ranges from 0.010 to 0.015 in. The flex life of the coined hinge is not as good as the integral molded hinge, but the ability to resist tearing is greater. Fig. 8–13 illustrates a coined hinge made from acetal resin. Successful coined hinges are made from nylon, acetal, polypropylene, and polyethylene materials.

Ball grip hinge. Fig. 8–11 shows a ball grip hinge design. This type of hinge is used on small boxes and is an accepted standard in the box industry. The two balls are approximately 0.125 in. in diameter and are molded into one half of the box. The balls snap into the two depressions on the other half of the box container. Although the two depressions are undercuts in the mold, the parts are stripped out. The depth of the depression is approximately 0.018 in.

Clasps

Fig. 8–14 displays many types of clasps that are used on plastic containers or boxes. It will be noted that most of the clasps work on the friction hook principle.

Snap-fits

The strength of a snap-fit comes from mechanical interlocking, as well as from friction. Snap-fits require a tough, stiff, plastic engineering material. ABS, nylon, acetal, polycarbonate are used mostly in this type of fastening assembly. Fig. 8–15 illustrates some of the many types of integral designs that make use of the snap fit.

Miscellaneous type fasteners

Hi-Lo. The hi-lo screw has two sets of threads, one high and one low (Fig. 8–16a). This gives greatly reduced radial pressure and minimal boss cracking. The high threads increase the axial shear area for

Figure 8–13. A mechanics rule that has a case injection molded of acetal resin. The hinges were coined after molding. The coined hinges are strong and durable, reduce the number of parts, and simplify assembly operations.

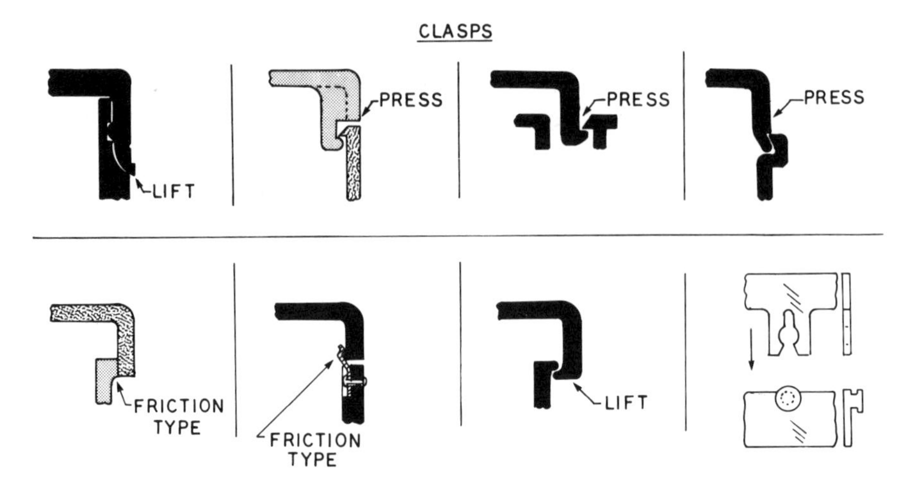

Figure 8–14. Different types of clasps that have been used on plastic boxes or containers.

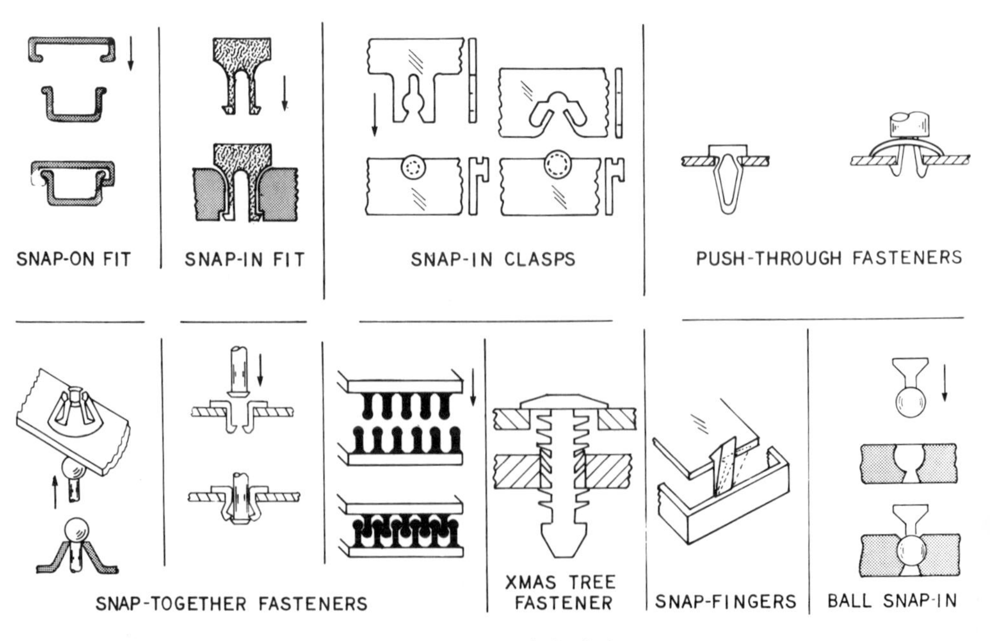

Figure 8–15. A group of many different types of plastic fasteners.

higher stripping torque. The low threads are used for good driving stability. This type screw is used mostly on thermoplastics.

Drive studs. Drive studs are used for fastening where the plastic assembly is not taken apart (Fig. 8–16b). The holding power of this type fastener is not equivalent to that of metal inserts or metal screws.

Unthreaded fasteners. The three classes of fasteners for unthreaded studs and bosses are push-on, turn-on, and cap (Fig. 8–16c). These special designed fasteners are used to minimize the danger of loose assemblies from bosses and increase the pull-out strength. If the push-on fastener is used the stud must be stiff enough to resist the push-on force. After the fastener has "bottomed," the surface from which the stud protrudes must be strong enough to withstand the pressure exerted on it. A turn-on fastener is essentially a self-threaded nut. The nut exerts a fair amount of torque as it is turned down and seated. The stud must be firmly anchored so it will not spin with the nut. This type of fastener is used for light load bearing assemblies.

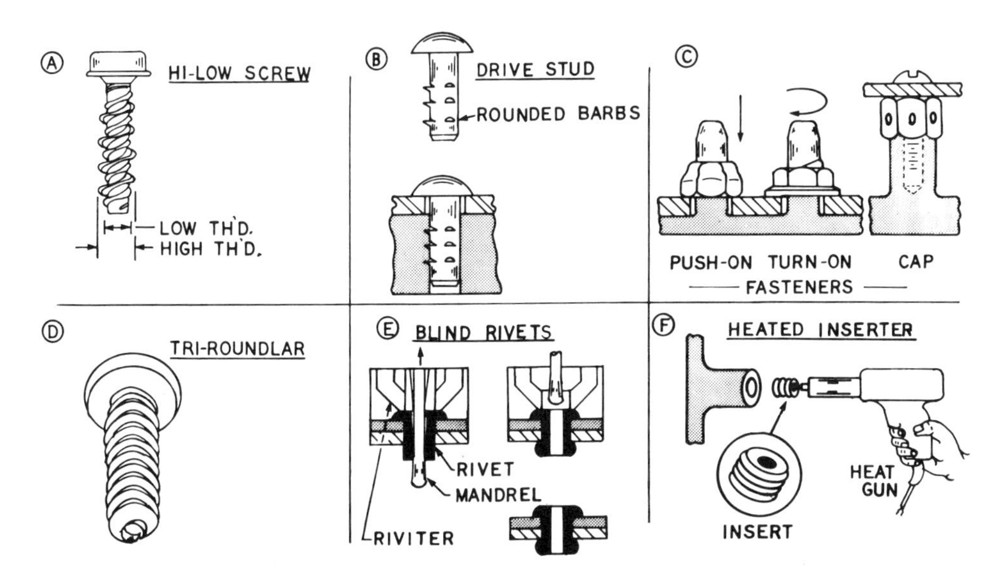

Figure 8–16. This illustrates a group of different type metal fasteners.

Cap type fasteners are made to fit over bosses with cored holes. This type fastener has a threaded receiving end and ears that cut into the plastic boss for anchorage.

Tri-roundlar. The tri-roundlar is designed so that cold flow of the thermoplastic material can occur as the thread is being rolled by the insertion of the screw (Fig. 8–16d). This is done to minimize radial stresses.

Blind rivets. In many applications, metal blind rivets are faster to apply, easier to use and provide a better joint than nuts and bolts, self-tapping screws or adhesives (Fig. 8–16e). Sometimes the rivets can serve as a clamp while an adhesive, intended for the permanent bond, cures. Joining plastics with most blind rivets is not always successful. Cracking, crazing, and delaminating are likely occurrences. Each plastic must be checked out for the proper rivet hole diameter and relative rivet and mandrel size to get a good joint without cracking. Blind rivets are relatively inexpensive to use.

Heated inserts. Heated metal inserts are sometimes inserted into thermoplastics to melt the compound as the insert enters (Fig. 8–16f). A rounded type insert is recommended. This type of insertion can be used for prototype work or repair work.

Press or Interference Fit.

The press or interference fit is the forcing of a slightly oversize part into a standard hole or opening. This is generally used in pressing metal inserts into a plastic part. It is limited to the stiff and tough plastic engineering materials. The designer should be aware of the fact that the joint may loosen in time, due to creep in the plastic material. Fig. 8–17 shows many nylon electrical connectors that are hand pressed together.

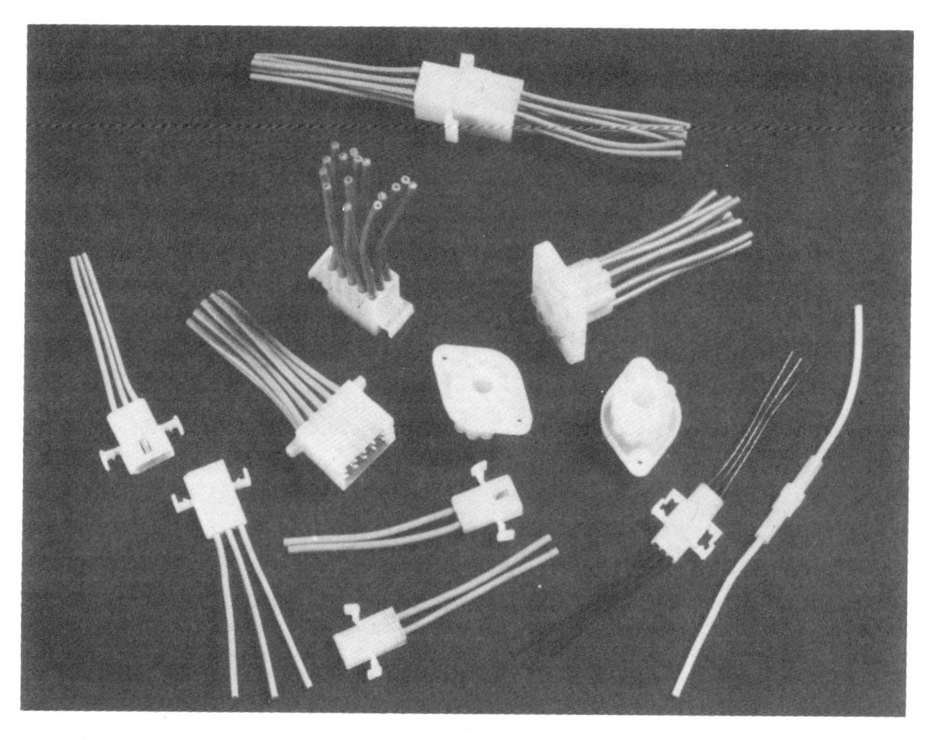

Figure 8–17. Injection molded nylon electrical connectors. The connectors are hand pressed together in assembly. The nylon used is self-extinguishing and is an important safety feature.

METHODS OF WELDING PLASTICS

Ultrasonic Welding

Ultrasonics is the science of producing and utilizing mechanical energy at frequencies above the range of human hearing. Ultrasonic welding is based on the induced heat generated in a thermoplastic by a high frequency electrodynamic field that causes the plastic surfaces to rub against one another and form a fusion bond. It is one of the fastest methods for joining thermoplastics. An ultrasonic tool, which is sometimes called a horn, is needed for this process (Fig. 8–18). The horn is vibrated by a mechanism that changes electrical energy into vibratory action by a transducer. A transducer converts electrical frequency into mechanical vibrations. The horn can be wedged or cone-shaped, and the narrow edge is pressed against the plastics to be

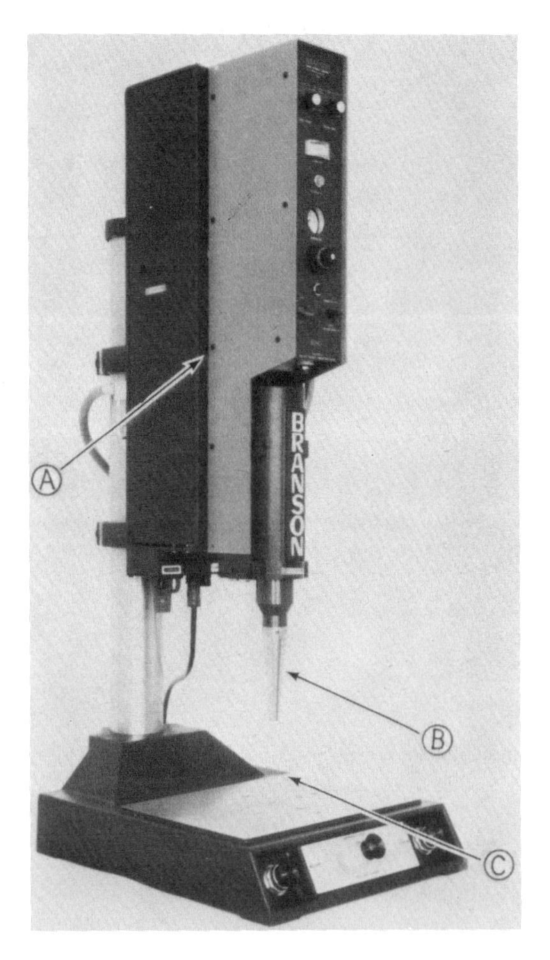

Figure 8–18. A typical ultrasonic machine. (A) Ultrasonic assembly stand. (B) Horn. (C) Work piece area. *(Courtesy Branson Sonic Power Co.)*

bonded. Intense vibration of low amplitude is transmitted from the tool to the plastics. Localized heat is generated by the friction of vibration at the surfaces to be joined.

The unusual feature of this bonding process is that no adhesive, no solvents, no heat, no surface treatments, or surface preparations are necessary. The plastic sections to be welded may be complex shapes. The distance that the vibrations travel in welding the plastic workpieces is very important. More energy reaches the workpiece if vibrations travel only up to 1/4 in. from the point of ultrasonic contact. This is called "near-field" welding. If the vibrations travel more than 1/4 in. from the point of ultrasonic contact, it is called "far-field" welding. Not all thermoplastic materials can be bonded by the ultrasonic method (see Table 8–1). Plastic material suppliers should be contacted for advice on what types of materials will bond.

TABLE 8–1. A GROUP OF THERMOPLASTIC MATERIALS THAT CAN BE ULTRASONIC WELDED AND STAKED. *(Courtesy Branson Sonic Power Co.)*

	ULTRASONIC WELDING					
	Ease of Welding*					
Material	Near Field†	Far Field†	Swaging and Staking	Inserting	Spot Welding	VIBRATION WELDING
Amorphous Resins						
ABS	E	G	E	E	E	E
ABS/polycarbonate alloy (Cycoloy 800)	E-G	G	G	E-G	G	E
Acrylic[a]	G	G-F	F	G	G	E
Acrylic multipolymer (XT-polymer)	G	F	G	G	G	E
Cellulosics – CA, CAB, CAP	F-P	P	G	E	F-P	E
Phenylene-oxide based resins (Noryl)	G	G	G-E	E	G	E
Poly (amide-imide)	G	F				G
Polycarbonate[b]	G	G	G-F	G	G	E
Polystyrene, GP	E	E	F	G-E	F	E
Rubber modified	G	G-F	E	E	E	E
Polysulfone[b]	G	F	G-F	G	F	E
PVC (rigid)	F-P	P	G	E	G-F	G
SAN-NAS-ASA	E	E	F	G	G-F	E
Crystalline Resins[c]						
Acetal	G	F	G-F	G	F	E
Fluoropolymers	P					G-F
Nylon[b]	G	F	G-F	G	F	E
Polyester (thermoplastic)	G	F	F	G	F	E
Polyethylene	F-P	P	G-F	G	G	G-F
Polymethylpentene (TPX)	F	F-P	G-F	E	G	E
Polyphenylene sulfide	G	F	P	G	F	G
Polypropylene	F	P	E	G	E	E

Code: E = Excellent, G = Good, F = Fair, P = Poor

*Ease of welding is a function of joint design, energy requirements, amplitude, and fixturing.

†Near field welding refers to joint ¼ in. (6.35 mm) or less from area of horn contact; far field welding to joint more than ¼ in. (6.35 mm) from contact area.

[a]Cast grades are more difficult to weld due to high molecular weight.

[b]Moisture will inhibit welds.

[c]Crystalline resins in general require higher amplitudes and higher energy levels because of higher melt temperatures and heat of fusion.

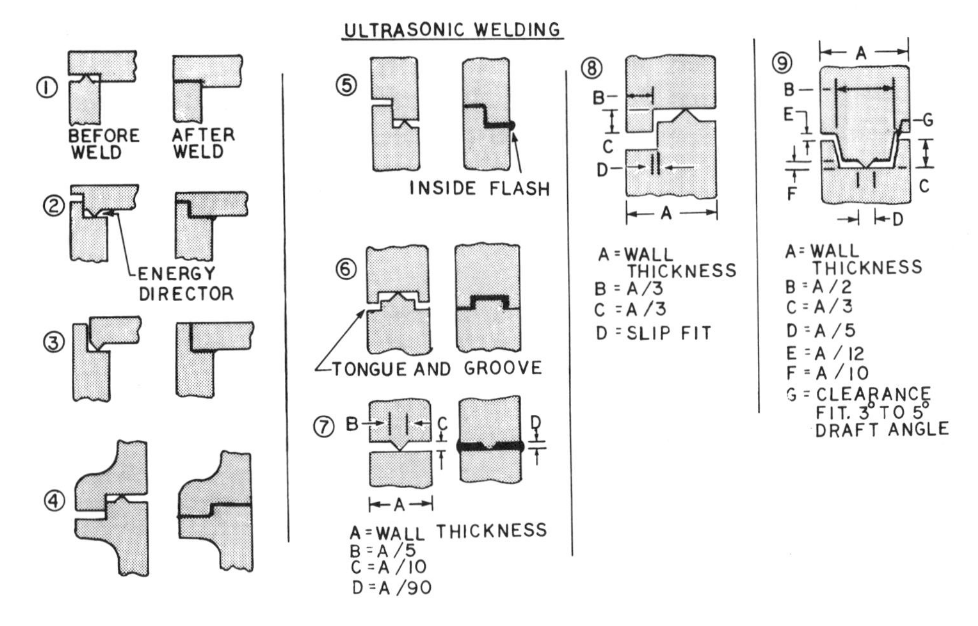

Figure 8–19. Some of the joint designs used in ultrasonic welding. *(Courtesy Branson Sonic Power Co.)*

Fig. 8–19 illustrates some of the joint designs used in ultrasonic welding:

1. A right-angle corner design with an inside step.
2. A right-angle corner design with an internal step.
3. A right-angle corner with a stepped lap joint.
4. A half lap joint with increased bonding area for additional strength.
5. A half lap joint with the flash going to the inside of the part.
6. A tongue and groove joint with the flash contained in the bonded area.
7. Recommended proportions for a simple butt joint.
8. Recommended proportions for a half lap joint.
9. Recommended proportions for a tongue and groove joint.

Ultrasonic Staking

This is similar to heat staking on thermoplastic parts, except that the plastic stud or protrusion is mushroomed over by an ultrasonic staking horn (Fig. 8–20). Ultrasonic staking is much faster than heat staking. Also, the head of the ultrasonic staking tool is cool and eliminates the sticking and cleaning problem that is always present in heat staking. The assemblies are tighter than those formed by heat staking, due to the fact that there is no recovery memory in the staked thermoplastic material. Also, there is less chance of material degradation. Fig. 8–21 shows the ultrasonic staking of a metal part onto a plastic part.

Ultrasonic Inserting

This is a new process of incorporating metal inserts in thermoplastic parts by means of ultrasonic heating. A molded or drilled hole is made slightly undersize compared to the size of the metal insert that is to be inserted into the hole by ultrasonic heating. Ultrasonic vibrations are applied to the metal insert, which causes the displaced plastic to flow into the knurls or undercuts on the insert. This mechanically locks the insert in place (Fig. 8–22).

A special designed metal insert has been made for this type of insertion. It is called an "Ultrasert" (Fig. 8–23 and 8–24). This same insert can be applied by either the ultrasonic method or by "spinserting." This is a frictional melting process where friction, due to relative motion between the metal insert and the thermoplastic, creates heat sufficient to melt the thermoplastic. While the insert is spinning, the installing tool is also pressing the insert into the molten plastic. The heating occurs at the interface between the insert and the plastic and only where the insert is in contact with the plastic. There is no danger of overheating the plastic. This type of insert has been designed to provide optimum transfer of frictional energy from the insert to the plastic. The hole, which is drilled or cored in the plastic, may be either tapered or cylindrical. The tapered hole has, however, three advantages: (1) it provides a good release of the plastic part from the mold; (2) it permits prepositioning of the insert into the hole and in-

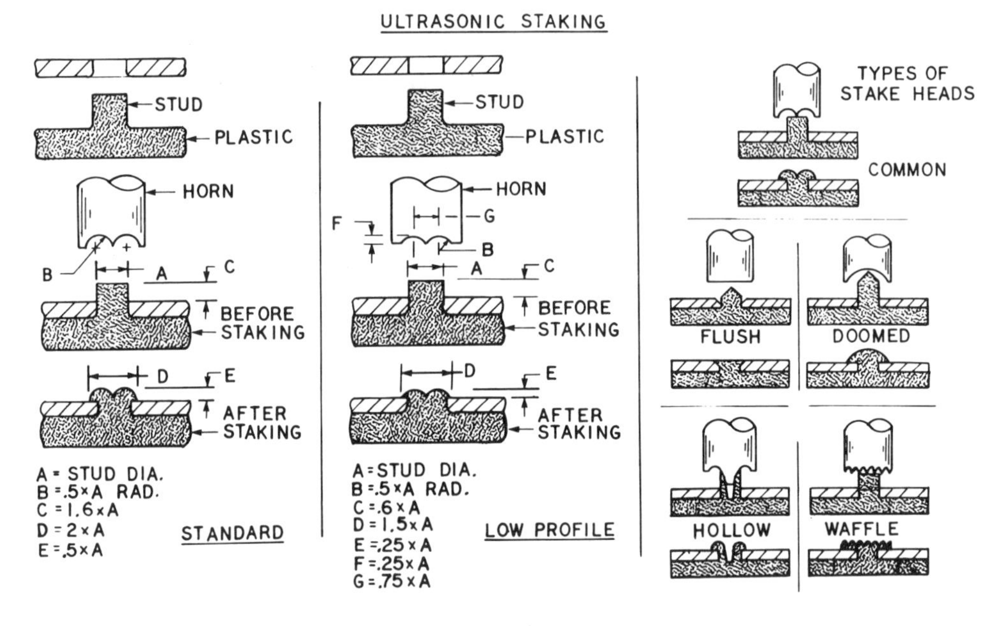

Figure 8–20. The design used for standard ultrasonic staking and low profile staking.

Figure 8–21. Ultrasonic staking metal onto a plastic part. (A) Sonic power supply. (B) Ultrasonic assembly stand. (C) Horn. (D) Workpiece. *(Courtesy Branson Sonic Power Co.)*

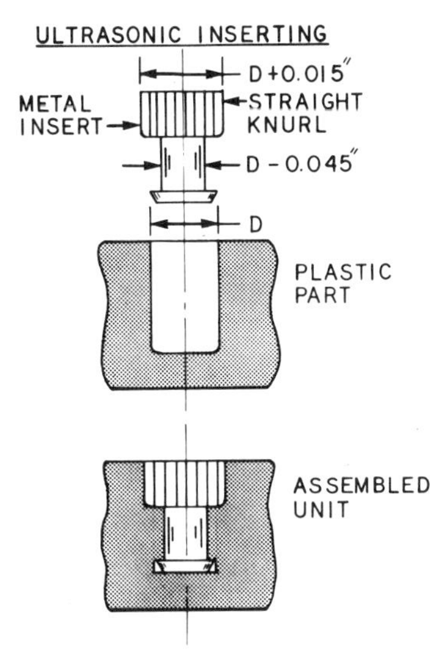

Figure 8–22. This illustrates a typical assembly of a metal insert in a plastic by ultrasonic inserting. *(Courtesy Branson Sonic Power Co.)*

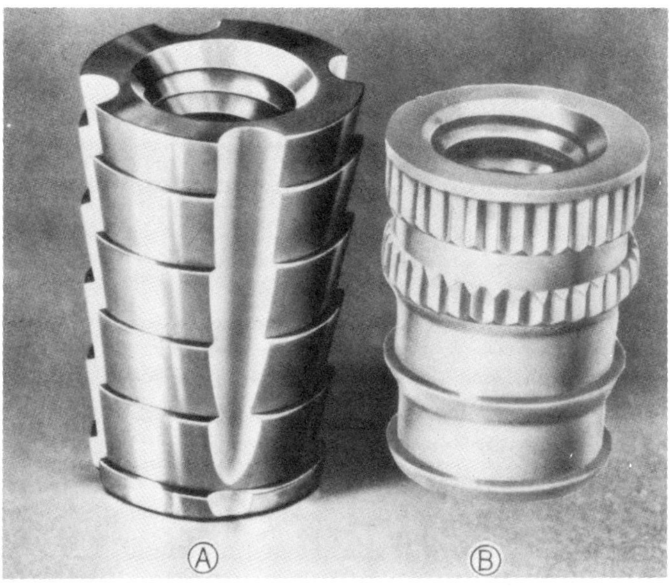

Figure 8–23. The "Ultrasert" is designed to be installed in thermoplastic by the frictional melting method of spinning or by ultrasonic vibration, after molding. (A) Ultraset 1. (B) Ultrasert 2, a later design. *(Courtesy Insert Products Div. of Heli-Coil Corp.)*

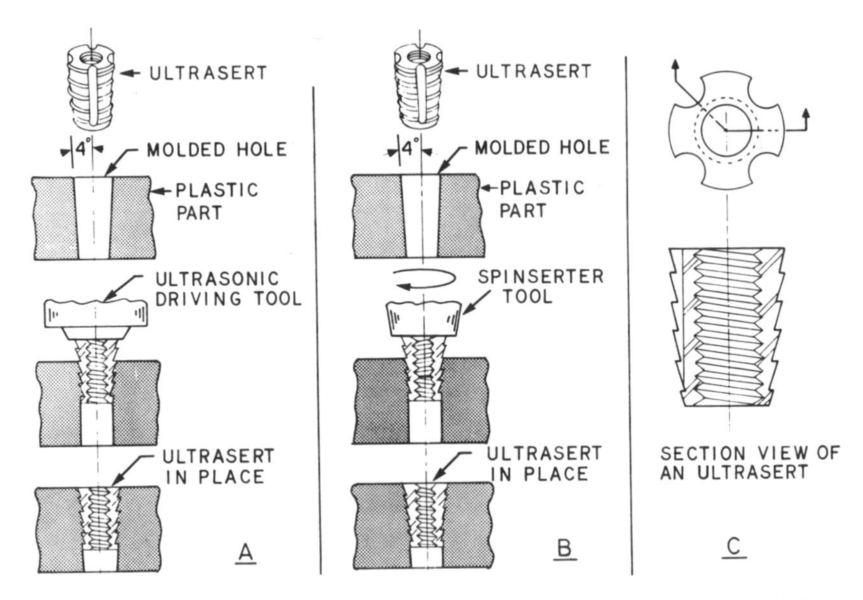

Figure 8–24. This illustrates two methods in which the "ultrasert" may be imbedded into thermoplastic parts. (A) The ultrasonic method. (B) The frictional melting process of spinning the insert. (C) A sectional view of the ultrasert.

sures alignment; and (3) it decreases the volume of plastic that must be displaced by the insert, thereby permitting a faster installation. This type of insert molds itself into the hole during installation, and its performance is like that of a molded-in insert.

Vibration Welding

Vibration welding is based on the principle of spin welding. In vibration welding the heat necessary to melt the plastic is generated by pressing one of the parts against the other and vibrating it through a small relative displacement in the plane of the joint (Fig. 8–25 and 26).

Vibration welding operates at a frequency of 120 Hz. The amplitude of displacement ranges from 0.120 to 0.200 in. The joint pressure is in the range of 200 to 250 psi. The weld time is 2 to 3 seconds. The complete cycle time is 5 to 8 seconds for most parts.

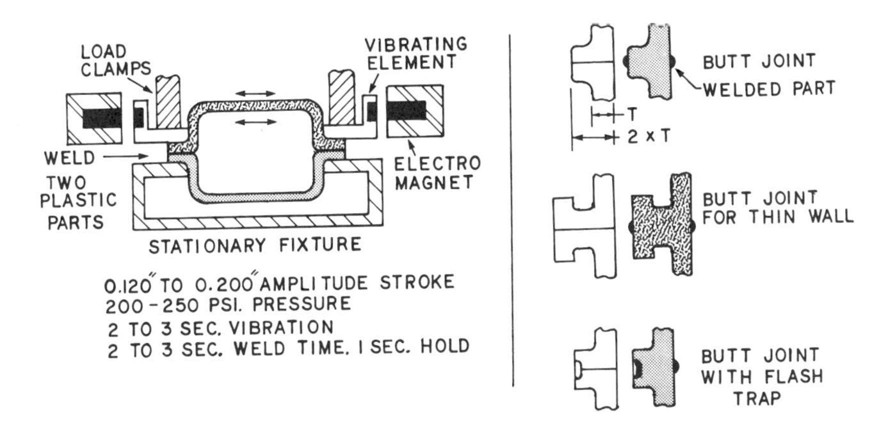

Figure 8–25. This illustrates the principle of vibration welding and three of the joint designs that are used.

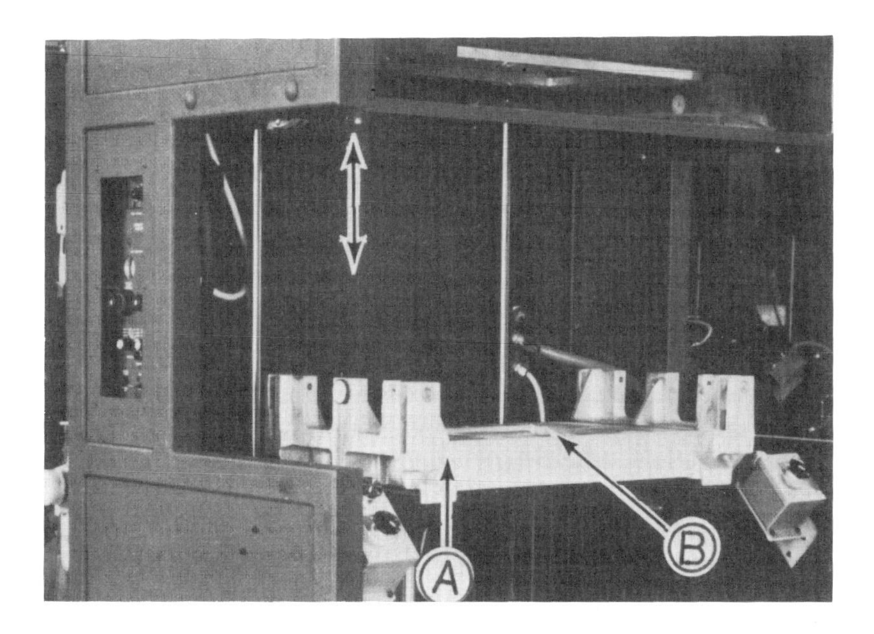

Figure 8–26. Vibration welder. (A) Stationary element. (B) Work area. The vibration element comes down over the work piece from the top. *(Courtesy Branson Sonic Power Co.)*

Joint Design

A flange on the parts to be welded is highly desirable. The flange makes it easier to grip the part in the fixture and also prevents flexing during welding. A minimum wall thickness of 0.030 in. is recommended, along with a flange at the weld surface. The welded joint should be in a single plane. However, parts with more than one plane can be welded. Melt traps can be molded into the flange. The flatness of the part should be 0.005 in. or less. During the welding process, the melted plastic flows at the joint area. The height of the welded assembly is lowered approximately 0.020 in.

Almost all thermoplastics can be vibration welded (Table 8–1). This includes crystalline and amorphous materials. Also cellulosics and fluoropolymers can be bonded by vibration welding. Applications for vibration welding include battery cases, emission control canisters and tail light assemblies.

Spinomatic Headforming

Spinomatic or orbital headforming is accomplished by pressure and a radial spinning or rotating tool. The tool is brought to bear against the part to be upset. This creates a line of pressure from the center point out, displacing a small quantity of plastic with each turn. Fig. 8–27 illustrates some of the various types of heads that can be produced.

Fusion Bonding

This process has been used successfully as an assembly technique for bonding almost any thermoplastic material. It has often been referred to as heat welding or hot plate welding. In this process the two surfaces to be joined are first held lightly against a heated metal surface to melt or until there is sufficient flow of molten material. After the part surfaces have been melted, they are quickly brought together and held under a slight pressure, usually from 10 to 30 psi. (Fig. 8–28). Positive stops are used in order to maintain tighter tolerances and assure dimensional accuracy of the bonded part. The two positive stops

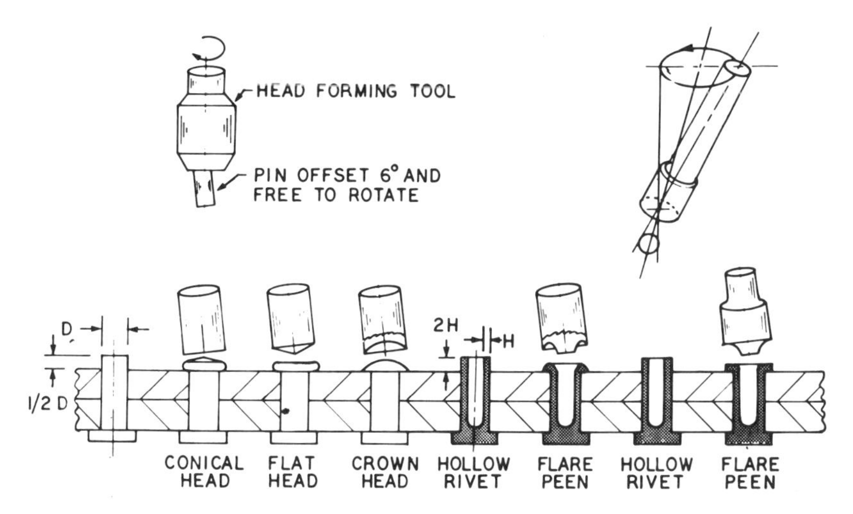

Figure 8–27. This illustrates the principle and some of the various types of heads that can be produced in headforming. *(Courtesy of VSI Automation Assembly, Inc.)*

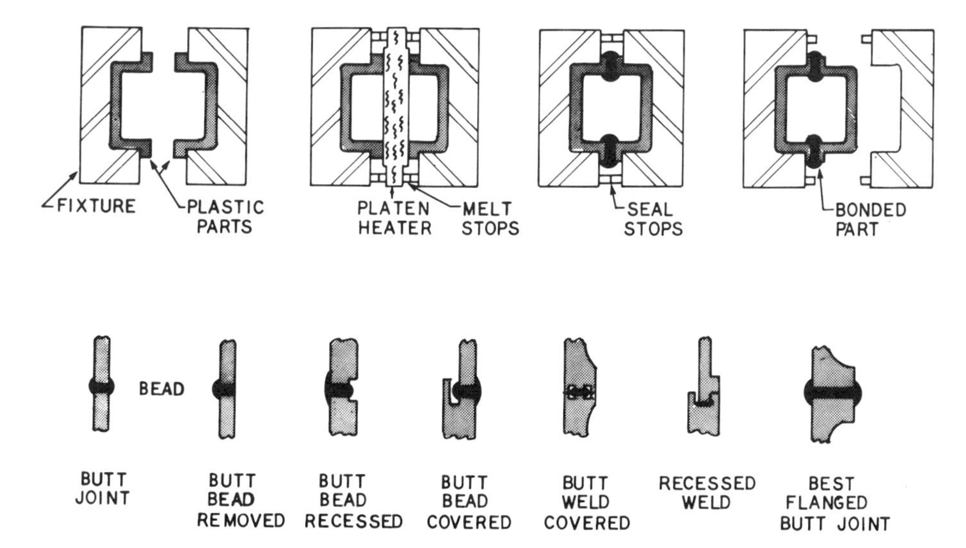

Figure 8–28. This illustrates the process of fusion welding or bonding. Also the many different types of welded joints.

are the melt stop and the seal stop. The heated platen surfaces should be covered, for release purposes, with a thin sheet of Teflon (fluorocarbon resin) impregnated fiber glass cloth. The heated platen can be made from aluminum or bronze. The normal heating temperature range of the platen is 300 to 700° F. The average overall cycle time for this type of bonding is approximately 30 seconds.

Joint design

Figure 8–28 illustrates some of the various joint designs that can be used. The flanged butt joint is the most desirable. The flange gives additional support for bonding and the bond area is increased. Also the flange section will help to reduce warpage on long sections. The straight butt joint is generally used on thin wall parts with straight wall sections. If the flash created by fusion bonding is objectional, flash traps can be added to the joint design. Objectional flash can also be removed by using a cutoff knife to remove the bead.

Design considerations. Excess plastic material must be provided for on the joining surfaces. The amount of excess material will vary from .005 to .050 in. The warpage and shrinkage of the molded part will determine the excess material that should be added. Thin-walled parts may require support by nesting in the holding fixture.

Flat parallel parting lines on the plastic molded part are highly desirable. Contoured parting lines require increased costs and tighter controls in the design of the heating platen. Unfilled plastic molded parts bond very well and show high tensile strengths at the bonded joint. Various fillers, such as glass fibers, talc, and pigments will reduce the tensile strength of the bonded joint by as much as 50%

Magnetic Heat Bonding or Induction Welding

This process is based on the induction heating principle. A thermoplastic electromagnetic compound containing metal particles is placed at the bonding interface of two thermoplastic materials. The composite is then subjected to a high frequency alternating current (Fig. 8–29).

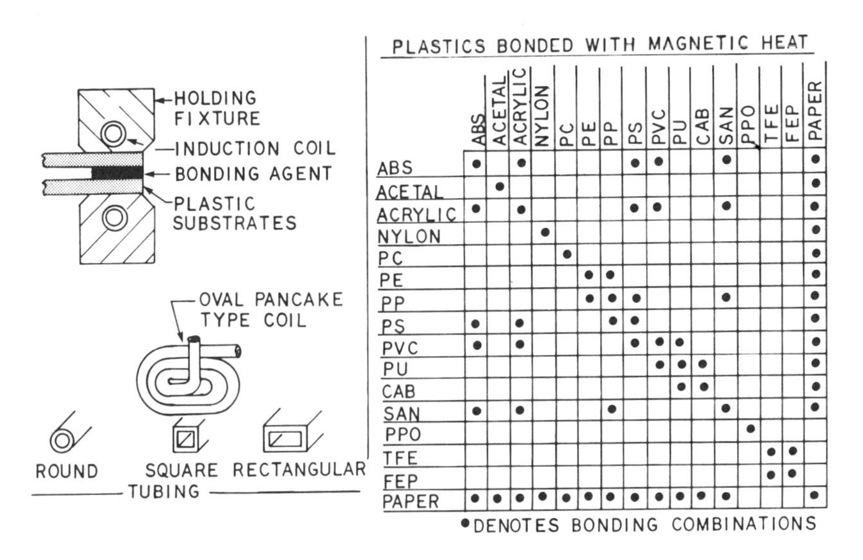

	ABS	ACETAL	ACRYLIC	NYLON	PC	PE	PP	PS	PVC	PU	CAB	SAN	PPO	TFE	FEP	PAPER
ABS	•		•					•	•			•				•
ACETAL		•														•
ACRYLIC	•		•					•	•			•				•
NYLON				•												•
PC					•											•
PE						•	•									•
PP						•	•	•				•				•
PS	•		•				•	•								•
PVC	•		•					•	•	•						•
PU									•	•	•					•
CAB										•	•					•
SAN	•		•				•					•				•
PPO													•			
TFE														•	•	
FEP														•	•	
PAPER	•	•	•	•	•	•	•	•	•	•	•	•				•

Figure 8–29. This illustrates the principle of magnetic heat bonding and the different combination of plastic materials that can be welded together.

The metal particles are heated through eddy currents and hysteresis losses, bringing the two bonding surfaces to the fusion temperature. The heating cycle is accomplished in one to two seconds. Contact pressure applied to the bonding surfaces assures uniform heat transfer and subsequent welding.

A frequency in the range of 2 to 7 megahertz is applied to the bond area by means of conventional induction coils. Iron oxide powder of submicron particle size is dispersed in the bonding agent. Approximately 6% by volume of iron oxide powder is used. The bonding agent can be a water or solvent-based liquid system, or in tape form, or incorporated in a molded component. The bonding agent is applied from a few mil thickness to as thin as 1/2 mil.

The process is very versatile and can bond plastic materials to many different substrates, such as sheet, film, fabrics, paper board, wood, etc. Almost all thermoplastic materials can be bonded by this process.

Hot Gas Welding

This is a welding process used for thermoplastic materials similar to that used for metals, except that an open flame is not used, since an open flame would burn the organic plastic material. The heated gas is directed at the joint to be welded, while a filler rod of the same material as the thermoplastic being welded is applied to the heated area (Fig. 8–30).

The welding gun blows a jet of air or nitrogen gas at temperatures of 400 to 600° F with a pressure of 3 to 4 lbs. per sq. in. If higher temperatures are needed for some thermoplastic materials, nitrogen gas is generally used. The nitrogen gas does not oxidize the plastic material like hot air. Welding guns for plastics contain an electrically or gas heated chamber for heating the gasses.

Proper joint preparation is essential. The ends to be joined in butt welding should be beveled to include an angle of 60°. All surfaces should be solvent cleaned or mechanically roughened. This will insure that all the surfaces are free from oil, grease, and mold release.

Hot gas welding is slow. A high degree of skill and familiarity with

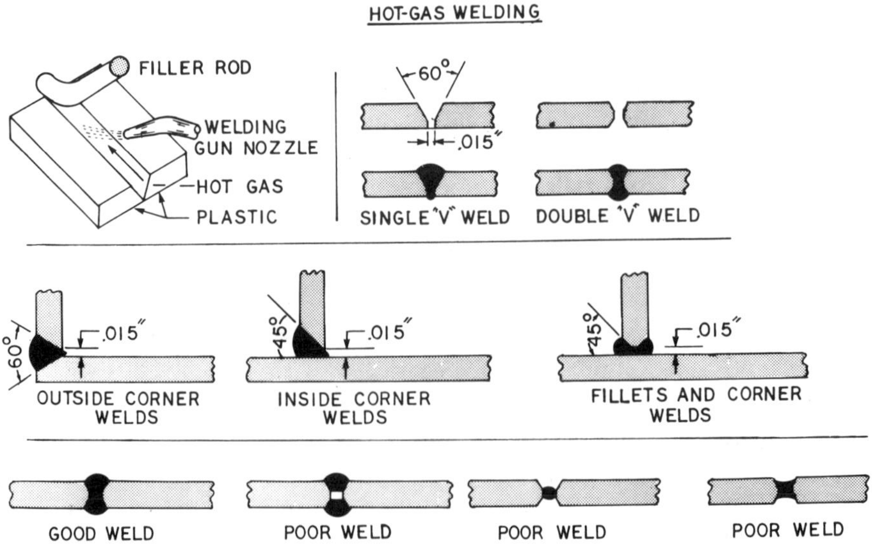

Figure 8–30. Position of the filler rod and the welding gun in the hot-gas welding process. Also shown are a few of the many different types of welded joints.

the materials and techniques are necessary before successful and consistent joints can readily be made with this method.

In hot gas welding of thermoplastics, complete fusion does not take place. The welds are never as strong as the parent material. High density polyethylene has a weld strength of about 60% of the original material, PVC 80%, polypropylene 75%, and acrylics 80%.

Spin Welding

Spin welding is accomplished by frictional heat developed at the interface of two thermoplastic parts. One part is rotated in contact with the mating stationary part to produce frictional melt. The frictional heat is of such intensity that it produces almost instantaneous surface melting, without actually affecting the temperature of the two plastic materials being welded. After the melt has been formed, the relative motion is stopped, and the weld is allowed to solidify under pressure (Fig. 8–31).

Spin welding is limited to parts with circular joints. The geometry of the joint is the most important factor influencing weld quality. The joint configurations shown in Fig. 8–32 are representative of many types used and are not specifically recommended for all thermoplastic materials. It is suggested that the designer contact the raw material supplier for specific types of welded joints that can be used with a specified material. The material supplier can also give the designer the rotational speed (rpm) and force (lbs) which should be used for spin welding a particular joint.

In Fig. 8–32 the following types of joints are illustrated:

1. Plain butt joint on a solid part.
2. A shallow "V" shape on a flat. This type joint helps in the speed-flow of melted material during the spin welding.
3. A shallow "V" shape. It is used on a solid part.
4. A convex on a flat. This type joint is used on a solid part.
5. A square tongue and groove. It is used on a solid part.
6. A groove on a flat. This helps in the speed-flow of melted material during the spin welding and is normally used on a solid part.

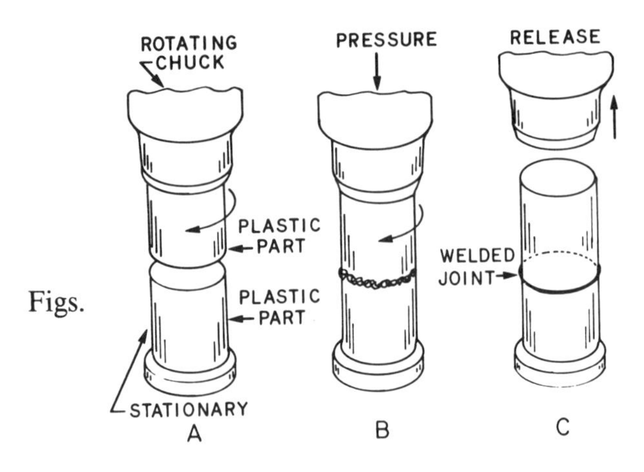

Figure 8–31. An illustration showing the principle of friction or spin welding.

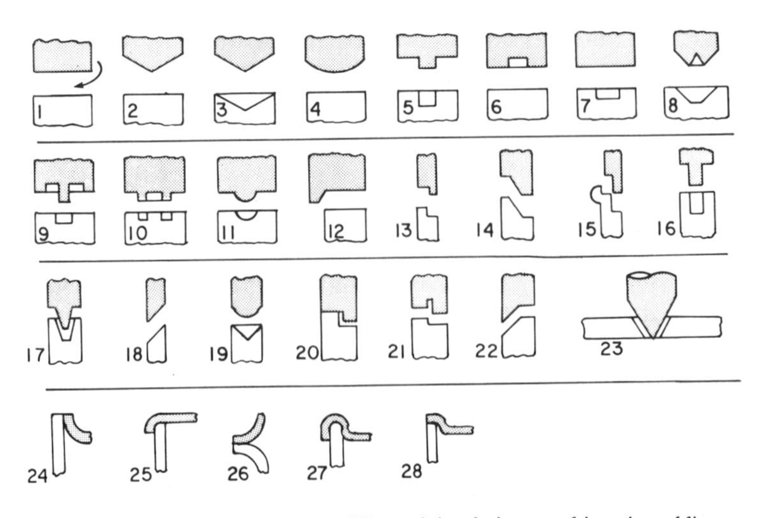

Figure 8–32. A few of the many different joint designs used in spin welding.

7. A flat on a groove. This helps in the speed-flow of melted material during the spin welding and is recommended for solid parts.
8. An angle joint. It is used on a solid part.
9. This design is used for internal flash retention on a solid part.
10. Double tongue and groove is used for flash retention and is recommended for use on solid parts.
11. A round tongue and groove. This type of design is used on a solid or hollow part. A hollow part will be considered like a bottle.
12. A butt joint with a locating ring. This design is used on a solid or hollow part.
13. A square step joint. This type of joint is used on a hollow part.
14. Matched taper. A joint that is used on a hollow part.
15. Outside flange. It is designed to be used on a hollow part.
16. Long square tongue and groove. This is recommended for use on a hollow part.
17. A taper tongue and groove joint. This type of joint is used on a hollow part.
18. A 45° butt joint. It is recommended for use on a solid member.
19. A round tongue with a sharp "V". This is used as a flash trap.
20. A step joint with a flash trap. The flash will be forced to the inside. A 0.010 in. clearance is left between the parts for this trap. It is used on hollow parts.
21. A step with a flash trap. This is used on a hollow part.
22. A 45° butt and a flat.
23. The fastening of a solid pin or stud.
24. An inside sleeve joint with a thin wall. Used on hollow parts.
25. An outside sleeve joint with a thin wall. This is used on a hollow part.
26. An outside butt joint with a thin wall. This is used on a hollow part.
27. An outside overlap joint with a thin wall. This is used on a hollow part.
28. A partial overlap joint with a thin wall. This is used on a hollow part.

Internal stresses are produced in spin-welded parts. The stresses are highest at the center of the weld and lowest at the outer periphery. If the internal stresses become objectionable, annealing of the spin-welded joint may be desirable. Actual service condition testing of the welded joint should be carried out to determine the desired strength that will be required.

A large variety of thermoplastic parts are being assembled by spin welding. A few examples are: fuel filters; spools; marking pens; bottles; blower fans (Fig. 8–33).

Figure 8–33. Plastic molded parts that have been friction or spin welded together. (A) Cross-section of a polypropylene drinking glass. (B) A polypropylene squirrel cage blower fan. (C) An automobile in-line gasoline filter made from nylon. (D) A nylon fastener that has been spun welded onto a velcro fastener. (E) An acetal auto windshield washer suction type valve.

Hook and Loop Fasteners

Hook and loop fasteners consist of two woven strips, one of which is covered with tiny stiff hooks and the other with soft pliable loops as shown in Fig. 8–34. When the two woven strips are pressed together, the hooks and loops engage to provide a secure fastening. The woven materials are nylon fibers. There are over 400 per square inch. The loop component has a dense napped pile which affords a continuous mass of loops designed to engage with the hook. The hook component is also made of solid molded nylon with double-barbed rigid hooks (Fig. 8–35). The rigid hooks provide for a better locking force. The holding strength of the hook and loop fastener is somewhat dependent upon the closure pressure. Actual holding strength tests should be conducted on each application.

This type of fastener can be attached to other substrates by sewing, adhesive-bonding, heat bonding, stapling, etc. Since the fastener is flat it can be easily die-cut to almost any shape. The fastener has

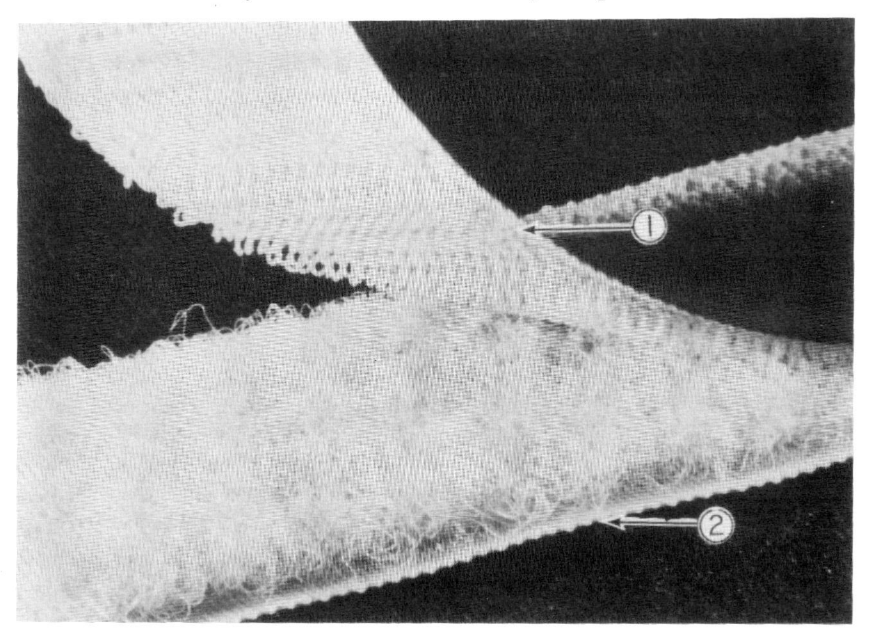

Figure 8–34. This shows the hook and loop fastener in the tape form. (1) Hook. (2) Loop. *(Courtesy Velcro Corp.)*

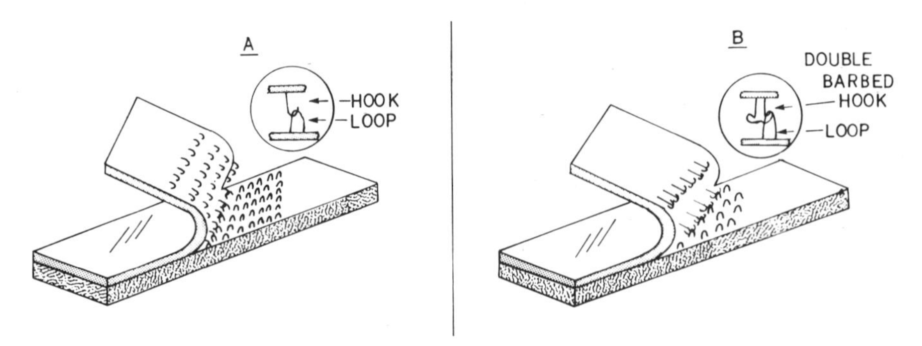

Figure 8–35. (A) the hook and loop fastener consists of two woven strips, one of which is covered with tiny stiff hooks and the other soft pliable loops or pile. (B) This illustrates a solid molded nylon double-barbed rigid hook and loop. The rigid hooks provide for a better locking force.

thousands of uses in many industries including automotive, aircraft, sporting goods, medical equipment, etc.

Flip-lok Bushing

This is a very unique bushing that has the ability to flare back upon itself and is permanently locked in place (Fig. 8–36). The bushing is

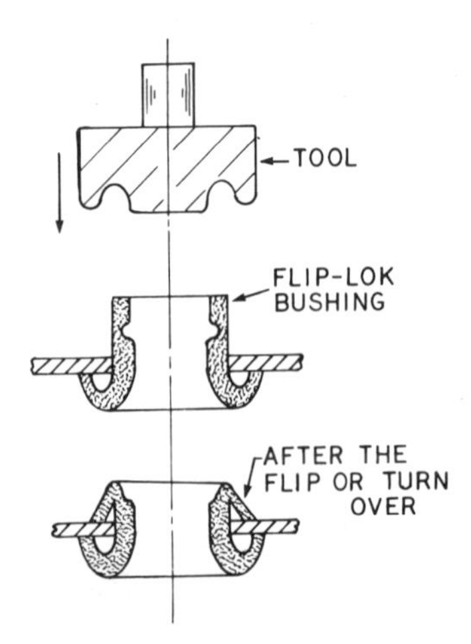

Figure 8–36. A flip-lok bushing. The nylon bushing is flared back upon itself and permanently locked in place. It can be used as a bushing, insulator, liner fastener, spacer and glide or foot. *(Courtesy Western Sky Industries)*

flipped over by a hand tool or a die. There is no heat applied or any secondary operations. It can be removed by cutting out the bushing. The bushing has many applications such as, a bushing, an insulator, a liner fastener, a spacer and glide or foot.

Adhesive Bonding

An adhesive is a substance capable of holding materials together by surface attachment. There is no one universal cement or adhesive that will bond all types of plastic materials. Adhesives are highly specialized and require technical knowledge in selecting and using the proper bonding material. There are essentially five different types of adhesives used in cementing or bonding plastics: (1) a solvent cement; (2) a bodied adhesive; (3) a monomeric cement; (4) elastomeric adhesives; and (5) reactive adhesives. Adhesives come in many forms such as, liquid, paste, powder, mastic, and film or tape. Figure 8–37 illustrates adhesives classified by form and the types of stresses that occur in adhesive bonds.

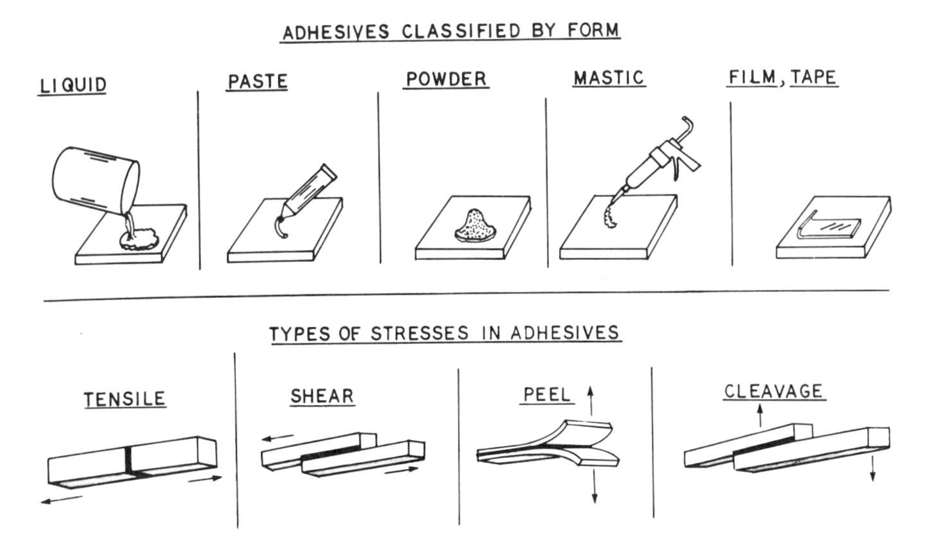

Figure 8–37. This illustrates adhesives classified by form and the types of stresses that occur in adhesive bonds.

Solvent cement. Solvent cementing is a process of joining thermoplastic parts by applying a chemical solvent capable of softening the surface to be bonded and then pressing the softened surfaces together. Solvent cementing is based on the fact that some thermoplastics are soluble in common solvents. The best thermoplastic resins suited for solvent cementing are the amorphous or less crystalline resins, such as ABS, acrylics, cellulosics, polycarbonates, polystyrenes, polyphenylene oxide, and vinyls. Solvent cementing cannot be used for thermosets.

The crystalline thermoplastics such as nylon, acetal, polyethylene, polypropylene, and fluorocarbons are less soluble and are joined by other methods.

In solvent cementing, the solvent is applied to the edges of the two pieces to be bonded. Sometimes the pieces to be bonded are held closely together, and the solvent flows between them by capillary action. The bonding takes place very rapidly, but the full strength of the joint is not reached until the solvent has completely evaporated. If the mating surfaces to be bonded are inconsistent with each other and gaps exist, it may be necessary to use a dope-type cement. A dope or bodied cement consists of solvents in which small quantities of the parent resin have been dissolved.

The joint design should be kept as simple as possible and the bonding surfaces should be smooth, clean and dry, and well aligned in order to obtain complete bonding across the edges of the two pieces. A light pressure should be applied until the joint has set enough so that there is no movement when the pressure is released. Extreme care should be taken so that the solvents do not attack or etch the plastic surface other than the bonding area.

Figure 8–38 illustrates a solvent cemented plastic part. It is a toy teapot that has been injection molded out of polystyrene. The two halves are solvent bonded together with a suitable solvent such as benzene, xylene, or toluene.

Body adhesives. This is sometimes called a dope or bodied adhesive. The cement consists of solvents in which small quantities of the parent resin have been dissolved. A bodied adhesive is compounded so that it can be "spread-on" the joint to be bonded and not run. In this

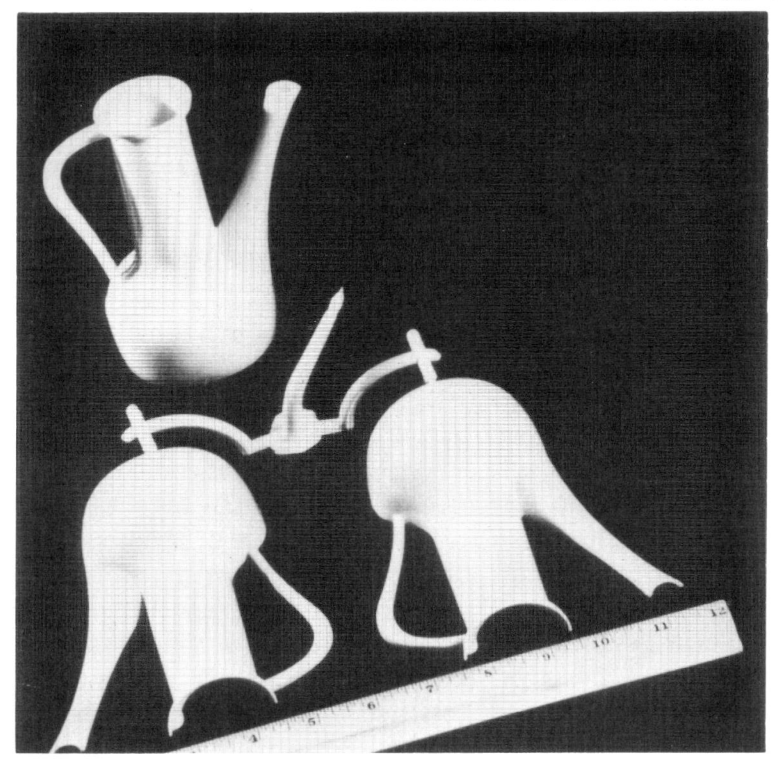

Figure 8–38. A toy teapot injection molded from polystyrene. The two molded halves have been solvent cemented together.

process, the adhesive is frequently applied on the surfaces to be joined by brush, spatula, trowel, flow gun, or flow brush.

Monomeric cement. This is a cement made from the same type of polymer plastic that is to be cemented together. The monomer is catalyzed so that a bond is made by polymerization rather than by solvent evaporation. A typical example of this type of cement is that of polymethyl methacrylate (a solid acrylic plastic). The liquid monomer (methyl methacrylate) is the starting material used in the manufacture of solid acrylic plastic. This liquid monomer may be used as a solvent cement. This liquid monomer may also be catalyzed and made to polymerize and become hard. Before it becomes hard, it may be used as a cement.

Elastomeric adhesive. Elastomeric adhesives, generally know as rubber adhesives, are made from natural, synthetic, or reclaimed rubber. The rubber may be dissolved in solvents, suspended in water, or other liquids. Adhesives containing rubber are the most versatile adhesives in use today.

Reactive adhesives. These adhesives are thermosetting materials and are generally used to bond thermosetting molded plastic parts. This is because thermosetting plastics, when cured, are insoluble in most organic solvents. The bonded joints between thermosets will be only as good as the bond formed by the thermosetting adhesive and the surface of the thermosetting plastics.

Epoxy resin based adhesives have 100% solids and contain no solvents that will attack the surface of any plastic. When the epoxy resin polymerizes or becomes hard, it has very little or no shrinkage. This makes it excellent for an adhesive, because very few stresses are set up in the adhesive joint. This is not true with some of the other type resins such as polyesters, phenolics, etc. Fig. 8–39 illustrates a good application of a bonded epoxy assembly. Three separate phenolic transferred molded parts, that make up a washing machine water mixing chamber assembly, are bonded together with epoxy resin. It

Figure 8–39. A washing machine water mixing chamber. Three separate phenolic molded parts are assembled and cemented together with epoxy resin. (Photo by Arthur Hollar)

would have been impossible to have made this part in one piece, because it is hollow and the geometric configuration would not allow for any type of core pins to be used.

Hot-melt adhesives. Hot-melt adhesives are 100% solids. They are thermoplastic adhesives that are intended for light-duty jobs in assembling, bonding, sealing, and gap-filling. Hot-melt assembly is a simple, two-step process. First the hot adhesive is applied to the joint line. Second it holds the parts together for a few seconds while the hot melt cools to its solid adhesive state. The hot-melt adhesive temperatures range from 300° F to 550° F. The material can be rigid or flexible.

Cyanoacrylate adhesives. The cyanoacrylates are thermoplastic monomers which polymerize in an extremely short period of time if confined in a thin film between close-fitting non porous parts. Polymerization is initiated by a trace amount of moistures on the bonding surfaces. This adhesive is easy to apply and sets or cures in a matter of a few seconds to 5 minutes. The adhesive is high in material costs, has a poor shelf life, and is considered hazardous.

Anaerobic adhesives. This adhesive is thermosetting and cures to a 100% solid. It is applied in the liquid form. The adhesive cures or polymerizes by the exclusion of air. It will not cure where air contacts it. It is a one-component system and is used for bonding or holding bolts, nuts, and static joints. The cure time can range from 15 min. to several hours.

Polyurethane adhesives. This adhesive is thermosetting and cures to a 100% solid. It is applied in the liquid form. It comes in one or two components and cures at room or oven temperatures. It is an excellent adhesive and bonds to almost any type of surface.

Table 8–2 and 8–3 will help to guide the designer in selecting the proper types or methods of joining or bonding different plastic materials. There is an overwhelming number of adhesive formulations available today. In selecting and specifying an adhesive for a specific application, it is advisable to consult your adhesive supplier and the

TABLE 8–2. A REFERENCE CHART TO HELP IN SELECTING THE PROPER METHOD OF FASTENING THERMOPLASTIC MATERIALS.

THERMOPLASTICS	MECHANICAL FASTENERS	ADHESIVES	SPIN AND VIBRATION WELDING	THERMAL WELDING	ULTRASONIC WELDING	INDUCTION WELDING	REMARKS
ABS	G	G	G	G	G	G	BODY TYPE ADH. RECOMMENDED
ACETAL	E	P	G	G	G	G	SURFACE TREATMENT FOR ADHESIVES
ACRYLIC	G	G	F-G	G	G	G	BODY TYPE ADH. RECOMMENDED
NYLON	G	P	G	G	G	G	
POLYCARBONATE	G	G	G	G	G	G	
POLYESTER TP	G	F	G	G	G	G	
POLYETHYLENE	P	NR	G	G	G-P	G	SURFACE TREATMENT FOR ADHESIVES
POLYPROPYLENE	P	P	E	G	G-P	G	SURFACE TREATMENT FOR ADHESIVES
POLYSTYRENE	F	G	E	G	E-P	G	IMPACT GRADES DIFFICULT TO BOND
POLYSULFONE	G	G	G	E	E	G	
POLYURETHANE TP	NR	G	NR	NR	NR	G	
PPO MODIFIED	G	G	E	G	G	G	
PVC RIGID	F	G	F	G	F	G	

E-EXCELLENT, G-GOOD, F-FAIR, P-POOR, NR-NOT RECOMMENDED

TABLE 8–3. A REFERENCE CHART TO HELP IN SELECTING THE PROPER METHOD OF FASTENING THERMOSETTING PLASTIC MATERIALS.

THERMOSETS	MECHANICAL FASTENERS	ADHESIVES	SPIN AND VIBRATION WELDING	THERMAL WELDING	ULTRASONIC WELDING	INDUCTION WELDING	REMARKS
ALKYDS	G	G	NR	NR	NR	NR	
DAP	G	G	NR	NR	NR	NR	
EPOXIES	G	E	NR	NR	NR	NR	
MELAMINE	F	G	NR	NR	NR	NR	MATERIAL NOTCH SENSITIVE
PHENOLICS	G	E	NR	NR	NR	NR	
POLYESTER	G	E	NR	NR	NR	NR	
POLYURETHANE	G	E	NR	NR	NR	NR	
SILICONES	F	G	NR	NR	NR	NR	
UREAS	F	G	NR	NR	NR	NR	MATERIAL NOTCH SENSITIVE

E-EXCELLENT, G-GOOD, F-FAIR, P-POOR, NR-NOT RECOMMENDED

plastic material supplier. They can both offer valuable information in solving the extremely complex problems in adhesive bonding.

In Fig. 8–40 the following types of adhesive joint designs are illustrated:

1. A butt joint.
2. A scarf joint.
3. A square tongue and groove joint.
4. An angled tongue and groove joint.
5. A half-lap joint.
6. A "V" joint.
7. A "V" type joint with a flat.
8. A round tongue and groove joint.
9. A double-scarf lap joint.
10. A simple lap joint.
11. A tapered simple lap joint.
12. A joggle-lap or off-set joint.
13. A double-lap joint.
14. A double-strap joint.
15. A beveled double-strap joint.
16. A "T" section joint.

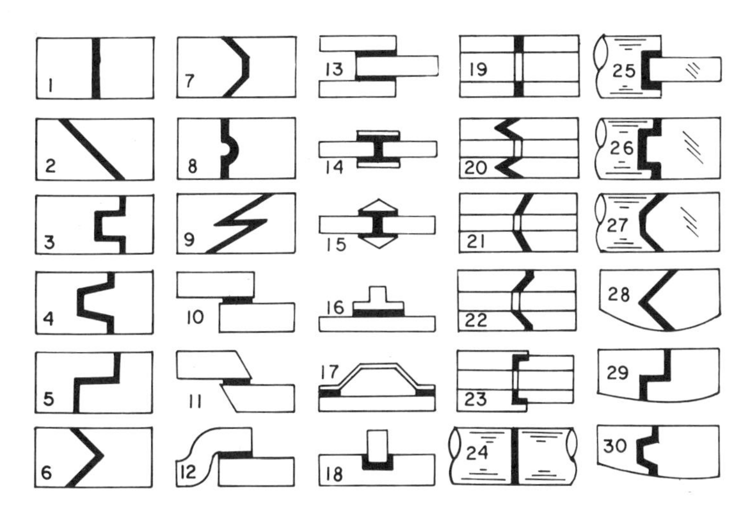

Figure 8–40. A few of the many joint designs used in plastic adhesive bonding.

17. A hat section joint.
18. A recessed right angle joint.
19. A tubular butt joint.
20. A tubular "V" joint.
21. A tubular half-lap joint.
22. An angled, tubular, half-lap joint.
23. A tubular lap joint.
24. A rod-butt joint.
25. A tongue and groove joint in a solid rod.
26. A landed tongue and groove joint in a solid rod.
27. A scarf tongue and groove joint in a solid rod.
28. A "V" type joint with increased bonding area for additional strength.
29. A half-lap joint with increased bonding area for additional strength.
30. An angled tongue and groove joint with increased bonding area for additional strength.

9

Decorating Plastics

Molded plastic parts are generally decorated for "eye appeal." However, other reasons for decorating can be functional as well as decorative. Functional improvements include resistance to wear, scratching, and marring, also increased resistance to heat, light, chemical and electrical exposure. A molded plastic part can be decorated to transmit part identification and product information. This is done by trade markings, stencilling, lettering, and numerals. Most plastic materials are decorative in themselves and can be molded with raised or depressed designs that may be accented with paint or lacquer. Decorative overlays provide contrast and novelty. All decorating depends on applying a surface marking or paint coating permanently to the plastic surface, without any damage to the original shape or appearance. Virtually every type of molded plastic product can be decorated in one way or another.

Decorative finishes can be applied to plastic parts by a variety of methods. However, only the following are considered herein: spray painting; metallizing; metal plating; printing and marking; labels and decals; screen processing; roll leaf hot stamping; and two-color molding. The majority of all decorating on plastic parts is done with vacuum metallizing, spray painting, silk screening, and hot stamping.

Surface Finish of the Plastic Part

The decorating of any plastic part is a surface treatment, and the first prerequisite is that the surface be clean and prepared in the right manner for decorating. Some of the common causes of failure in decorating plastic parts are: mold lubricants; dust; natural skin greasiness; excess surface plasticizers; surface moisture and humidity conditions; and strains locked into the molded part.

Almost all plastics can be decorated successfully, but some are more difficult than others. The thermoplastic materials are prone to solvent attack while the thermosetting plastics have excellent solvent-resistance characteristics. It becomes obvious that each family of plastic materials must be considered separately when selecting coatings, thinners, foils, tapes, etc.

It should always be remembered that a plastic molded part is no better than the mold that produced it. Because the decorative part surface is reproduced by the paint, film, or foil overlay, the finished decorative item is no better than the initial molded part. Molded parts with the best finish are always produced from highly polished molds. It is desirable to mold for high gloss and a low stress level in the part. The use of fillers in the plastic material causes some difficulty. Fiberglass does not always give a good luster finish, nor do cloth-filled thermosetting materials. A clear thermoplastic material may show flow marks. Flow marks are caused by welding or knitting of the material as it enters the mold and by the tumbling motion of the material as it passes along the mold surfaces.

Buffing of thermosetting parts produces a highly polished surface, but a buffed surface does not wear as well as a non-buffed surface, because buffing removes a small portion of the outer resinous skin. Buffing helps, however, to disguise flow marks on the thermosetting materials. Buffing of the thermoplastic materials does not disguise flow marks, but it helps to remove outside blemishes. Flow marks may be disguised nicely by using a mottled two to three-color material.

If a large, flat area is required, flow marks may be hidden by graining the surface. Since graining of the part is done by the mold, the grained surface always should be made on a plane perpendicular to the draw, or adequate taper should be allowed so that the molded part can be removed from the mold. A grained surface may cause undercuts in the part.

Frequently, it is necessary to pretreat a plastic surface before it can be painted or decorated. This is true with polyolefins. A flame or chemical pretreatment is required to ensure good bonding on the slippery surface. Static electricity on plastic surfaces has always caused problems. Airborne impurities become attracted to the plastic surface and settle on the part. Wiping or rubbing with a cloth may only increase the static charge and aggravate the problem. There is equipment available on the market that will remove the static charge from the plastic surface. The decorating of plastics should be carried out in clean areas and controlled room conditions. A "white room" is desirable, but not always necessary.

SPRAY PAINTING

A plastic substrate may be coated by spray painting. Two types of coating systems are in use. They are enamels and lacquers. The enamels are coatings containing thermosetting resins dissolved in a solvent. Lacquers are coatings containing thermoplastic resins dissolved in a suitable solvent.

Enamels have good properties of high gloss and hardness, but require curing temperatures at which most thermoplastics warp or distort. Lacquer coatings are used more extensively to decorate thermoplastic molded parts. Lacquers usually dry at room temperatures by solvent evaporation and are easy to apply. The plastics that are most often coated are shown in Table 9–1. It will be noted that urethane is very versatile and can be considered for coating many plastics.

Almost all industrial paint coatings are solvent-based. In order to avoid the solvent-emission problem, water-based paint systems have been introduced. These paints are based on such resins as acrylics and alkyds. The same coating methods used for solvent-based paints are used. The water-based paints offer particular advantages for plastics because the absence of a solvent eliminates attack on the plastic substrate.

Mask Spray Painting

A paint spray mask is used to provide sharp lines of demarcation where different colors are designated. The spray mask works best with a step at the paint line. If the mask does not have a step, it is nothing more than a stencil and will not give good sharp lines. Masks are generally made out of electroformed nickel. Sometimes paper, rubber, and plastic can be used for masks for short production runs.

There are three types of spray masks in use today (Fig. 9–1). The plug mask is used to fill and keep depressed areas clean while the surrounding area is painted. The plugs are suspended from wire bridging and must fill the depressed sections to prevent paint leakage. Plug masks are also used for protecting areas to be vacuum plated.

The block or cutout mask is utilized to confine paint to a shallow area that has no paint steps. A secondary wiping is usually required after removal of the mask. This type of mask is used where small lines and lettering are too small for practical individual masking.

The lip mask caps raised areas and vertical sidewalls to allow painting of recessed areas with a fine line of definition and without a secondary wiping step. The lip keeps the top clean and all, or as much as may be desired, of the sidewalls. The mask must have a "lip" of metal that extends down the sidewalls.

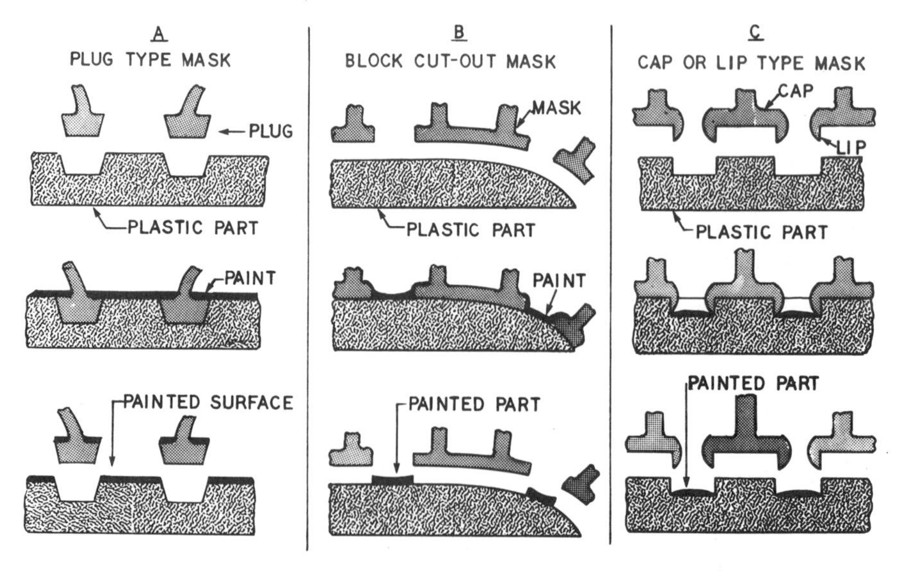

Figure 9–1. (A) The plug mask is used to fill and keep depressed areas clean while the surrounding area is painted. (B) The block or cut out mask is used to confine paint to a shallow, recessed area that has no paint step. (C) The cap or lip mask covers raised areas and/or vertical sidewalls to allow painting of recessed areas with a fine line of definition and without a secondary wiping operation.

If spray masks are going to be worthwhile, the molded part must be held to close tolerances. This calls for good mold design and molding technique. Slight variations in dimensions from part to part will allow paint to blow by the mask and cause rejects. Good consistent molded parts are a must in spray mask painting. One molder has found that by weighing each injection molded shot as it comes from the mold it can be determined at once whether the spray mask is going to give a precision fit when the part reaches the paint room a few hours later. Different lots of the same material vary and moisture conditions can cause trouble.

Mask Spray Painting Equipment

There are five types of spray gun machines (Fig. 9–2). (A) The stationary gun is used to spray coatings on large parts or only one side of a part. The spraying may be done manually or automatically. (B) The reciprocating gun or guns are mounted on a traveling bar that moves back and forth across the part to be painted. (C) The rotating guns actually oscillate or move back and forth above or beneath the spray mask. The work piece moves back and forth. (D) The rotating guns revolve below or above the part to be painted. Sometimes the

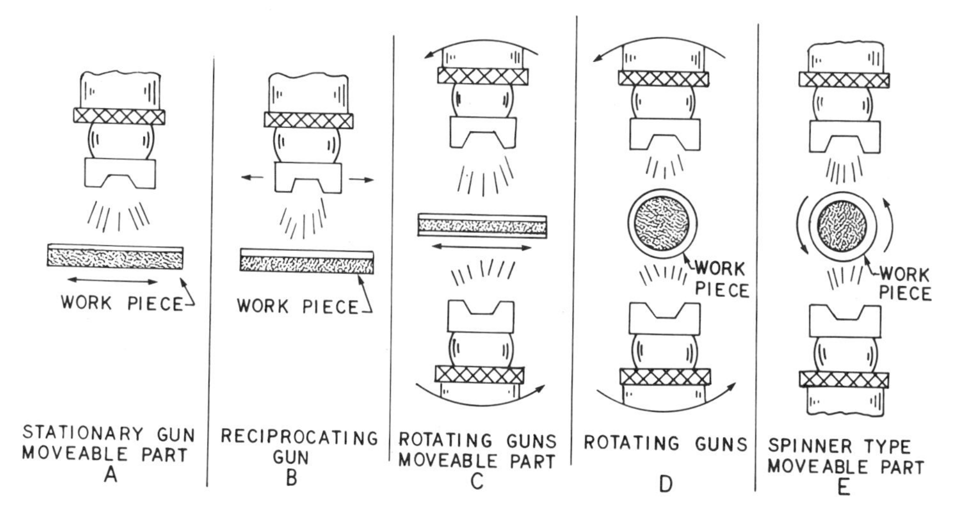

Figure 9–2. The five basic types of spray gun machines.

part itself may revolve. (E) In the spinner gun-type, guns are generally stationary, and the part or workpiece moves.

Flow Coating

This is a painting process where the parts to be painted are drenched under a curtain of lacquer. After painting, the parts are slowly rotated while the coating dries. The excess paint drains off and is collected in a reservoir and filtered. Plastic parts with sharp corners and edges should be avoided in flow coating. It is difficult to get a coating build up on these sharp edges. The cost of flow coating is low, since very little waste occurs. Also, the equipment costs are generally low.

Dip Coating

The dip coating is a system wherein the part is dipped into a tank of lacquer, withdrawn, and allowed to dry. The coating thickness is controlled by automatic methods of lowering and raising the parts into and out of the lacquer tank. If parts are withdrawn from the dip tank at a rate faster then they drain, the film will be thicker at the bottom of the part. The process is economical and the equipment costs are low.

Roller Coating

This is a process of coating plastic surfaces with fluid paint by contacting the surface of the part with a roller on which the fluid paint is spread (Fig. 9–3). The roller is usually covered with a neoprene jacket. For hand rolling, a small amount of paint is placed on a smooth flat surface, such as a piece of plate glass. The paint on the glass plate is transferred to the hand tool by rolling it back and forth across the glass. The roller is then rolled back and forth over the raised areas to be coated. Close control of coating viscosity is required to allow flow-out, while avoiding run-off in raised areas. Small raised areas such as letters, figures, and designs may be decorated by roller coating. The raised area to be painted should have smooth flat surfaces, and the edges and corners should be sharp for better details.

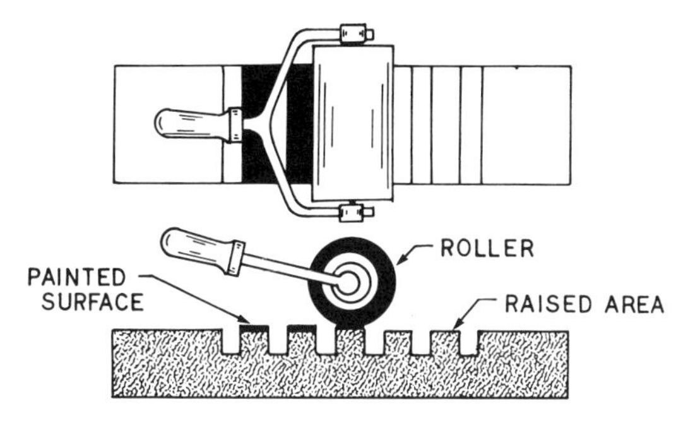

Figure 9–3. Small raised areas such as letters, figures, and designs may be decorated by roller coating.

Fig. 9–4 shows a name plate that has been painted by roll coating. The name plate was compression molded from phenolic material. The name "Castle" is molded in raised letters and was painted white by roller coating. The word "sterilizer" is molded through letters. The unique design of this molded through letter permitted a small red translucent piece of plastic film stock to be placed in back of the letters. A light from the back of the name plate shows the word "sterilizer" in red.

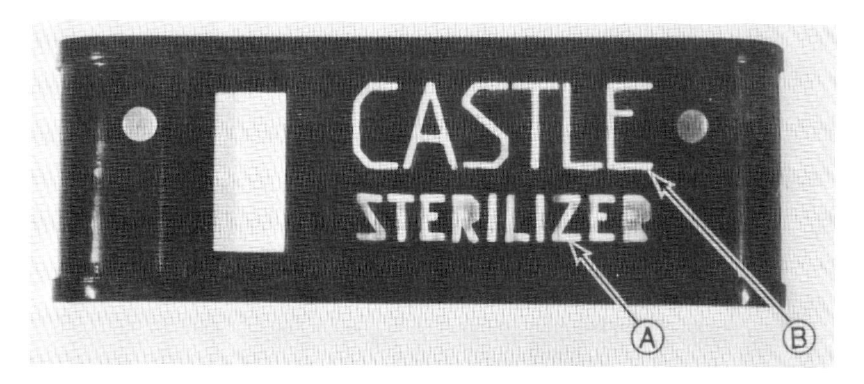

Figure 9–4. A name plate compression molded from phenolic material. The word STERILIZER (A) is molded through letters. The name CASTLE (B) is raised letters and decorated by roll coating.

Spray and Wipe

In this process, paint is sprayed over a recessed area in the plastic part, and the excess paint is wiped away. Wiped-in letters are indented on the part, and either first or second surfaces can be sprayed and wiped.

Two types of spray and wipe coatings are used—wet and dry wipes. The wet wipes are usually air-dry spray paints that are removed from the plastic surface with a suitable solvent-soaked cloth that is blended to remove the paint without damaging the plastic. The dry wipe coatings are sprayed upon the plastic part in the same manner as the wet wipe paints. After spraying, the paint soon drys to a fine powder, and the excess is removed with a dry cloth.

Recess or surface depressions for wipe-in decorations should have a depth to width ratio of one and one-half to one (Fig. 9–5a). Recesses are seldom over .031 in wide. Indentations or recesses should be sharp and abrupt where they meet the surface of the part. This prevents streaking of the paint across the surface as it is wiped off (Fig. 9–5b). Wiped-in letters or decorations will not chip or rub away

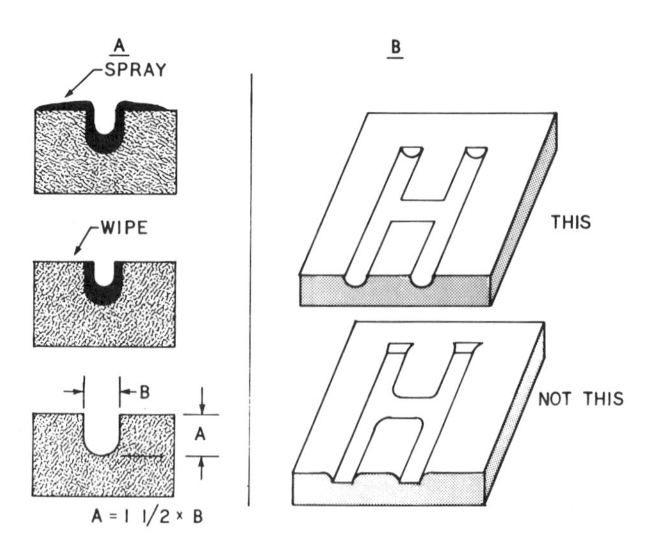

Figure 9–5. (A) The proportions of depth to width for letters or depressions in the spray and wipe paint process. (B) Wiped-in letters are depressed and should be sharp where they meet the surface of the part.

as easily as stamped letters. This is a more expensive method, but it does improve visibility of letters and decorations. Dry wipe coatings are limited to black, white, and pastel colors. Fig. 9–6 shows an antenna rotator knob and a washing machine control knob that has black paint wiped into depressed letters and decorations.

An injection molded part that has depressed letters, which are to be filled in with paint, should be gated in such a manner that weld or knit lines do not form between the letters. If weld or knit lines are present, paint may run into them when the letters are being filled, thus resulting in a scrap part.

Screen Printing or Decorating

Screen printing or decorating is a process used to force paint or ink through a stencil fabric, commonly called a "silk screen," onto the plastic part that is being decorated (Fig. 9–7). The screen consists of a taut woven fabric securely attached to a rectangular frame and carefully masked with a stencil in a manner that allows the paint to be pressed through the screen only at areas where the stencil is open. For most applications a screen with 230 perforations per sq. in. is adequate. The stencil fabric is generally nylon or stainless steel. The

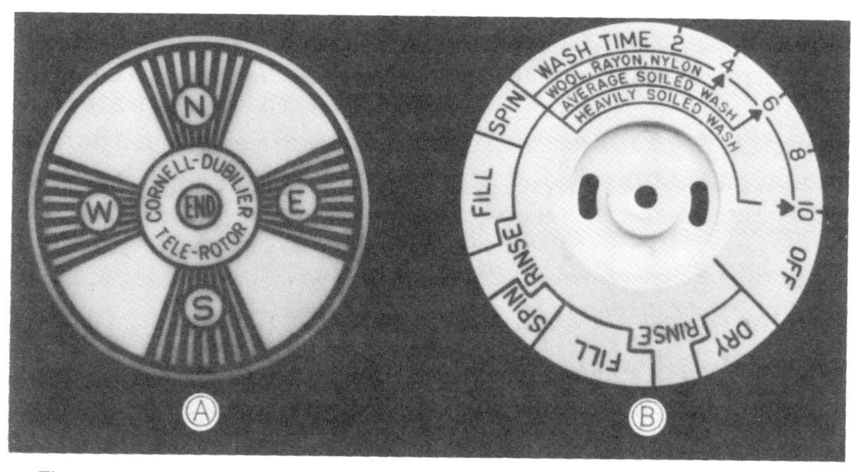

Figure 9–6. Injection molded knobs that have black paint wiped into depressed letters and decorations. (A) Antenna rotator knob. (B) Washing machine control knob.

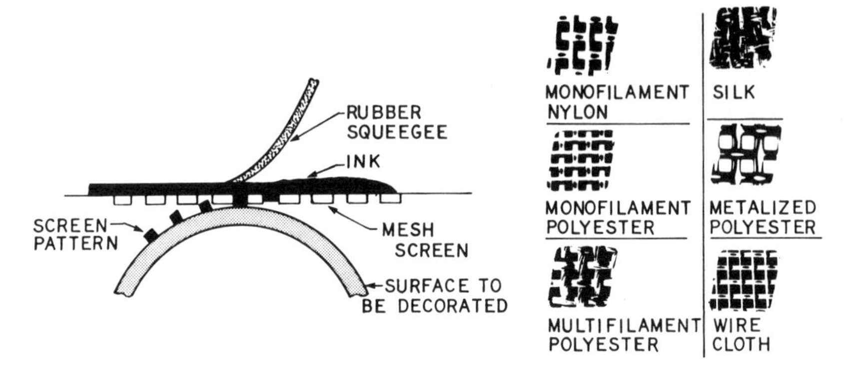

Figure 9–7. Screen printing or decorating is a process of transfering paint through an open screen onto a flat or gently rounded surface.

screen is placed above the plastic part to be decorated, and a flexible rubber squeegee forces the ink or paint through the openings in the screen onto the surface of the plastic part. The rubber squeegee ranges in hardness from 50 to 80 durometer, Shore type "A".

Multicolor decorating and intricate designs can be used in this process by using a series of different screens. Silk screen decorating is done on a flat or curved surfaces and does not give the three dimensional depth effect. Although the inks used for this process are expensive, the overall costs are generally economical. Fig. 9–8 shows a lamp shade that has been silk-screen decorated. The injection molded shade is made from high-density polyethylene, and the surface is flame treated before being screen painted.

Painting Plastic Materials

Painting plastic molded parts is the last step in a series of operations including product design, mold design, the molding process, and finishing operations. A paint adheres to plastics through intermolecular attraction, solvent etching, or a combination of both. Painting is a special case of adhesive bonding, requiring a chemically cleaned sur-

Figure 9–8. An injection molded high density polyethylene lamp shade. The decoration on the lamp shade has been screen painted.

face. The two types of paints used on plastic surfaces are lacquers and enamels. The lacquers are generally air-dried, and the enamels require a baking operation. Each type of plastic material requires a different type of paint formulation. (Table 9–1). This is not a difficult problem, provided the plastic and paint coating system is understood.

Annealing the plastic before painting is a good policy, since it removes surface strains which, if present, might develop a pattern of tiny fissures. This effect is called "crazing." The following plastic materials indicate some of the important paint characteristics in painting.

TABLE 9–1. THIS TABLE ILLUSTRATES THE MANY TYPES OF PAINTS THAT ARE USED TO PAINT PLASTIC MATERIALS.

A GUIDE TO PAINTS FOR PLASTICS

PLASTIC	PRETREATMENT	COATABILITY	TYPES OF COATINGS USED
ABS	SOLVENT WIPE	GOOD	VINYL, MODIFIED VINYL, URETHANE
ACRYLIC	SOLVENT WIPE	GOOD	ACRYLIC MODIFIED ACRYLIC, NITROCELLULOSE, URETHANE, EPOXY
CAB	SOLVENT WIPE	GOOD	ACRYLIC, VINYL, NITROCELLULOSE
NYLON	DETERGENT WASH	FAIR	ACRYLIC, VINYL, URETHANE
PHENOLIC	SOLVENT WIPE	DIFFICULT	ALKYDS, URETHANE, EPOXY
POLYCARBONATE	PRIMER	FAIR	URETHANE, ACRYLIC
POLYESTER	SOLVENT WIPE	GOOD	URETHANE, EPOXY
POLYESTER T.P.	PRIMER	GOOD	ACRYLIC
POLYOLEFINS	FLAME, ETCH	DIFFICULT	URETHANE, NITROCELLULOSE
POLYSTYRENE	PRIMER	GOOD	URETHANE, NITROCELLULOSE
PPO	PRIMER	FAIR	URETHANE, ALKYD, ACRYLIC
POLYURETHANE	PRIMER	GOOD	URETHANE
PVC	SOLVENT WIPE OR PRIMER	FAIR	URETHANE, NITROCELLULOSE

ABS. This thermoplastic material is opaque and can be painted. It is made in many formulations and each paint application must be considered separately. There is a large broad range of paint systems available for ABS plastic materials. The stresses in the molded part can cause solvent etching when it is painted. Annealing is best accomplished by heating the mold part for two hours at a temperature of 10° F (5 to 6° C) below its distortion temperature.

Acetals. This crystalline thermoplastic exhibits poor adhesion to most paints. A surface treatment is required before painting. An acid-etched surface is recommended for good paint adhesion. A bake-type primer, similar to that used on metals, is generally recommended. The bake temperature should be below 315° F (157° C). Temperatures above 320° F (160° C) adversely affect the physical properties of the acetal resin. A standard commercial color is applied over the primed surface.

Acrylics. This thermoplastic is subject to crazing caused by stress. Most coating systems used on acrylics are lacquer types. Safe solvent systems are ketones, esters, and alcohols. Checks for stress may be made by immersing the part in ethyl acetate. Crazing may be corrected by annealing the parts before painting (150 to 160° F, (66 to 74° C).

Cellulosics. This family group of plastics contains substantial amounts of internal plasticizers that cause problems in painting. The plasticizer migration to the surface of the painted part causes tackiness in the film after a period of time. Adequate testing and aging of the painted plastic part should be done to ensure good paint adhesion. The cellulosics are generally coated with an air-dry lacquer.

Ionomer. This type of plastic resin material can be successfully painted. An epoxy enamel is used for a prime coat. The prime coat can be applied by standard spray technology or a wipe-on coat. The paint coat is a polyurethane enamel. The enamel is cured for approximately 15 minutes at 145° F (63° C).

Nylons. This crystalline thermoplastic can be painted with many coating systems. If necessary, it may be baked in order to cure enamel-type paints. Nylon is hydroscopic and under normal conditions absorbs as much as two and one-half percent moisture. This moisture absorption problem should be considered in painting and baking.

Polycarbonate. Many solvents craze or etch polycarbonates. Paint coatings that are to be used on polycarbonates should first be tested for crazing by painting sample molded parts. Molded parts can be painted with either air-dry or baked coatings. A baking temperature up to 250° F (120° C) can be tolerated. If molded parts show stress crazing, they can be annealed at 250° F (120° C).

Polyester Thermoplastic. This plastic material can be painted with a lacquer or an enamel. It requires a alkyd primer that is baked at 330° F (166° C).

Polyolefins. This group of plastics include polyethylene, polypropylene, polyallomers, etc. Polyethylene requires a surface treatment before it can be painted. This may be in the form of chemical oxidation or flame treatment. Flame treatment is the most widely used and involves a highly oxidizing flame applied directly to the plastic. Polypropylenes generally do not require a pretreatment, and several paint systems are available for the untreated plastic.

Polyphenylene Oxide. This thermoplastic material requires a primer. It is painted with an enamel or lacquer and is baked at 250° F (121° C).

Polystyrene. This thermoplastic is extremely sensitive to solvents. Air-dry type coatings are used, and the solvents in the paint coating must be blended to prevent solvent attack. Base coatings of acrylic, alkyd, and urea resins are used. If necessary, the part may be annealed at 145° F (63° C).

Vinyls. A wide range of paint coatings are available for this family of materials. A special paint formulation is required for each type of plasticized vinyl. The main problem is the plasticizer migration in the vinyl formulation.

Painting Thermosets

Thermosetting plastic materials are generally painted with enamels the same as in painting metals. Most of the paints used for thermosetting plastics are of the baking type, because the heat distortion point of the thermosets is not critical.

Phenolics. This thermosetting plastic is painted with enamels based on alkyds, ureas, and melamines. Almost any type of commercial paint solvent may be used. The baking temperature is between 275 and 300° F. (135° C and 149° C).

HOT STAMPING

This is a process of marking plastics by transferring the decorative coating (paint or bright metalized) from a carrier foil to the part by heat and pressure. The foil is pressed against the plastic surface by means of a heated die, thus transferring and welding selected areas of the foil to the plastic surface (Fig. 9–9). This method is sometimes called hot-roll leaf stamping. In hot stamping, the pressure on the die creates a recess for marking, which protects the stamped letter from abrasion, and heat causes the marking medium to adhere to the plastic.

Block lettering can be hot stamped either on raised or flat areas. A flat metal plate or silicone rubber can be used to transmit heat and pressure to the foil for stamping. Silicone rubber dies are used to cover irregular surfaces. The flexibility of the die allows good transfer over erratic surfaces (Fig. 9–10). Silicone rubber does not conduct heat as well as metal. If a silicone rubber die face must have a temperature of 275° F (135° C), the back up metal will be heated to 325° F (191° C). Surface pretreatment of the plastic part is not a requisite in hot stamping.

The roll leaf tapes and foils that are used in hot stamping are made specially for each type of plastic. Simple, pigmented foil is made in a three-ply laminated roll form. The film carrier is generally Mylar. On one surface of the Mylar, the pigmented coating is placed with a suitable binder and an adhesive or release agent. During the stamping process, the heated die activates release agents in the foil and causes the coating to transfer to the plastic surface. Depending on the plastic and the tape mark-up, either thermal bonding or activation of the adhesive occurs. Metallized foils are much more complicated, having as

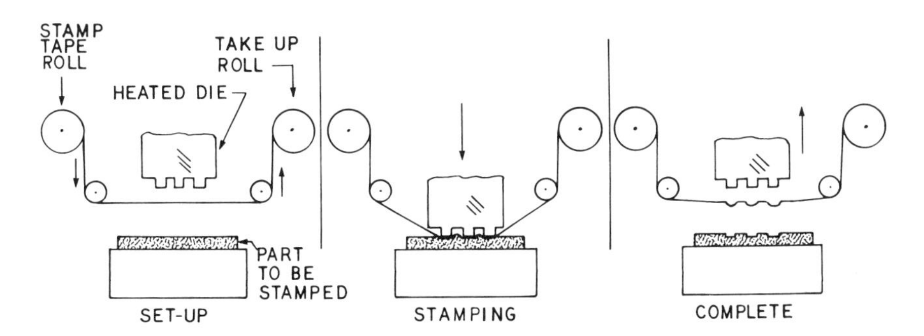

Figure 9–9. This illustrates schematically the process of hot roll leaf stamping by using a heated metal die.

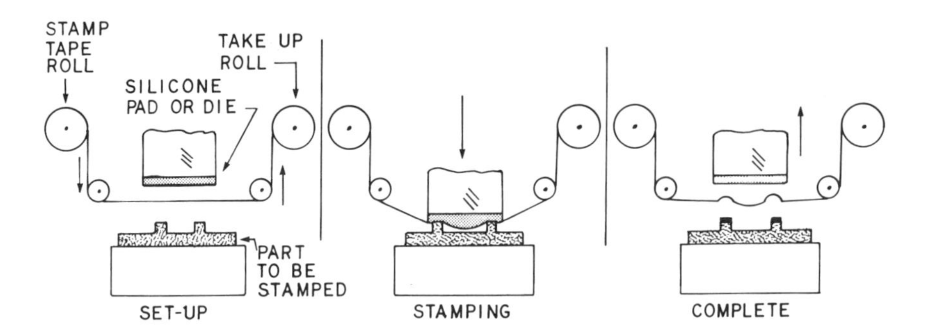

Figure 9–10. This illustrates schematically the process of hot roll leaf stamping by using a hot silicone pad or die.

many as five layers. Fig. 9–11 illustrates a typical engraved hot stamp die. Note the intricate details that have been engraved into this die.

Dies For Hot Stamping

Hot stamping dies are made from magnesium, brass, steel or silicone rubber bonded to metal. Inexpensive photoetched dies work well for prototype or short runs on flat areas. However, production stamping dies should be pantographed and deep cut from either brass or steel. Dies that require contours must always be pantographed and hand benched to match the radii of the imprint area. Improperly fitted dies produce poor quality work. The improper fit cannot be overcome by increasing the load on the die.

Magnesium dies. Magnesium dies are made in a photoetched process. The metal is rather soft and is used only for flat prototype work on rigid plastics or for medium production runs on flexible plastics.

Brass dies. Brass dies can be either photoetched or pantographed. Photoetching produces a rather shallow die and, for this reason, it is not desirable. Brass dies are acceptable for medium production rates of 25,000 to 50,000 pieces, but are more easily damaged in production if parts are not located properly on the fixtures by the operator. These

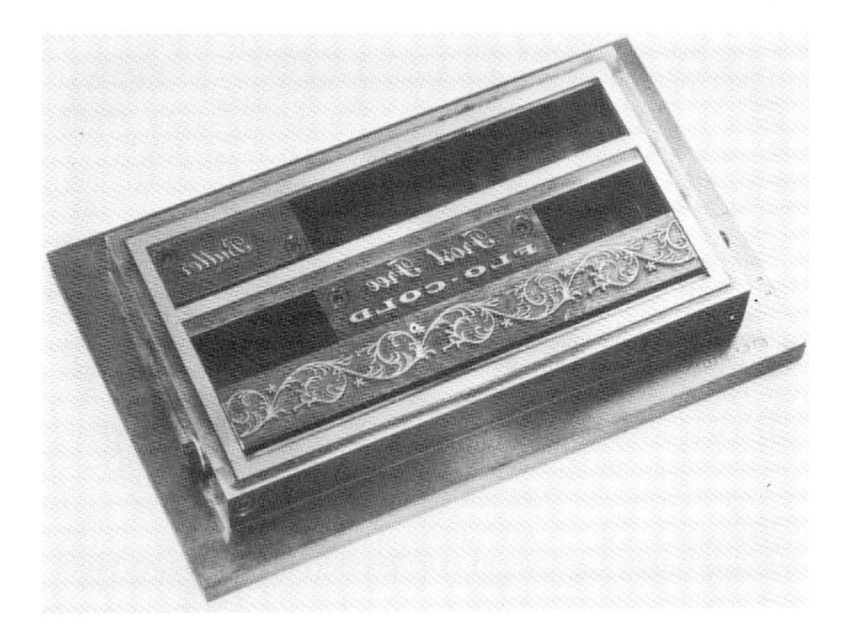

Figure 9–11. An engraved hot stamp die. The die will be heated to the transfer temperature of the foil (about 275° F or 135° C). When it is pressed against the part, an exact duplicate of the die pattern will transfer. Note the intricate detail this process will transfer.

dies also wear more readily, and letters tend to round off, thus producing marginal imprints on rigid plastics.

Steel dies. Steel dies, made in a pantograph method and hardened, are the best dies for long production runs on rigid plastics. These dies are rarely damaged. Occasionally, where fine-line detail is involved, breakage can occur, but repair is relatively easy by welding on metal and recutting. Should a deep-cut pantographed steel die need to be resurfaced, this also is a simple job that requires only a touch-up on a surface grinder to remove high spots and even out the surface of the die.

Silicone rubber dies. Silicone rubber bonded to metal is the die material used on all raised applications, as well as for covering large opaque areas. Silicone rubber is available in 50 to 90 durometer

(Shore A). The flexibility of silicone compensates for variations in the molding of plastics where raised letters and borders are involved. This material also lends itself well to forming where contours are involved, such as concave or convex parts. Fig. 9–12 illustrates some design recommendations for hot stamping raised letters, borders, etc.

1. The raised section to be hot stamped should be at least 0.030
2. The width of a section to be hot stamped should be at least 0.010 in. wide.
3. Raised sections to be hot stamped should have a slight crown, if possible.
4. The side walls or raised sections cannot be hot stamped.
5. Raised sections located at the bottom of a straight wall of a part should be at least 0.250 in. from the side wall.
6. Raised sections located at the bottom of an angled side wall should have a minimum of 0.060 in. from the side wall to allow for roll leaf clearance and silicone rubber squeeze out.
7. In raised sections that are to be hot stamped with several colors, the sections should be separated by a minimum of at least 0.040 in. to avoid overlap of the hot silicone stamped dies.

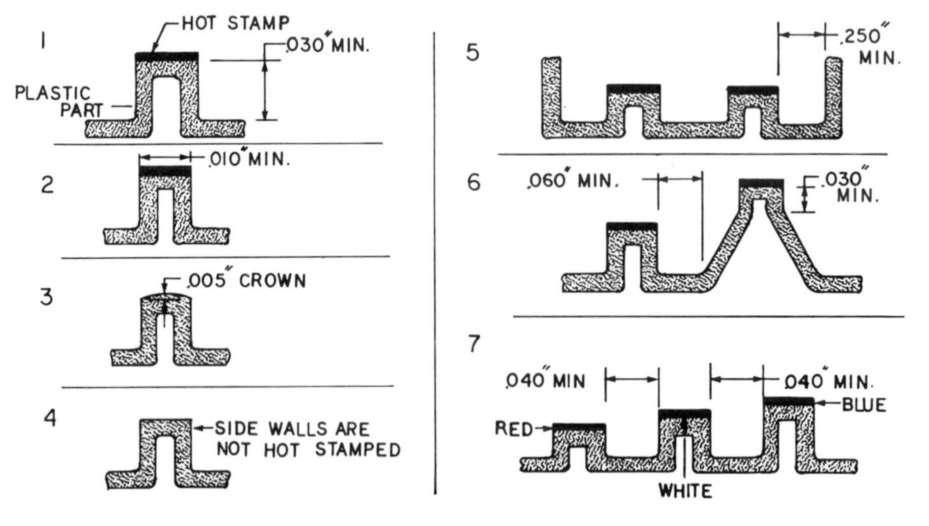

Figure 9–12. Design recommendations for hot stamping. Raised sections on plastic parts can be hot stamped by using silicone rubber.

Hot Stamping Multicolor Transfers

This plastic decorating method may soon make several existing decorating processes obsolete, because of its low cost. The process is very simple and produces a high quality, multicolor effect in a one-step operation. The printed transfers are made by printing on a lacquer film bonded to a (paper or plastic) backing. The printed film is brought into contact with items to be decorated, and a transfer of image is achieved by means of heat and pressure. The equipment used in this process is either similar to or the same as that used in hot stamping metallic foils.

High speed processes are used to print the decoration of the carrier, and all colors are later transferred to the plastic part. Several transfer systems are available. They vary by the type of carrier used and the equipment required to effect the transfer. Fig. 9–13 illustrates three methods of heat-transfer decorating: Electrocal, Therimage, and Di-Na-Cal.

Electrocal. The Electrocal hot-stamping transfer process is a combination of screen printing, hot stamping, and transfer decorating. The design is screen printed on a film carrier in roll form. The transfer of the design from the film carrier to the plastic part is accomplished

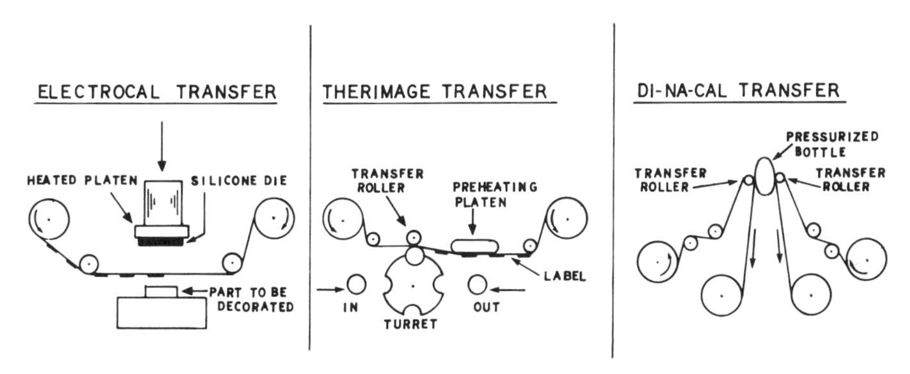

Figure 9–13. Three methods of heat-transfer decorating. The Electrocal Transfer has a printed design on a roll-fed release-coated paper and is transferred to the part by heat and pressure. The Therimage Transfer method has a design printed on roll-release-coated paper and is preheated before it is transferred to the part under heat and pressure. The Di-Na-Cal Transfer method has parallel feed and take-up lines with printed transfer designs on roll-fed release-coated paper. Both sides of the article to be decorated are done at one time.

with regular hot-stamping equipment. The transfer takes place in one cycle using heat and pressure. Silicone rubber stamping dies that conform to the shape and surface of the plastic part to be decorated are generally used. This process is most effective on plastic parts that are rigid, instead of the thin walled, softer plastic materials.

Therimage. The Therimage process of heat-transfer decorating is used mostly on round tubes, bottles, vials, and jars. The printed design is transferred from the carrier foil to the round plastic part by a special machine that preheats the carrier foil and then presses it against the surface to be decorated. The transfer roller provides additional heat and pressure. This system of transfer decorating is especially effective on plastic bottles of high-volume production rates.

Di-Na-Cal. The Di-Na-Cal process of transfer decorating is similar to the Therimage process except that two different multicolored designs can be applied to both sides of a plastic part simultaneously. This process is used mostly on decorating plastic bottles.

In this process printed transfer design rolls are fed to both sides of the part to be decorated. The maximum decorating area on cylindrical bottles is approximately 135 deg. per side or about 75% of the total coverage. The best design use of this process is when two different designs are required (front and back) on a flat or oval bottle. Thin-walled bottles are generally internally pressurized by compressed air to provide the support necessary for the decorating process.

Tampo-Print

This is a relatively new method of decorating plastic parts. Contoured parts that were previously considered impractical to decorate are now made rather simple. This process (Fig. 9–14) consists of a design or decoration etched into a metal plate. The design is filled with pigment by the wiping action of a moving brush and doctor blade. The pigment is picked up by a soft silicone rubber transfer pad. The pad is pressed against the etched plate, picking up the pigment from the recessed area. The transfer pad, with the pigment pick-up is then pressed against the part to be decorated. The pad presses against the

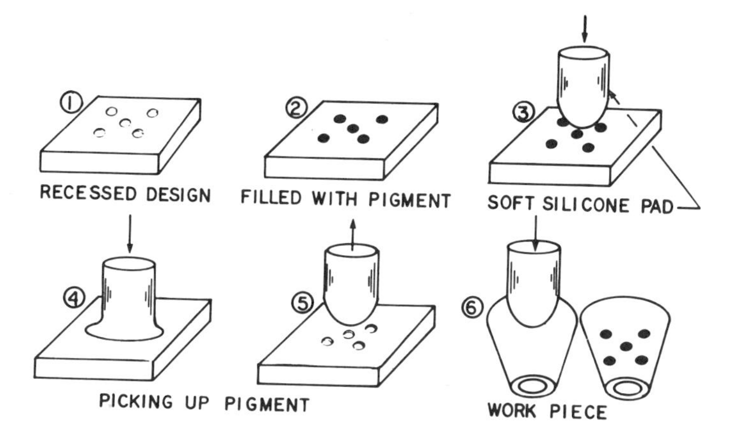

Figure 9–14. This illustrates schematically the Tampo-Print method of decorating.

part and distorts to conform with the decorating surface. The entire pigment or ink is carried by the transfer pad and is deposited on the part, leaving the pad clean of any residue. The pad distorts to conform to sharp ridges and deposits a perfect reproduction on every decorated part.

The process uses very light pressures and can be used to print on highly sensitive thin-walled products. It also can apply to multicolor designs. The transfer pad applies each required color separately. It is a rapid operation that is done without drying between color applications.

The cost of tooling for this process is normally less costly than those involved with other decorating methods. The acid-etched chrome-plated copper plates normally cost around $200.

Vacuum Metallizing

This is a process of depositing a thin layer of metal onto a plastic surface by vaporizing the metallic filaments (usually pure aluminum) and condensing it while under a high vacuum. Various metal films including silver, gold, nickel, chromium, and aluminum can be used to metallize plastic surfaces. The film thickness is three to five millionths of an inch, and the film appears opaque and shiny.

The actual vacuum metallizing step is carried out in a steel vacuum chamber. The parts are first prime-based coated. The purpose for the base coat is to bond the aluminum layer to the substrate.

The parts to be metallized are placed in the vacuum chamber, and aluminum staples are hung on tungsten wire filaments near the center of the chamber (Fig. 9–15). The filaments are connected to an electrical power source, and the pressure within the chamber is reduced to 0.5^{m} or below. The temperature of the filaments is raised to 1200° F. At this temperature, the aluminum melts and spreads over the tungsten filaments. Increasing the temperature to 1800° F (982° C) evaporates or "flashes" the molten aluminum.

The metal atoms travel in a straight line from the filaments and condense on all surfaces they encounter. The plastic molded parts must be rotated in the chamber to insure complete surface coverage. If entire coverage is not desirable, masks can be used to protect specific areas. Metallizing is done on the first surface of opaque moldings and the first or second surface of transparent moldings. Fig. 9–16 illustrates the first and second surface metallizing of plastic parts.

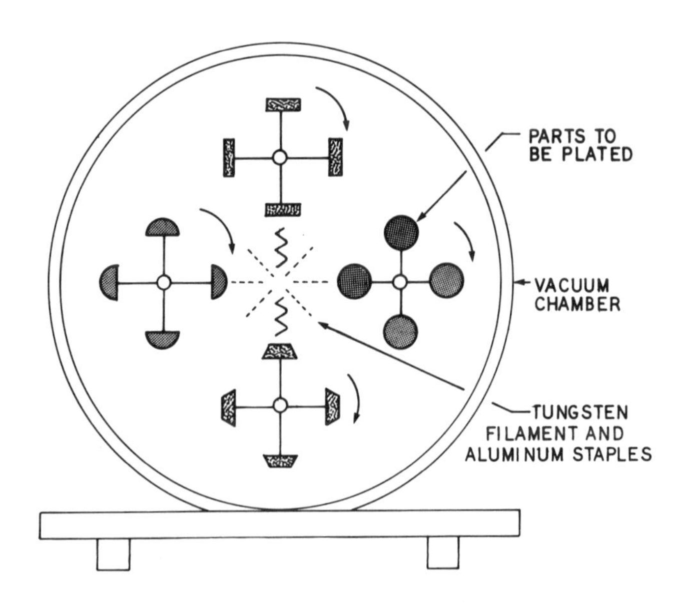

Figure 9–15. An open end view of a vacuum metallizing chamber. The plastic parts are rotated to ensure complete metal coverage.

The melting and flashing takes only a few seconds, and the complete cycle may take as little as 15 mins. After flashing, the vacuum chamber is vented to the atmosphere, and the metallized parts are removed and given a protective or top coat of lacquer.

In designing plastic parts that are to be metallized, it should be noted that large flat surfaces and sharp corners or edges should be avoided. They should be replaced with a patterned, textured, or crowned surface and rounded edges. It is very difficult to get a molded surface that is optically flat. If a surface is not flat, the surface distortion is magnified and becomes plainly visible after it is metallized.

Almost any plastic can be metallized, provided a suitable base-coating system is available that will adhere to the plastic part and will accept the metallic film. Polystyrene, methyl methacrylate, ABS, SAN, urea, melamine, polyester, and epoxy resins can be metallized. Polyolefins can be metallized, but they require a surface pretreatment similar to that needed for printing. Plasticized vinyl and cellulosics are not recommended.

Sputtering

This process of metallic coating is sometimes called cathode sputtering. The sputtering process involves the use of an electrical dis-

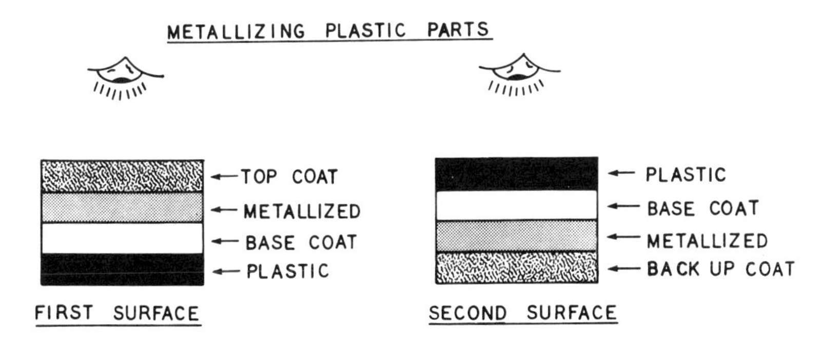

Figure 9–16. The difference between first and second surface metallizing on plastic parts.

charge between two electrodes in order to condense metal atoms from the cathode onto the plastic surface. This is done in a partial vacuum and an atmosphere of argon gas.

Sputtering is generally done in a vacuum chamber. The metal used in this process is physically broken away, atom by atom, by the impact of ionized gas molecules. The inert gas used is argon. The dislodged metal atoms are deposited as a coating on the part surface. In sputtering the metal is not melted but is mechanically chipped away. The metal atoms form a cloud inside the chamber that fogs the parts with a thin layer of metal. The typical thickness of sputtered coatings is 300 to 500 angstroms (1 to 2 millionths in.).

The sputtering process uses a two-coat system along with the metallized layer. The base coat is applied to the surface of the plastic part. It helps to level the surface and increases the brightness of the metal deposit. It also increases the adhesion and stress-resistance of the metal layer. The top coat, applied over the metallized layer, protects the metal deposit from abrasion and weathering.

Sputtering is used in applying special coatings to electronic parts and optical lens. In automobile applications, the metal of primary interest is chromium because of its good appearance and high corrosion resistance. Almost any metal or alloy can be sputter deposited.

Composite Decorating

A plastic molded part may be decorated by using a combination of methods. An instrument control panel may be vacuum metallized, spray painted, and hot stamped. Table 9–2 demonstrates the possible combinations that can be used to decorate a molded plastic part. The chart is based on the use of air-dry lacquers on thermoplastics. The horizontal columns indicate the method by which a coating is first applied, and the vertical columns indicate the second method of decoration.

It will be noted that spraying over a sprayed coating can be done as long as the coatings are compatible and not intermixed. In the spray and wipe method, caution should be taken with the other processes of decorating. Also, flow coating and dip coating are generally not applicable with other decorating methods. Spray painting over silk

TABLE 9–2. A METHODS CHART SHOWING THE POSSIBLE COMBINATIONS THAT CAN BE USED TO DECORATE A PLASTIC PART.
It is based upon the use of air-dry lacquers. Horizontal columns indicate method of application of first coat, vertical columns indicate second method of decoration.
(Courtesy Bee Chemical Co.)

SECOND COAT	FIRST COAT	*Spray*	*Spray and Wipe*	*Silk Screen*	*Roll Coat*	*Hot Stamp*	*Vacuum Metal-lizing*	*Flow Coat*	*Dip*
	Spray	Yes	Caution	Yes	Yes	Yes	Caution	Not applicable	Caution
	Spray and wipe	Yes (2nd surface)	Not recommended	Not applicable	Not applicable	Not applicable	Yes (2nd surface)	Not applicable	Not applicable
	Silk screen	Yes	Caution	Yes	Yes	Yes	Caution	Not applicable	Not applicable
	Roll coat	Yes	Caution	Yes	Yes	Yes	Caution	Not applicable	Not applicable
	Hot stamp	Caution	Not recommended	Caution	Caution	Yes	Caution	Not applicable	Not applicable
	Vacuum metallizing	Caution	Caution	Caution	Caution	Yes	Not applicable	Not applicable	Not recommended
	Flow coat	Yes	Caution	Yes	Yes	Yes	Caution	Not applicable	Caution
	Dip	Yes	Caution	Yes	Yes	Yes	Caution	Not applicable	Caution

screening is frequently done on second surfaces, such as washer or dryer control panels.

Lettering Plastic Parts

Numerous types of lettering are used on plastic parts, generally for purposes of identification. Raised letters are often less costly than depressed letters, because the raised letters are molded from letters recessed in the mold cavity, while depressed letters require the mold steel to be cut all around them (Figs. 9–17 and 9–18). Hobbing of the mold, wherever possible, will often aid in reducing the total cost of making raised letters on the mold. The letters in this case would be cut into the hob, and the hob itself would be forced into a softer steel which, after hardening, would become the mold. Raised decorations generally allow some savings in molding. Many types of letters and numerals, as well as some types of decorative designs, can be raised in the cavity by hobbing.

A raised letter is visible if it is only 0.003 in. high. Letters normally 0.015 in. high are easily read, because they catch sharp highlights. Letters that are over 0.030 in. high should be tapered and have fillets at the base. Raised letters can be decorated or painted by roll coating.

The cost of applying molded-in letters to a part will vary. Table 9–3 gives comparative costs for raised, depressed, or photoetched letters that are made by either a machined or hobbed cavity.

Electroplating Metal on Plastics

Many of the common plastics can be electroplated by a process similar to that of plating metal. The plastic surface is made conductive by the precipitation of a conductive metal such as copper. A bright chrome is plated over the copper. The key to successful electroplating of these polymers is a very thin conductive film of copper deposited on the plastic. The copper film is bonded chemically and physically to the plastic and provides a base for electroplating with chrome, silver, gold, nickel, and other metals. Chrome is of the greatest interest, because of its weatherability and excellent abrasion resistance.

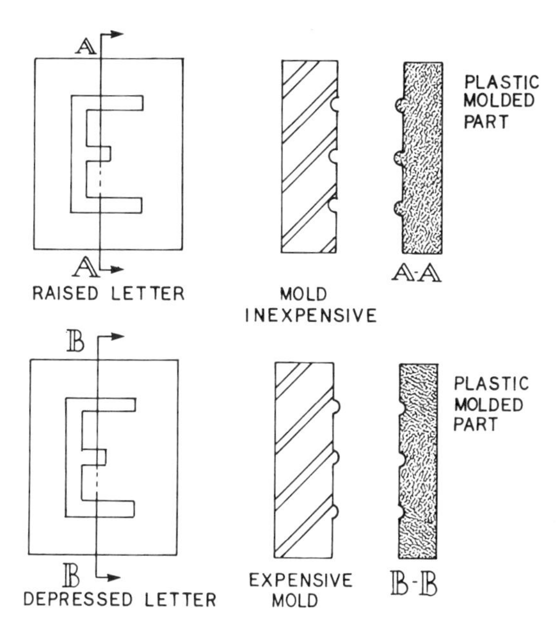

Figure 9–17. Raised letters on the molded plastic part are depressed letters on the mold. This allows engraving the mold which is a cheap procedure. Depressed letters on the molded plastic part are raised letters on the mold. This means cutting away the mold around the letter which is an expensive operation. If a hob is used in making the mold, the reverse is true.

Figure 9–18. Molded plastic parts can have raised or depressed letters. Depressed letters on the plastic part are more expensive. (Photo by Arthur Hollar)

TABLE 9–3. COMPARATIVE TOOL COST OF PRODUCING LETTERS FOR MOLDED PLASTIC PARTS.

Letters	*Machined Cavity*	*Hobbed Cavity*
Raised on molded surface	Low	Medium
Raised on depressed pad	Medium	Medium
Depressed on molded surface	High	Low
Photoengraved letters 0.003 in.	Low	Low

Since plastic parts are nonconductors, they require an electrically conductive surfaces on which the electroplate can deposit. The initial conductive coating, which is applied by chemical deposition, must adhere throughout the electroplating operation and service life of the part. All traces of grease, oil, mold release, and other film residues must be removed.

Generous radii should be used. It is difficult to obtain good plates on parts having sharp edges and corners. Blind holes are to be avoided, and a smooth mold is needed for maximum plate adherence to the plastic surface. Fig. 9–19 illustrates good design practice for metal plating on plastic. (A) Large flat surfaces to be metal plated should be made convex by crowning 0.010 to 0.015 in. This tends to disguise any surface imperfections. Also, areas can be textured or

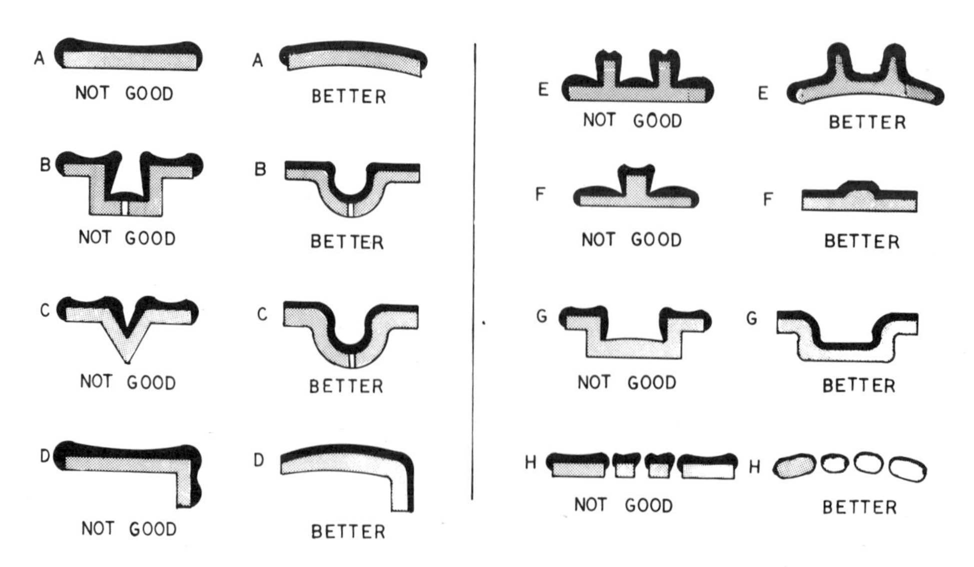

Figure 9–19. Good design practice for metal plating on plastic surfaces. *(Courtesy Marbon Chemical Co.)*

grained to produce a patterned finish. (B) Deep holes or recesses should be rounded, and a drain hole should be provided so that plating solutions are not carried from bath to bath. (C) Avoid deep "V" shaped grooves, as they are almost impossible to metal plate. (D) Sharp angles should be avoided, and all inside and outside corners should be well radiused. (E) The corners of all ribs and bosses should be rounded. Adequate draft angles should be provided so that mold lubricants do not need to be used. Mold lubricants must be removed from the part before good plate adhesion can be obtained. (F) The height of letters and embossed decorations should be no more than half their width, with a maximum height of .187 in. (G) Grooves and indentions should be at least three times as wide as they are deep. All corners should be rounded (H) Louvers and slots should be crowned and as widely spaced as possible.

Electroplating is primarily suited to ABS plastic, but can be used on other thermoplastics. In addition to appearance, the electroplating of plastics yields increased advantages in tensile strength, impact resistance, modulus, and heat distortion temperatures.

Two-Color Molding

This is sometimes called double-shot molding. It is an injection molding process for making two-color molded parts by means of successive molding operations. This is accomplished by first molding the basic case or shell and then using this as an insert, molding numerals, letters, and designs in and around the insert. There are two methods of two-color injection molding (Fig. 9–20). The "A" part of the drawing illustrates the method of first molding a shell and then transfering the shell to another mold with a different cavity and core pin. The second color is injected into the shell and around the second core pin. In the "B" part of the drawing, the inner plug is molded first and then transferred to another cavity, where the second color is injected around the plug.

Fig. 9–21 illustrates the full "shot" of two-color adding machine numerals. The "A" part of the picture shows the front view of a complete "shot" of finished control keys. The "B" part of the picture

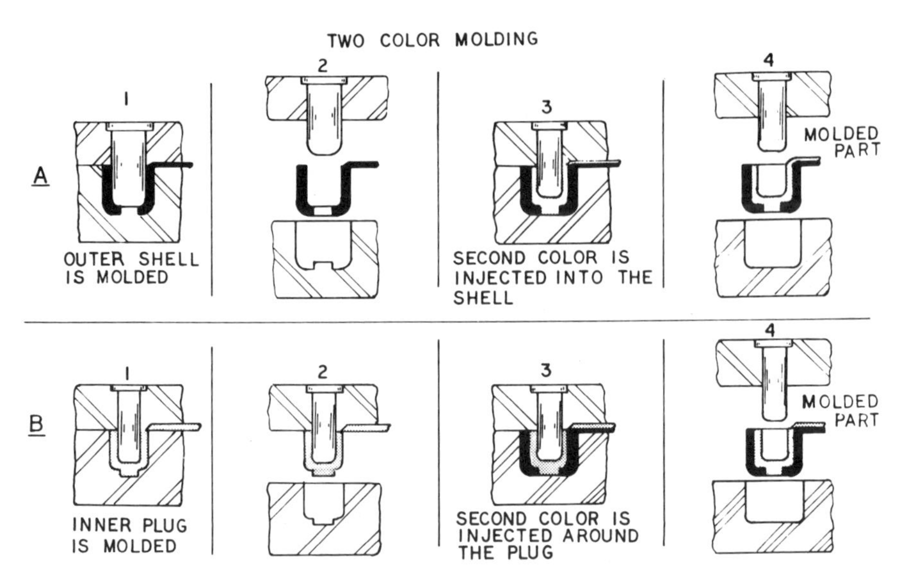

Figure 9–20. This drawing demonstrates two methods of making two-color molded parts.

Figure 9–21. Two-color injection molding of adding machine keys. (A) Front view of a complete shot of finished control keytops. (B) Rear view of a complete shot of finished control keytops.

shows a rear view of a complete "shot" of finished control keys. Fig. 9–22 shows a few of the many items that can be molded by two-color molding. It should be noted that it is possible to mold three-color molding by first inserting one component part into the die before molding the other two colors. There are special injection molding machines made for three-color molding. The machines and dies are complicated and expensive.

In this type of molding, mold design is very critical and gating is specialized. Typical parts currently being molded include caps for typewriter and business machine keyboards, automotive tail light lenses, pushbuttons, and telephone dial components.

Insert Mold Decorating

The insert mold decorating process begins with a predecorated sheet (.020 in. ABS). The ABS sheet is thermoformed and die cut to con-

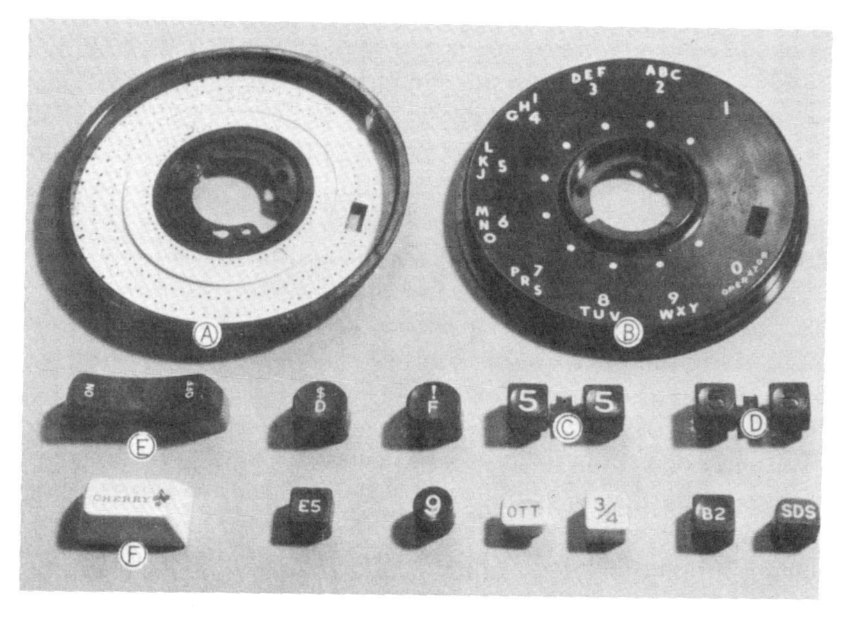

Figure 9–22. This illustrates parts that have been made by two-color injection molding. (A) and (B) show the front and back of a telephone dial plate. (C) and (D) show the shell and the filled number "5". (E) and (F) shows three color molding, one colored part was inserted into the die before molding the other two colors. (Photo by Arthur Hollar)

form to the die cavity of an injection mold (Fig. 9–23A). The formed insert is placed in the injection mold and molten plastic (ABS) is injected into the cavity where the back surface of the thermoformed insert under heat and pressure becomes an integral part of the molding by fusion with this melt.

Special grades of ABS, PVC, and polystyrene are suitable for this type of molding. The stretching or deformation of the part should not be more than 250%. A minimum radius of .062 in. is recommended. The formed part should have at least a 2° draft angle. Fig. 9–24 shows an automobile door arm rest insert that has been insert-molded.

Inmold Decorating

This process consists of placing in a mold an overlay of thermoplastic film of the same material as the plastic to be molded. A design has previously been printed on the transparent thermoplastic film (Fig. 9–23B). The transparent film containing the design, is placed against the mold with the printed side exposed to the molten plastic. When the molding cycle is begun, the molten polymer enters the mold and flows over the film, and fuses it to the molding.

Decorating in the mold utilizes the heat and pressure involved in normal molding to bond or laminate the decorated foil to the part being molded. Clear film of the same plastic material as the part to be molded is used. Rotogravure printing, screen printing, offset printing, and hot stamping are used to decorate the film or foil. The thickness of the foil varies from .003 to .010 in. The foil must have sufficient body to be easily inserted into the mold. This type of decorating works best on flat surfaces. The foil will not go around an angle that is more severe than 45°.

Locating of the gating is the most important consideration in designing for in-mold decorating. The gate will determine the point where the plastic will hit the foil and the location of the weld line. The gate should be at least .500 in. away from the decorating surface.

The best method to anchor overlays to the mold surface is by static electricity. The electrostatic charge on the film causes it to adhere to

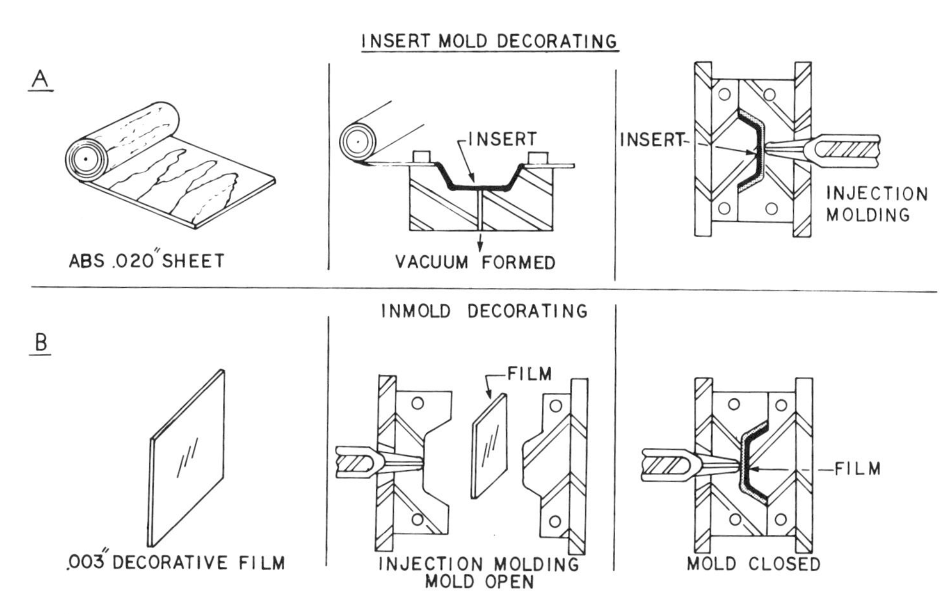

Figure 9–23. (A) This illustrates insert mold decorating by using a .020 in. sheet. (B) This illustrates inmold decorating by using a decorative .003 in film.

Figure 9–24. This illustrates an automobile door arm rest insert that has been insert-molded (A). The .020 in. ABS predecorated sheet is shown in (B). (Photo by Arthur Hollar)

vertical surfaces. The film is given a high charge of static electricity, then is placed in the mold where it holds smooth and tight against the mold surface.

Decorating Molded Melamine Parts

Melamine dinnerware (Fig. 9–25) is made by molding a color printed foil pattern on the top surfaces of the part. In making a decorative melamine part, the mold is first charged with melamine powder and is then closed until the material has cured sufficiently so that the material is blister-free when the mold is opened. The part must also be rigid enough to stay in the cavity as the mold is opened. The printed foil is inserted on the partially cured melamine molded part, and the

Figure 9–25. The materials that make melamine dinnerware are shown in this picture. (A) Melamine powder in a preform. (B) A molded melamine plate without the overlay sheet. (C) The printer overlay sheet. (D) A molded decorated melamine plate.

mold is again closed. The printed overlay is molded into the melamine part and becomes a part of it. No fillers or pigments are used in the foil paper, because the overlay must disappear into the molding, leaving only the design on the molded product.

The printed overlay is made of high-grade alpha cellulose or rayon paper, three to four mils thick. The paper foil is saturated with a melamine resin solution (65 to 67%). The dry saturated foil is printed by rotogravure, offset press, or silk screen in any variety of colors and designs. The offset printing method is the most common. In this process, the image is transferred from the printing plate to a rubber blanket that actually puts the ink on the melamine treated foil. The inks used in this printing process are made specially for this type of decorating. The overlay is generally inserted ink-side down in the molding process.

The adhesion of the printed foil to the base molding is a critical part of the decorating technique. The best adhesion occurs when the foil is inserted as soon as possible. The initial cure cycle on the molding compound is in the range of 50 to 80 secs. The final cure on the overlay is 50 to 60 secs.

Some applications for this type molding and decorating include dinnerware, wall plates for light switches, utensil handles, clock faces, instrument dial faces, and color-coded control knobs.

10

Extrusion Design and Processing

PRINCIPLES OF EXTRUSION

Extrusion is a molding process used to form continuous shapes by forcing a molten plastic material through a die. It is used to produce tubes, filaments, film, and shapes with a wide variety of profiles (Fig. 10–1). It is also used in preheating plastic materials in injection molding machines, producing parisons for blow molding, and for pelletizing.

In dry, hot extrusion, thermoplastic materials (in the form of powder, chips, or pellets) are fed from a hopper into a barrel of the extruder. A screw rotating inside the barrel kneads the plastic to a uniform consistency and impels it toward the die orifice. The barrel has closely controllable temperature zones that aid in producing a uniform melt of the plastic material. This molten plastic is forced through the die and the extruded form is carried through a cooling medium by a takeoff mechanism, which is generally either a conveyor belt or caterpillar type belt with variable-speed controls.

There are a number of different kinds of extruders in use today. The most common is the single-screw machine. The single-screw extruder is the type discussed in this book (Fig. 10–2A). Many adaptations are used with the single-screw extruder, such as cross head die extrusion, dial extrusion, metal embedment, and vented extruder. A cross-head die extrusion (Fig. 10–2B) is used mostly for wire or cable covering. Here the die is placed at right angles to the extruder. Dual

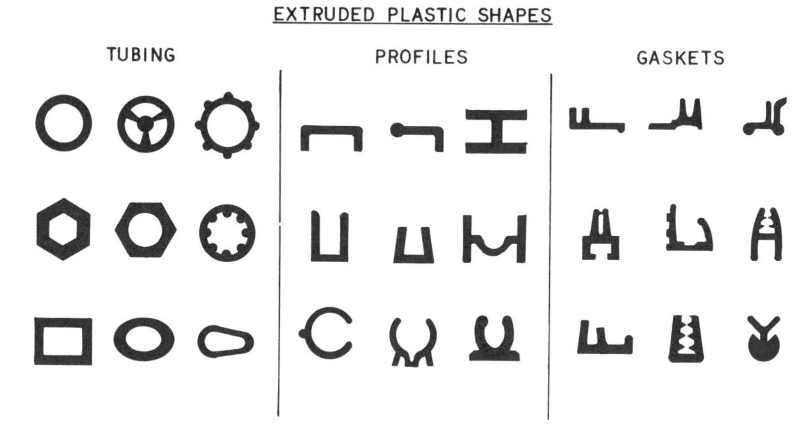

Figure 10–1. Plastic extrusions are produced in a variety of sizes and shapes.

extrusion uses two extruders, as shown in Fig. 10–2C. Metal embedment provides a means of incorporating metal strips, wires or roll-formed shapes in the plastic profile extrusion (Fig. 10–2D). The metal components are fed through an offset extrusion die where they are partially or completely embedded into the extrusion. Venting an extruder is the process of removing air, gases, and volatiles from the plastic after melting (Fig. 10–2E). The plastic softens, melts, and mixes in the first pump stage. It then passes into a deep channeled zone open to the atmosphere. Gases flash out of the hot plastic before it moves on to the second pump stage. The molten plastic then moves on to the delivery end of the extruder to form the final product.

The size of an extruder determines the rate at which a plastic material can be processsed. The size description refers to the outside diameter of the screw, which mates closely with the inside diameter of the barrel. Thus, a machine with a 3 in. diameter screw is referred to as a 3 in. extruder. The most common sizes of extruders used in profile extrusion are from 1-1/2 to 6 ins.

The extrusion die is a streamlined orifice that reduces the heat-softened plastic mass to a definite shape. A good alloy or tool steel is generally used to make the extrusion die. The die orifice size controls the thickness or width of any extended part dimension. In general, it is developed oversize to allow for draw and shrinkage during the conveyor cooling operation.

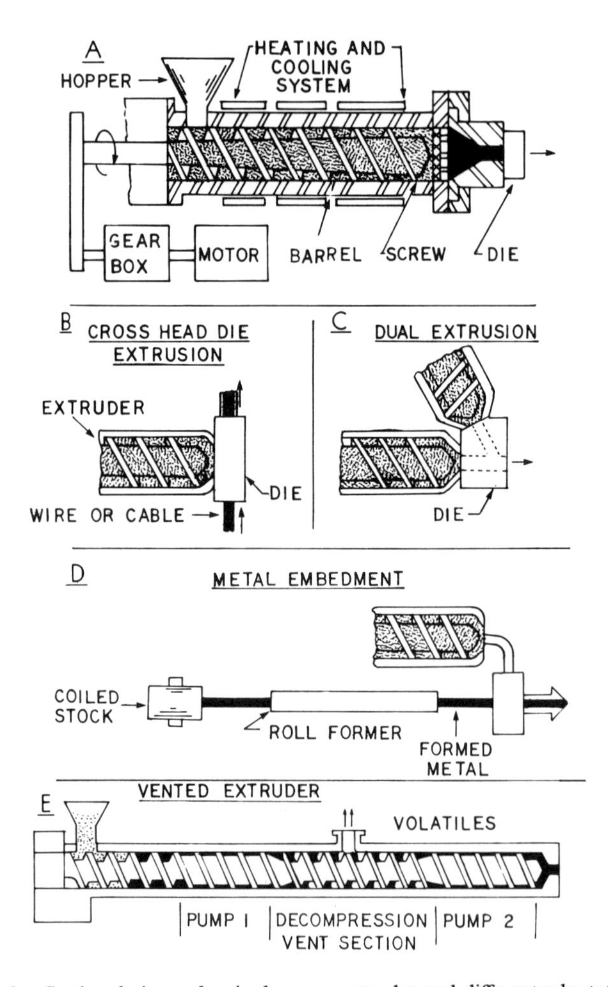

Figure 10–2. Sectional views of a single screw extruder and different adaptations such as dual extrusion, cross head die, dual extrusion, metal embedment, and vented extruder.

The rate of take-off influences the dimensions of the extruded shape (Fig. 10–3). In all cases, some type of take-off or pulling conveyor is used to pull the extrusion away from the die. This conveyor may be in the form of pinch rollers or opposed caterpillar belts. By changing the speed at which the take-off pulls the plastic, it is possible to control the dimensions of the part. Almost always, the size of

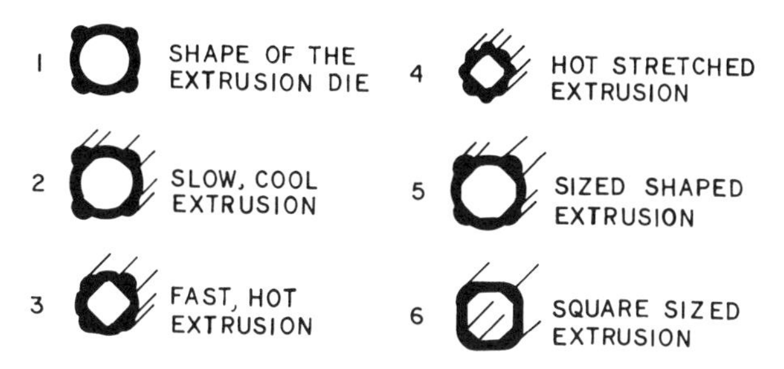

Figure 10–3. This illustrates the products that can be obtained from a single die by adjusting the take-off methods.

opening of the die will be larger than the finished section desired. The plastic is pulled away faster than it is extruded, thus causing it to draw down or get smaller as it leaves the die. This is called "drawdown," and it keeps the extrusion straight and permits size adjustments. The amount of "drawdown" is defined as "drawdown ratio" and is the ratio of the die size to the final cross-sectional area of the extruded part.

Cooling and Shaping

As the molten or fluid plastic material leaves the die it must be supported with fixtures to retain the desired shape during the cooling period. The simplest method is to draw the extrusion directly into a water bath, keeping it submerged so that the water can cool the extrudate. The temperature of the water bath may vary from cold to hot, depending on the material and section being extruded. The emerging shape also may be cooled with hot or cold air, depending upon the type of plastic and the viscosity of the compound used. Rigid materials, such as polystyrene, methyl methacrylate, rigid vinyls, and cellulose acetate are not generally quenched in cold water. This causes undesirable stresses and poor surface appearances. Some of the crystalline materials such as polyethylene, nylon, and polypropylene may be (and generally are) cooled in cold water. Small plates that have the shape of the finished extrusion are often used to hold the extruded plastic as it passes through the water. These are called sizing plates. Other types of holding fixtures may be used such as rollers, adjustable fingers, or blocks.

Tubing or Hollow Parts

Tubing is made by a die that forms the outside and a core or mandrel that forms the inside. The mandrel is supported by a spider. Air is introduced through a rib of the spider to the core. The air supports the inside of the tubing to keep it round until cooled. In all hollow extruded sections, some provisions must be made to admit air inside the hollow; otherwise, the vacuum generated would flatten the section immediately. After the tubing leaves the extrusion die, it is run through a sizing die or over a cooling mandrel, to maintain concentricity during cooling.

PROFILE EXTRUSIONS

Principles

Plastic profile extrusion is still somewhat of a trial and error process, because of many and interrelated reasons. First, each thermoplastic material has its own unique extrusion flow characteristics. Next, each combination of extrusion machine and screw has certain flow-generation characteristics. Each extrusion die usually requires testing and modification to perfect the profile being extruded. The skill of the operator is also important in balancing these variables to produce a satisfactory extrusion.

Profile Design

The range of extrudable profile shapes is practically unlimited. But to realize full design and economic potential from the extrusion process, particular attention must be given to wall balance, hollows and cores, legs and projections, and corners and radii.

Wall Balance

The most important consideration in profile design is the balance of various wall thicknesses. A profile with uniform wall thickness throughout the cross-section is easiest to produce. Uneven walls cause material flow variations between the large and small portions of the profile. Also, thinner sections cool faster that thick ones, causing a bow, or warpage, toward the heavy side. To compensate, it is necessary to provide external cooling for the bulkier sections. This requires additional equipment and reduces extrusion speed, resulting in higher production costs.

The penalty of an unbalanced wall is the loss of tolerance control. Tolerance limits are often double those of a balanced profile. With flow and warpage problems minimized through planned wall balance, it is possible to achieve more complex shapes, and still hold to reasonable tolerances.

Some thermoplastic materials are more difficult to extrude in uneven or unbalanced wall profiles. Vinyls, ABS, and polystyrene, for example, cool quicker and are easier to run through an unbalanced die than polyethylene and polypropylene. The last two materials have a low melt strength and, therefore, come out of the extrusion die in a more fluid condition that is difficult to control. Changing to uniform wall thicknesses improves their flow, cooling, and warpage characteristics.

Although the balanced-wall profile is ideal, it is not always possible. Often, an unbalanced profile can be slightly redesigned to eliminate a heavy area or to undercut or relieve a leg, thus providing an even wall thickness. Examples of balanced-wall profiles are shown in Fig. 10–4. Examples of design modifications to provide balanced walls are shown in Fig. 10–5. Successful unbalanced-wall profiles, shown in Fig. 10–6, have been made by using specially designed take-off fixtures.

As with injection molding, a sink mark almost always occurs in extrusion on a flat surface opposite and adjoining leg or rib. The greater bulk of material in this area retains its heat longer and causes post-extrusion shrinkage at the juncture. If appearance of the part is important, sink marks can be concealed by adding a design feature, such as a series of serrations on the area where they occur. Fig. 10–7 shows three ways that may be used to eliminate sink marks. Although sink marks can seldom be eliminated, such design treatments make the mark less obvious and often improve the appearance of the extruded section.

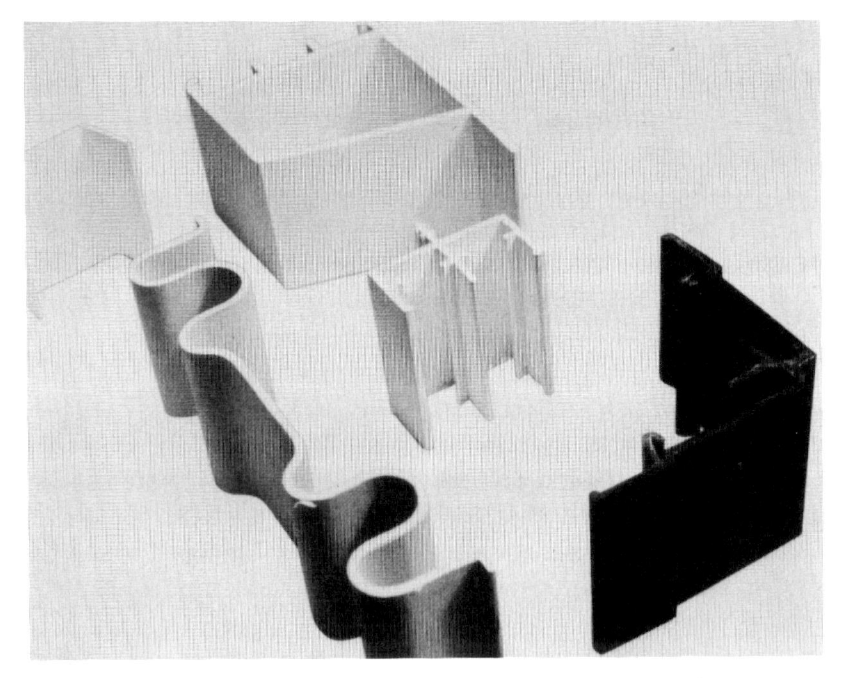

Figure 10–4. Typical examples of balanced-wall extruded profiles.

Hollows

Hollow sections are not as difficult to extrude in thermoplastics as they are in aluminum. However, hollows do increase die cost and increase tolerances. In addition, a hollow extrusion usually requires air pressure, internal mandrels, or vacuum sizing to maintain the hollow until the extrusion has cooled sufficiently to hold its shape. Unless a hollow is functionally necessary, it should be eliminated or minimized. Despite the added complexity of extruding hollows, a hollow cross-section is more desirable than one with an unbalanced wall, if the use of the hollow results in uniform walls. Fig. 10–8 illustrates extruded holes or hollows that will not be round, because of the unbalanced wall conditions.

Tolerances for hollow profiles cannot be held as tight as those for solid, uniform-wall cross-sections, but they can be improved if the shape has uniform walls. Fig. 10–9 illustrates several typical hollow profiles extruded from rigid vinyl.

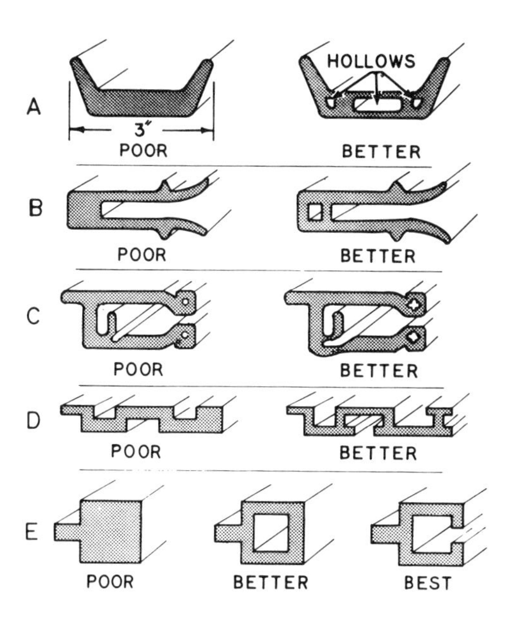

Figure 10–5. Design alterations to achieve balanced walls in extruded plastic profiles. Changing to a uniform wall shape (A and B), even if hollows are added, is preferred. Better wall balance (C). Functional surface are maintained in redesign to uniform wall profile (D). If function permits, eliminate the hollow as shown in (E).

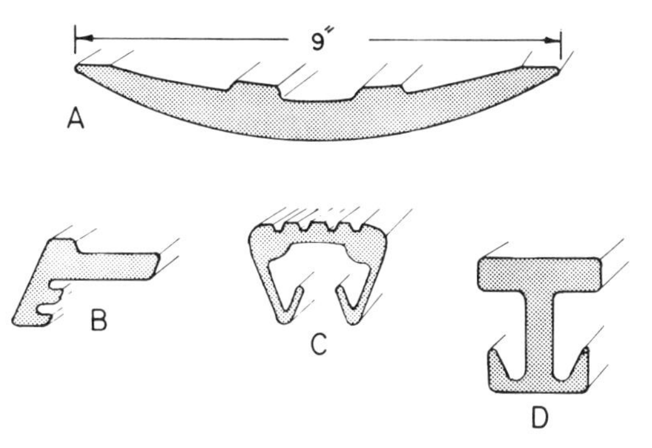

Figure 10–6. Example of successful profile shapes with unbalanced walls. (A) Rigid PVC wear shoe. (B) ABS house-trailer trim section. (C) Flexible PVC arm rest for bus. (D) Rigid PVC insulator for electrical bus bar.

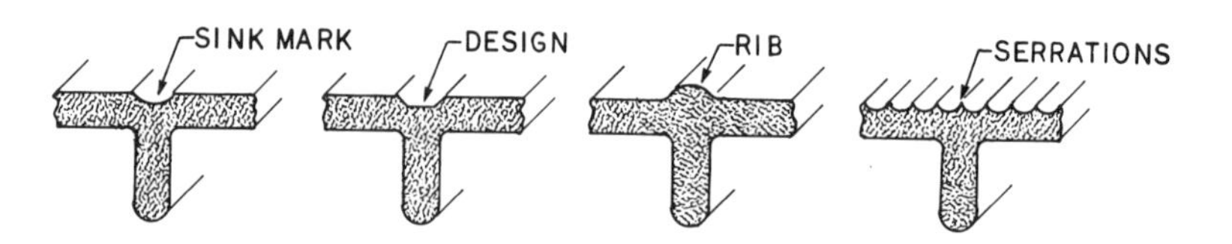

Figure 10–7. Sink marks can be eliminated by creating a design, a rib, or serrations.

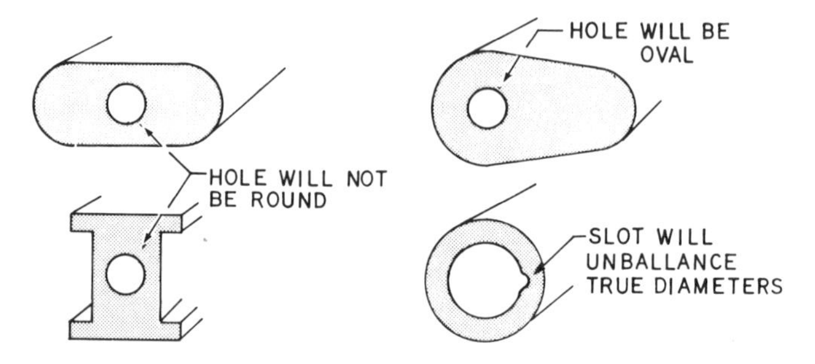

Figure 10–8. Unbalanced walls will not produce round holes.

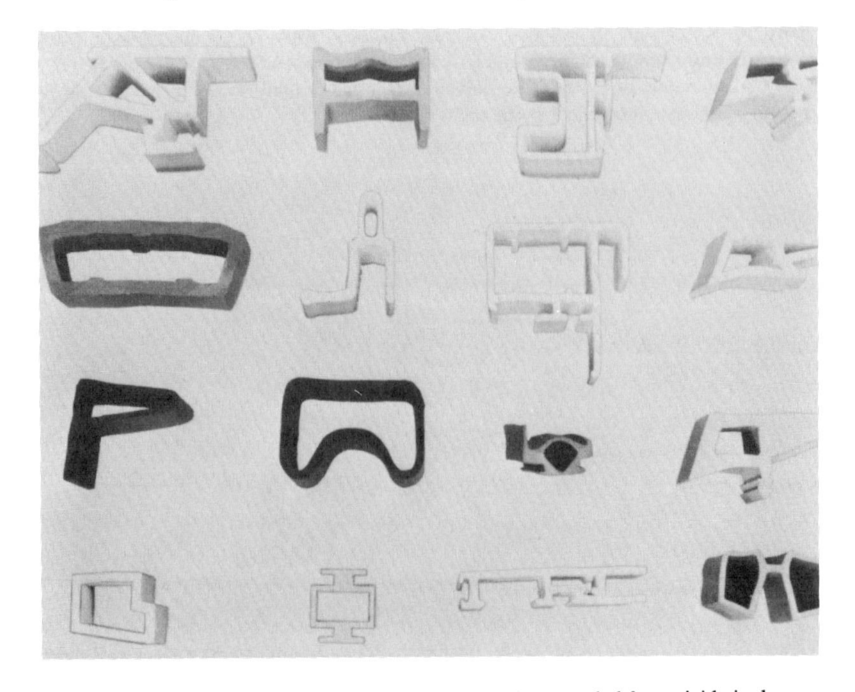

Figure 10–9. Examples of typical hollow profiles extruded from rigid vinyls.

Legs or projections inside a hollow profile should be held to a minimum or eliminated. Where projections are essential, generous tolerances must be allowed, because there is no practical way at present to hold the shape of these internal projections while they are cooling. A design rule of thumb is that the projection of a leg into a hollow be held to a maximum penetration equal to the thickness of the wall (Fig. 10–10). All of the problems of projections within hollows are compounded many times in a hollow-within-a-hollow design. Avoid this extremely costly and poor tolerance-control configuration.

Corners and Radii

Sharp outside corners are difficult to achieve with most thermoplastics, since the material tends to bridge across the sharp corner of the die and form a radius. The sharpest practical outside corner radius that can be extruded is 0.016 in. This is a fairly sharp corner that suffices for most applications. Fig. 10–11 illustrates outside radius guidelines. The larger radius aids material flow, minimizes warpage, and eliminates stress concentration at the corners.

Sharp inside corners should also be avoided. At least a 0.016 in. radius should be specified. This is particularly necessary in the more rigid materials to eliminate a natural notch or easy breaking point. Recommended design considerations for inside radii are shown in Fig. 10–12. Achieving the "sharp as practical" inside radius usually requires that the outside radius equal at least one-half of the wall thickness, although it is possible to extrude both sharp inside and

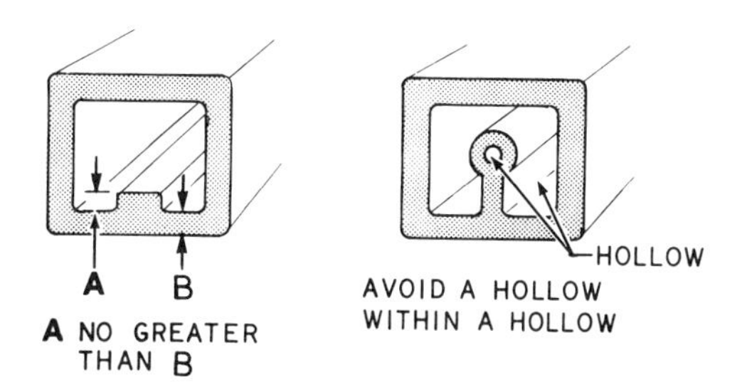

Figure 10–10. Design tips for hollow section profiles.

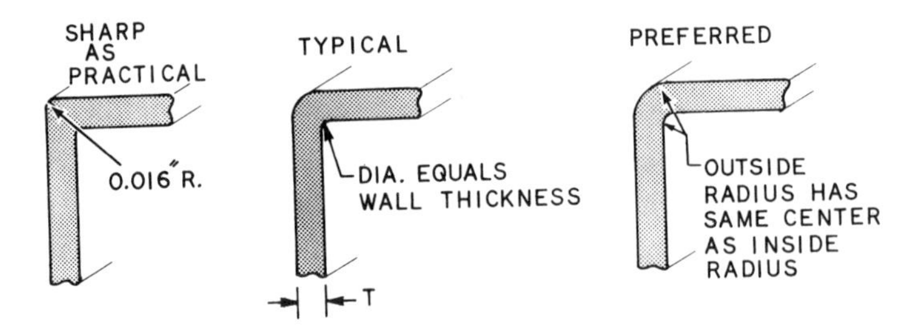

Figure 10–11. Recommended minimum radii for outside corners.

outside corners (0.016 in. outside radius and 0.010 in. inside radius) at the same juncture.

Corner radii also depend on the material being extruded. Sharp corners are relatively difficult to extrude in polyolefins and nylons. Rigid flexible vinyls, cellulosics, and ABS present no problems in this regard. Table 10–1 illustrates a tolerance guide for plastic profile extrusions. Fig. 10–13 is an example of a profile extrusion with tolerances taken from the table.

DUAL EXTRUSION

Principles

Dual extension is the combination of two dissimilar thermoplastics into a single homogeneous profile (Fig. 10–14). Combining different materials in a single extrusion can effectively reduce the number of parts in an assembly and improve product appearance.

The most common materials married in dual extrusions have been rigid and flexible vinyls. A positive, permanent, homogeneous bond

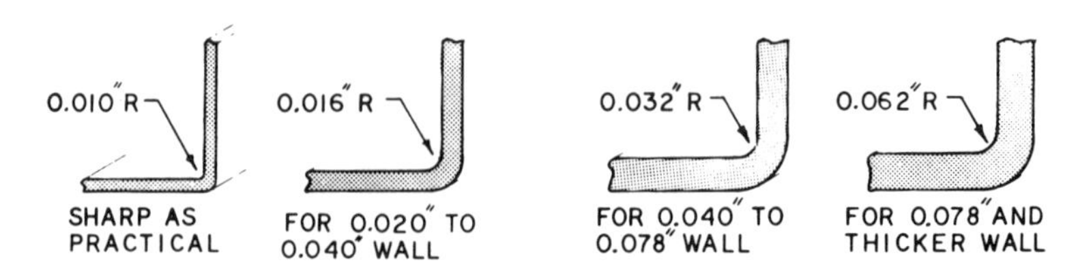

Figure 10–12. Recommended minimum radii for inside corners.

TABLE 10–1. TOLERANCE GUIDE FOR PLASTIC PROFILE EXTRUSIONS.

	Material					
	Rigid Vinyl	*Poly-styrene*	*ABS, PPO, Poly-carbonate*	*Poly-propyl-ene*	*EVA, Vinyl Flexible*	*Poly-ethy-lene**
Wall thickness (% ±)	8	8	8	8	10	10
Angles (Deg. ±)	2	2	3	3	5	5
Profile dimensions (Inches ±)						
To 0.125	0.007	0.007	0.010	0.010	0.010	0.012
0.125 to .500	0.010	0.012	0.020	0.015	0.015	0.025
.500 to 1	0.015	0.017	0.025	0.020	0.020	0.030
1 to 1.5	0.020	0.025	0.027	0.027	0.030	0.035
1.5 to 2	0.025	0.030	0.035	0.035	0.035	0.040
2 to 3	0.030	0.035	0.037	0.037	0.040	0.045
3 to 4	0.045	0.050	0.050	0.050	0.065	0.065
4 to 5	0.060	0.065	0.065	0.065	0.093	0.093
5 to 7	0.075	0.093	0.093	0.093	0.125	0.125
7 to 10	0.093	0.125	0.125	0.125	0.150	0.150

*Low density and regular grades.

between these plastics is achieved during a single-extrusion operation. Dual rigid and flexible vinyl extrusions offer advantages in both assembly and sealing applications. The rigid portion of the extrusion is used for shape retention or for attachment purposes, and the flexible portion, for sealing, cushioning, or absorbing impact. Any number of rigid and flexible members can be combined in a single profile.

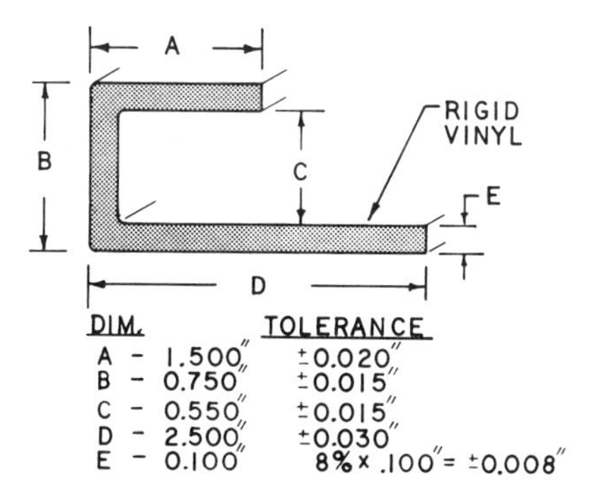

Figure 10–13. Tolerances for a plastic profile extrusion taken from Table 10–1.

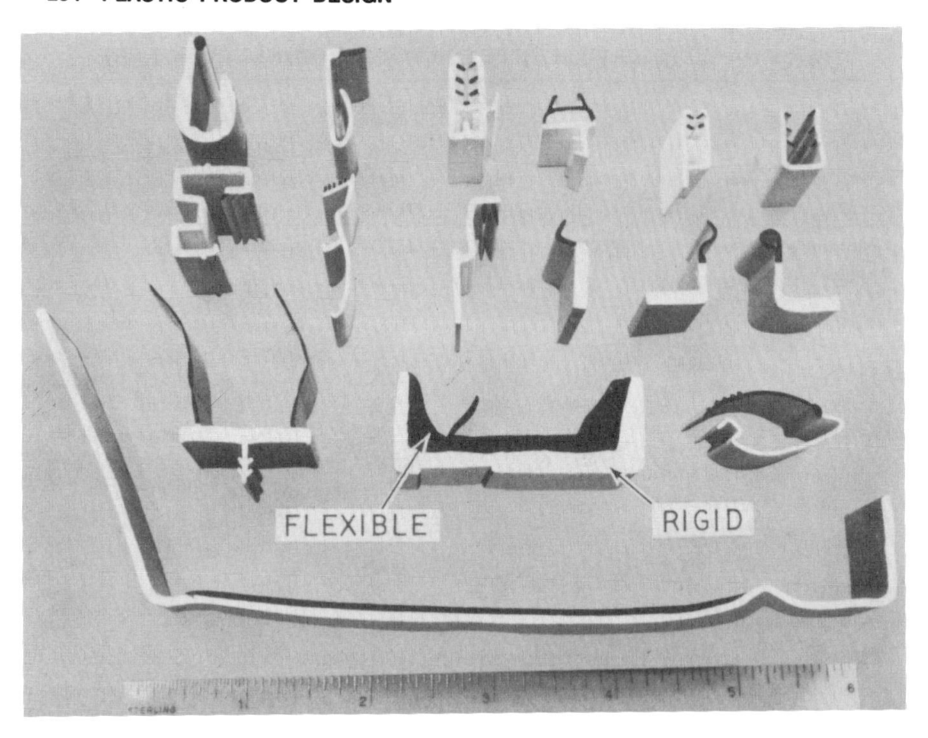

Figure 10–14. Dual durometer extrusions. The white sections of the extrusions are rigid plastic. The black sections of the extrusions are soft or flexible plastic.

Dual Color Extrusion

This is an extrusion process that produces articles similar in effect to the barber pole or striped toothpaste. The principle is rather simple in that two extruders are used. Each extruder has a different colored material and both materials are compatible. The two colored materials are brought together into a common die and extruded into the desired shape and color. The advantage of this process is that two colors are produced in one operation, eliminating any post-extrusion decorating.

Compatibility

Although vinyls are the most common combination, other dissimilar materials can be joined. Table 10–2 illustrates the compatibility of

TABLE 10–2. THIS TABLE ILLUSTRATES THE COMPATIBILITY OF DIFFERENT THERMOPLASTIC MATERIALS IN DUAL EXTRUSION

PLASTIC MATERIALS	RIGID VINYLS	FLEXIBLE VINYL	ABS	POLYETHYLENE	POLYPROPYLENE	EVA
RIGID VINYLS	YES	YES	YES	NO	NO	NO
FLEXIBLE VINYLS	YES	YES	NO	NO	NO	NO
ABS	YES	NO	YES	NO	NO	NO
POLYETHYLENE	NO	NO	NO	YES	YES	YES
POLYPROPYLENE	NO	NO	NO	YES	YES	NO
EVA	NO	NO	NO	YES	NO	YES

different thermoplastic materials in dual extrusion. Any two compatible thermoplastic materials can be extruded together. But with the exception of vinyls, the bond in most cases will not be complete, and the two materials can be pulled apart, unless dovetails, undercuts, or other mechanical joints are provided (Fig. 10–15).

Some thermoplastic combinations are not possible, because of incompatibility. Extrusion of flexible vinyl and polystyrene, for example, are not practical unless special polymeric plasticized flexible vinyl is used. Otherwise, the plasticizer in the flexible vinyl would attack the polystyrene.

Minimum Wall

Minimum wall thickness of the rigid portion of a dual extrusion is 0.020 in. Although there is not maximum thickness, parts over .250

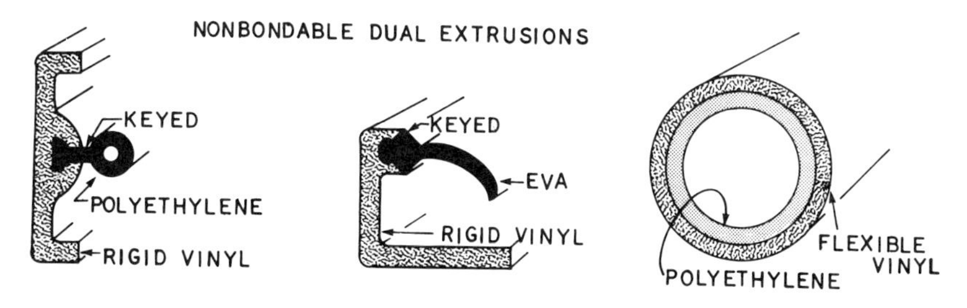

Figure 10–15. Non-bondable plastic materials can be joined by keying or fit.

in. thick cause flow imbalance during extrusion. The thinnest practical wall thickness for flexible material is 0.005 in.; the maximum thickness should not exceed .250 in. Any combination of colors can be used for the two materials in the extrusion.

Hollows

Hollows in a dual extrusion are easier to achieve, than in a standard flexible profile, when one part of the hollow is flexible material and the other part is rigid material. Fig. 10–16 shows how two materials can be used together to form a half-round shape. Note that it is difficult to extrude a half-round shape from flexible materials. The air pressure inside the half-round shape will tend to make the bottom flat section bulge.

Tolerances

Tolerances and dimensioning for dual extrusions follow the same rules as for standard extrusions. In a dual extrusion of rigid and flexible vinyls, however, there will be two sets of tolerances. Tighter tolerances can be held on the rigid material than on the flexible material.

Fig. 10–17 illustrates a dual extrusion for a modular cabinet-wall panel. The outer surfaces have a dual-extruded flexible vinyl coating with embossed pebble-grain finish.

Fig. 10–18 shows a dual extrusion replacement seal. If the flexible

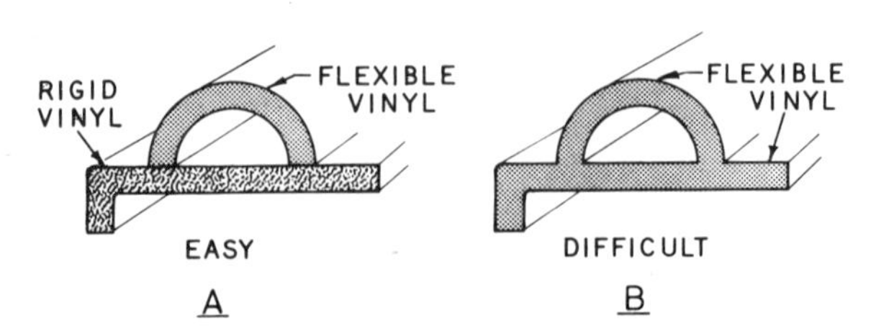

Figure 10–16. Non-circular hollows are easier to form if each part of the surrounding wall is made from the same family type of plastic material. (A) The rigid vinyl base will remain flat and not bulge. (B) The air pressure inside of the hollow will cause the flexible base section to bulge.

vinyl seal becomes damaged through wear or abuse, it can be trimmed off, and a replacement seal can be slid into the slot provided in the rigid vinyl section.

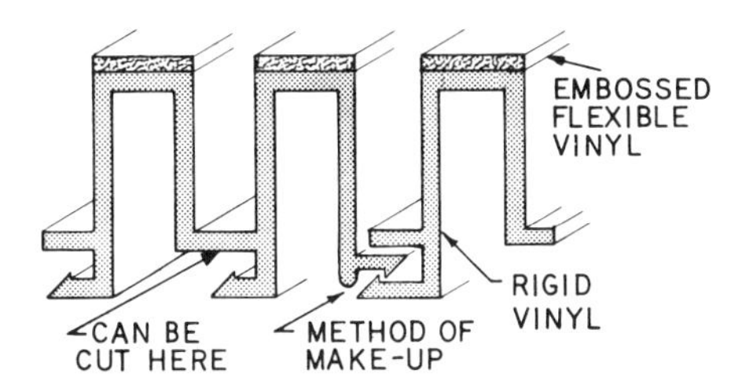

Figure 10–17. A dual extrusion for a modular cabinet-wall panel.

Fig. 10–19 shows a cross-section of a ball-return trough for a billiard table. The dual extrusion is made from rigid vinyl and EVA. The one-piece member replaces a seven-piece assembly of wood, hardboard, aluminum, and EVA.

Fig. 10–20 shows a tube and a bowling-ball return trough. The trough starts out as a standard six-in. diameter extruded tube of high-impact rigid vinyl. While the tube passing out of the die is still workable, it is slit and quided over a simple forming die where it cools into its permanent shape. Developed width of such parts can be as large as 18 ins.

Fig 10–21 shows typical rigid/flexible extrusions. The edge protec-

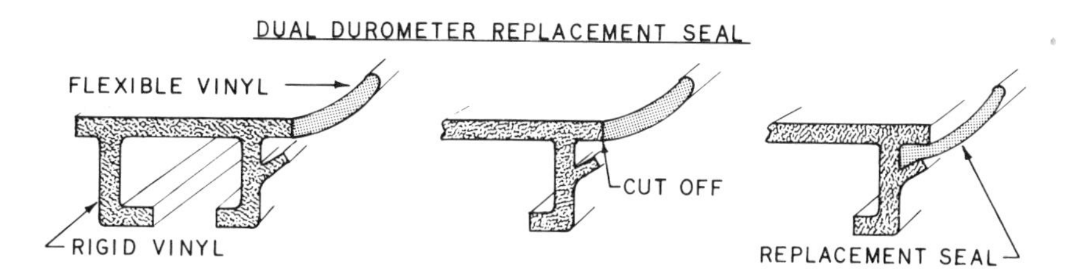

Figure 10–18. If the flexible sealing portion wears out from abrasion, a replacement flexible insert can be slid into the slot in the rigid portion.

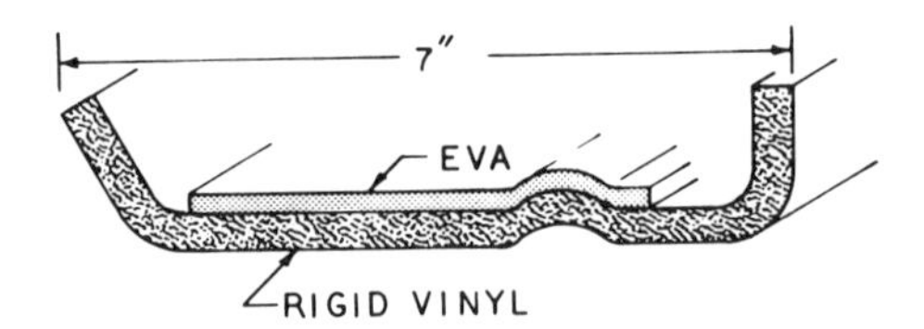

Figure 10–19. A cross section of a dual extrusion. A ball return trough for a billard table.

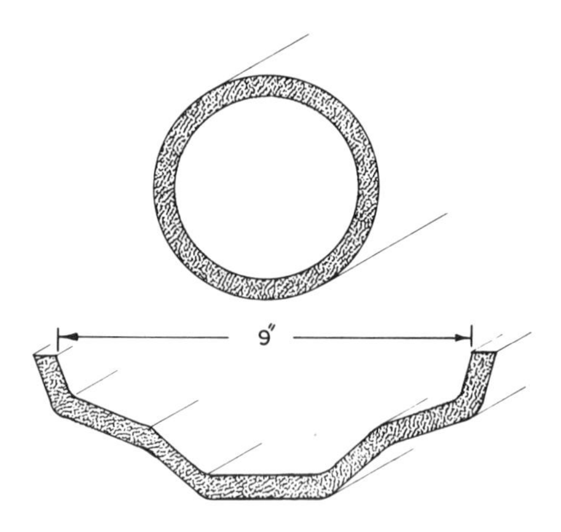

Figure 10–20. A bowling ball return trough made from a 6-inch diameter extruded tube. The extruded tube is slit while still workable and guided over a forming die.

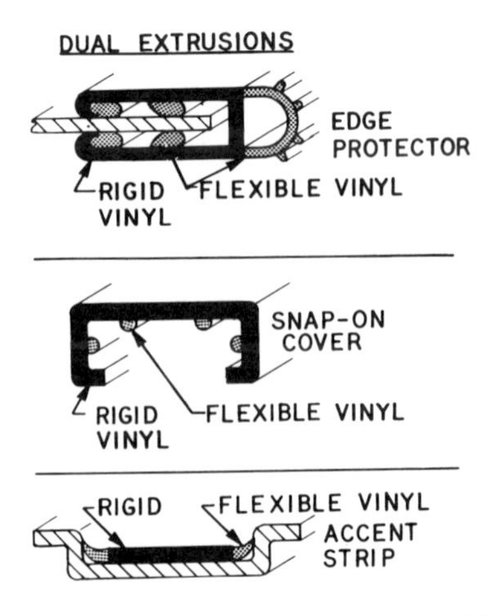

Figure 10–21. Typical dual extrusions of rigid and flexible vinyls.

tor for the sheel-metal panel is quickly applied in lengths to 10 ft. The snap-on cover has vibration-absorbing pads to prevent rattling. Pads on each side are of a different color for easy identification. The accent strip, with matching-color, flexible gripping fingers, installs easily and holds firmly without fasteners or adhesives.

Fig. 10–22 illustrates more uses for rigid-flexible extrusions, but for different applications. The seal and cap is used for sealing glass to a movable partition and is applied without fasteners. The protective bumper for a bus seat combines a resilient pad for passenger protection and a rigid member for attachment. The flexible compound used can be specified within the range from 60 through 90 durometer.

METAL EMBEDMENT

Metal embedment is the most recent addition to profile extrusion technology. With this technique, continuous lengths of wire, metal strips, or roll-formed shapes can be completely or partially embedded in the extruded profile. The metal-embedment technique can be used with both standard and dual-extrusion profiles. Any of the extrudable plastics can be used.

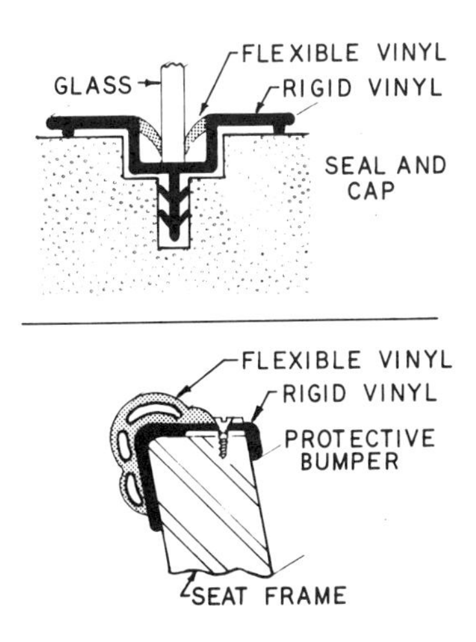

Figure 10–22. Typical dual extrusion of rigid and flexible vinyls showing different applications.

The primary advantages of embedding metal are increased rigidity and dimensional stability. The metal inclusion in the profile practically eliminates thermal expansion and contraction problems in plastics. Tests show that the thermal expansion and contraction characteristics or a metal-embedment are approximately the same as those of the metal itself. Other advantages include providing a means for rigid mounting or welding or other metal components.

With metal embedment, thermoplastic extrusions can be considered for use as structural components or for long, unsupported, load-bearing members. The combination provides a part that benefits from the strength of the metal, yet has the chemical resistance, warmth, and color of plastic. Die charges for metal embedment profiles are relatively high. This is because of the complexity of handling the metal feed during extrusion. The feed problem also is the reason for limiting embedding to wire, strip, or roll-formed metal shapes in the extruded profile.

Fig. 10–23 shows a window frame with a metal embedment extrusion. The metal and rigid vinyl provide a rugged member, while the flexible vinyl grips and surrounds the glass. The construction withstands outdoor exposure for many years.

Fig. 10–24 illustrates a storm-window frame of rigid vinyl with three 0.020 in. diàmeter metal wires to minimize thermal expansion and contraction in outdoor exposure. The tubular extrusion or welded members covered with rigid vinyl provide strong corrosion-resistant components for many applications. The metal used in the embedment process must be roll formed and continuous.

Metal foils are extruded with clear plastic materials to give a decorative effect (Fig. 10–25). Materials in a wide variety of colors are used to match metal finishes. The colors can be gold, copper, chrome, and brass.

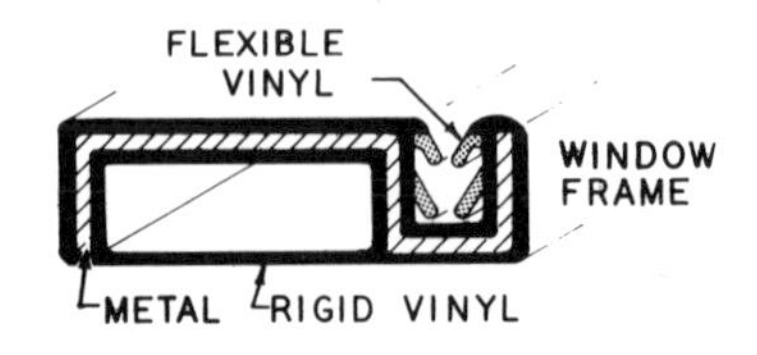

Figure 10–23. A cross-section of a window frame with a metal embedment.

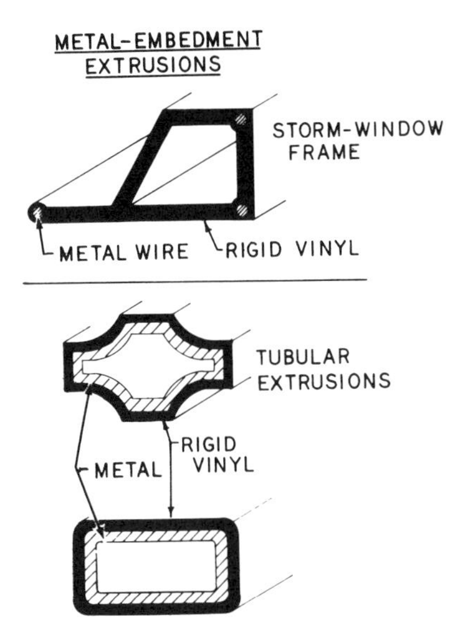

Figure 10–24. Different applications for metal-embedment extrusions.

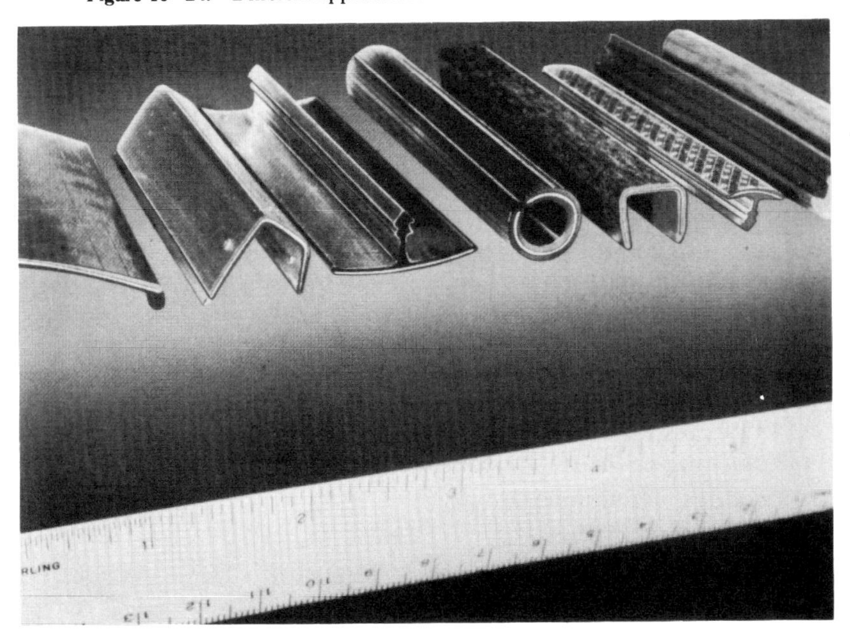

Figure 10–25. Extruded plastic shapes with metal embedment. Used mostly with clear plastic for decorative effects.

SECONDARY OPERATIONS

In-Line Operations

Rarely are plastic extrusions used in the as-extruded form. Secondary operations can be handled "in-line" (directly on the extrusion production line), or as separate operations. Typical in-line operations include:

1. Cutting-to-length (to almost any degree of tolerance, but not as tight as can be achieved in a separate operation).
2. Punching (special shapes, holes, notches, cuts, etc.).
3. Rotary hot stamping.
4. Heat sealing or heat forming.
5. Applications of films (wood grains, protective tapes, foam-rubber backing, etc.).
6. Printing (wood grain or other decorative pattern).
7. Embossing.

The most economical application of film substances, printing, and embossing, is handled directly on the extrusion production line. These techniques are limited, however, to profile shapes that cannot be supported on the surface opposite the printing or embossing roller. It is difficult to emboss or print on hollow profile shapes. Fig. 10–26 illustrates some of the many secondary operations that are performed on extrusions, both in-line and off-line.

Off-line Operations

Secondary operations that can be performed separately from extrusion include:

1. Cutting-to-length (to achieve extremely close tolerances, special angles, compound cuts, etc.).
2. Hot stamping
3. Silk screening
4. Decal labeling
5. Pressure-sensitive labeling
6. Punching (special shapes, holes, or notches)

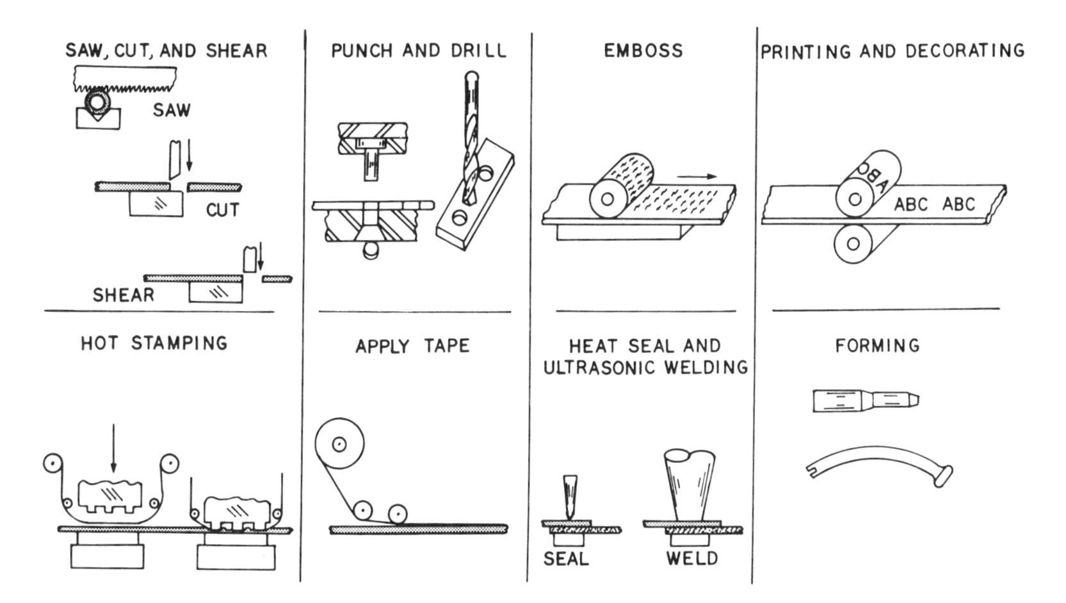

Figure 10–26. This illustrates the many off-line operations that are performed on extruded plastics.

7. Screw-machine operations
8. Broaching or milling
9. Heat sealing or heat forming

The dual extrusion technique has provided a simple means of adding a decorative embossed surface to a rigid extruded part. A flexible vinyl pad, for example, can be added to the exposed surface of a rigid profile and then embossed with a pebble grain or texture finish. This approach is much easier than embossing rigid profiles.

COMPARISON OF THERMOPLASTIC EXTRUSION MATERIALS

Acrylonitrile butadiene styrene (ABS). This material offers the best balance of properties of the common rigid thermoplastic materials. ABS is opaque and can be obtained in many colors. It can be extruded very easily into complex profiles and has above average tolerance control. This material is used in such applications as door jams, slides, breaker strips, housings, and handles. The extrusion die is built 30% oversize. Fig. 10–27 illustrates the die expansion of four different profiles. The parting line of the die should be designed to facilitate ease of machining and rework.

High-impact polystyrene. This material is the least expensive rigid material for extrusion. It has a translucent to opaque natural color and can be obtained in many color combinations. The material can be extruded in the above average complex profiles and has average tolerance control. It is used in such applications as nameplates, trim strips, and sliding door guides. The extrusion die is built 30% oversize.

General-purpose polystyrene. This material is the lowest-cost thermoplastic available. Polystyrene is a natural crystal-clear material that permits a full range of transparent and opaque colors. The surface finish of polystyrene is high gloss. It has limited use in profile extrusions, because of brittle characteristics. It is widely used for light-shield applications. The extrusion die is built 30% oversize.

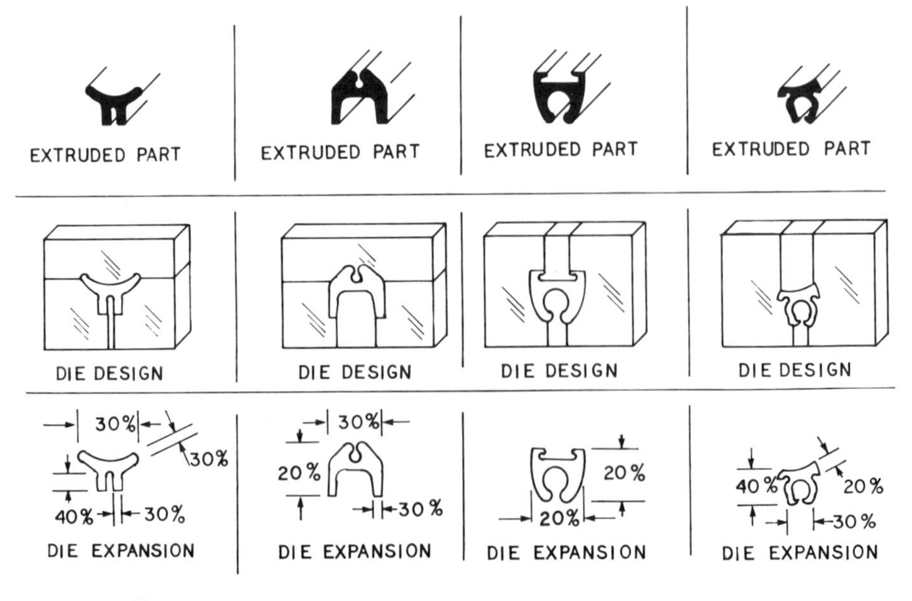

Figure 10–27. This illustrates the die expansion of four different profiles.

Low-density polyethylene. This is a flexible material with very good electrical properties and low moisture absorption. Low-density polyethylene is difficult to extrude into profile shapes and is also difficult for the control of tolerances. It has a milky-white natural color and is used for handles, straps, tubing, bumpers, edgings, and guides. The extrusion die is built 50% oversize.

High-density polyethylene. The high melt characteristics of this plastic material make it difficult to control during extrusion. It requires loose tolerances and can only be extruded in simple shapes. Polyethylene has a translucent, milky-white color and can be obtained in many color combinations. It is used in simple shapes such as straps, belting, and rods. The extrusion die is built 20% oversize.

Polypropylene. This material has excellent hinge properties and is one of the lightest thermoplastic materials. It is difficult to extrude this material in complex profiles. The material has only average tolerance control. It has a translucent milky-white natural color, but can be pigmented to almost any desired shade or color. Polypropylene is used for high-performance hinges, air-conditioner baffles, slide guides, and weather stripping. The extrusion die is built 25% oversize.

Polyphenylene oxide-based resin (modified PPO) can be extruded. It has a fast extrusion and through-put rate. The drawdown ratio is up to 50% and has virtually no effect on properties. Both single and two-stage screw designed extruders can be used. Slow cooling of the extrudate is preferred over quick quenching in order to minimize residual stress. The die expansion is 30%.

Ethylene vinyl acetate (EVA). This is a copolymer material and has good flexibility. It has a translucent, milky-white natural appearance and can be obtained in many colors. It has average tolerance control and can be extruded in simple profiles. It is used for seals, gaskets, and weather-stripping. The extrusion die is built 25% oversize.

Cellulose acetate. This is a semirigid to rigid material and can be extruded into extremely thin profiles. It has a clear natural color but can be pigmented to make many colors. The surface is high gloss.

Cellulose acetate extrusions are used for edgings, clear packaging, and tubes. The extrusion die is built 25% oversize.

Rigid vinyl. This is one of the best thermoplastic materials for extruding profiles. It has good tolerance control and complex profiles can be extruded. Rigid vinyl has a natural amber color and a matte surface finish. Because of the fillers used in the plastic, the colors are only fair. Rigid vinyls are used to make appliance breaker strips, building and transportation products, electrical conductor covers, and raceways. The extrusion die is built 10% oversize.

Flexible vinyl. This thermoplastic material is available in a wide range of compoinds and hardnesses. Its natural color is clear to opaque amber, depending on the fillers and extending agents used. It has good colorability. Flexible vinyls are extruded into only average complex shapes, and average tolerances can be used. It is used for gaskets, seals, strapping, and trim. The extrusion die is built 25% oversize.

Methyl methacrylate. This is one of the best outdoor weathering thermoplastics. It is a clear transparent plastic and has a high-gloss surface. Although the material can be made optically perfect, this feature is not possible by the extrusion process. It is used for signs, fluorescent light shields, and name plates.

Acetals. This crystalline thermoplastic material is not generally extruded into profiles, because the material's high melt temperature characteristics make it extremely difficult to control during extrusion.

Polycarbonate. This thermoplastic material has the most desirable overall balance of properties of any thermoplastic material available. It has a clear transparent natural color and has good colorability. Polycarbonate can be extruded into average shaped profiles and has good tolerance control. This material is high in cost; also, special extrusion handling requirements (because of moisture absorption characteristics) limit its use. Applications include lenses, tubing, and light shields.

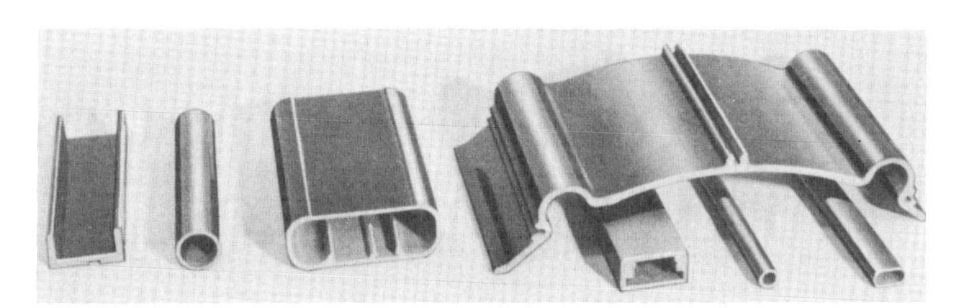

Figure 10–28. A group of different sizes and shapes of thermosetting extrusions. (Photo by G. N. Freund)

Nylon. This crystalline thermoplastic material requires loose tolerances and can be extruded in simple shapes. It has a natural milk-white color and is extruded with a high-gloss surface finish. Applications include tubing, guides, glides, and channels where low friction is required.

THERMOSETTING EXTRUSIONS

Thermosetting materials such as phenolics, melamines, epoxies, dially phthalates, polyesters, and alkyds are extruded into many shapes on special equipment. The equipment is not the same as used in the extrusion of thermoplastics. Three types of fabrication are used. The oldest method of fabrication is a ram-type extruder. It uses pressures as high as 50,000 psi. and the long land hot-die technique. The modern method of extruding thermosets is the cold-screw, hot-die process. The extrusion rates are slow and the cost of processing is high. The third technique of processing thermosets is the pultrusion method. Continuous reinforcements such as fiberglass with catalyzed resin is drawn through dies that are heated by both conductive techniques and the use of dielectric heat to cure.

Phenolic materials are extruded by the ram extruders with the hot-die technique and the cold screw extender with the hot-die technique. Fig. 10–28 shows many different sizes and shapes of phenolic thermosetting extrusions. Polyester resins can be extruded by the cold-screw, hot-die technique and the pultrusion method. Epoxy resins are fabricated by the pultrusion process and are more difficult to process than the polyesters.

11

Reinforced Plastics and Composites

The subject of reinforced plastics will be divided into three parts: (1) reinforced thermosetting plastics; (2) reinforced thermoplastics; and (3) laminated plastics.

REINFORCED THERMOSETTING PLASTICS

Reinforced thermosetting plastics sometimes referred to as reinforced plastics (RP), are plastic compositions in which fibrous reinforcements are imbedded in a thermosetting resin. The reinforcements are usually fiberglass in the form of rovings, mats, fabrics, chopped fibers, and short milled fibers. Other reinforcements include asbestos, chopped paper, macerated fabrics, jute, sisal, cotton, nylon, carbon, etc. The plastic resins most commonly used are polyester, phenolics, epoxies, diallyl phthalate, alkyds, melamine, and silicones. The five most important methods of fabricating thermosetting reinforced plastics are preform mat, premix, sheet molding, filament winding, and pultrusion (Fig. 11–1). Other methods include compression molding, transfer molding, screw injection molding, bag molding, and hand lay-up.

Preform Molding

In this process, a fiberglass preform is made over a perforated metal screen the same size and shape as the article to be molded. The screen is mounted on a rotating turntable of the preform machine. Chopped strands of glass fiber are blown onto the screen as it slowly revolves. There are two methods of making preforms–one is the directed fiber

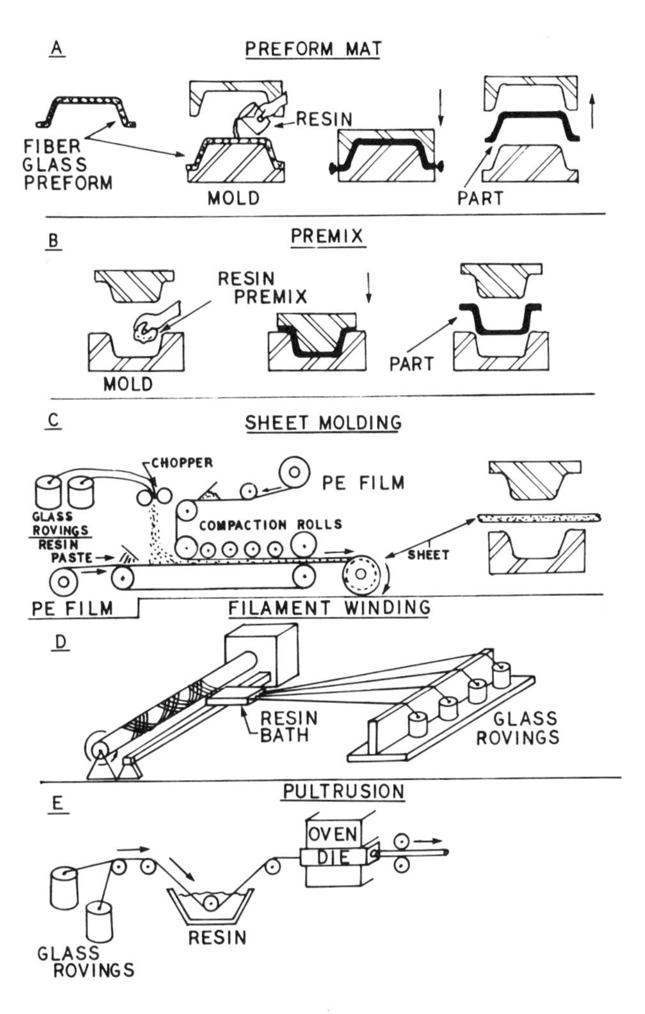

Figure 11–1. This illustrates the five most important methods of fabricating reinforced thermosetting plastic parts.

process, and the other is the plenum chamber process. In making the fiberglass preform by the directed fiber process, the roving is cut into one to two in. lengths of chopped strands that are blown through a flexible hose onto a rotating preform screen (Fig. 11–2).

Suction holds the strands in place while a binder is sprayed on the preform. The wet preform is heated in an oven to cure the binder.

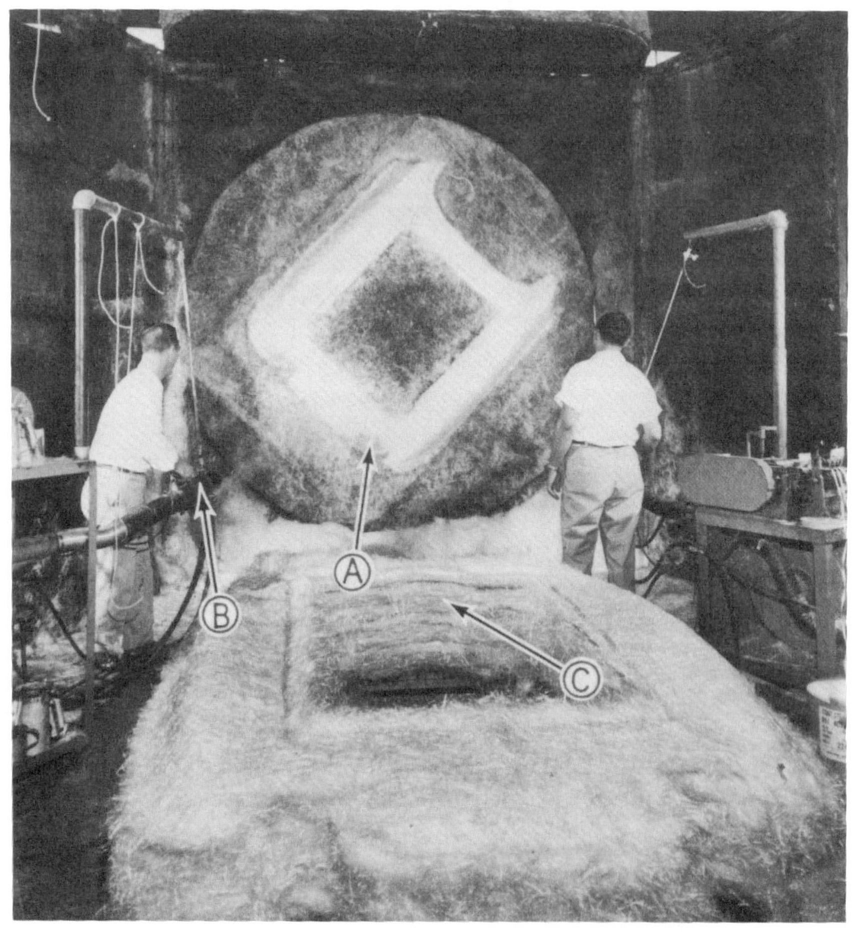

Figure 11–2. This picture illustrates the making of a fiber glass preform for the fenders and cowl of an automobile. A perforated screen, formed to the shape of the finished part, is attached to a spinning turntable. Long strands of fiber glass are blown against the rotating preform screen. A vacuum inside the screen draws the fibers down tight. (A) Preform screen mounted on the turn table. (B) Fiber glass being blown from a tube. (C) A stack of preforms.

In the plenum chamber process of making preforms, the fiber glass roving is fed into a cutter on top of the plenum chamber. The chopped strands are directed onto a spinning fiber distributor to separate the chopped strands and distribute them uniformly in the plenum chamber. From the plenum chamber, the fiber glass strands are

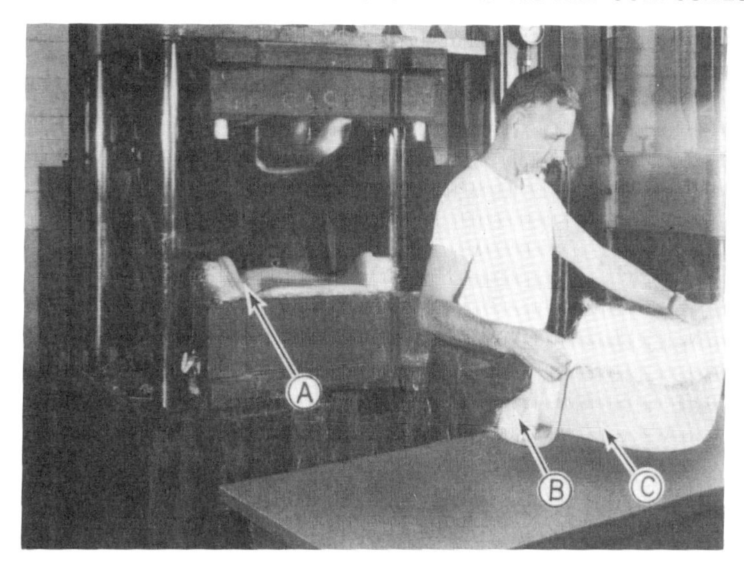

Figure 11–3. The preform molding of a reinforced fiberglass polyester resin chair seat. (A) The mold in the open position showing a molded part. (B) A fiberglass preform used in making the chair seat. (C) The molded chair seat.

sucked onto the rotating preform screen. As the strands are falling, a resinous binder is sprayed on them. The wet preform is heated in an oven to cure the binder.

The completed preform, which looks like a thick white blanket, is placed in the cavity or over the male portion of the metal die. A resin mix, comprised of the resin, catalyst, fillers, and pigments, is measured out and poured on the chopped strand fiberglass preform. The press is then closed, and under these conditions, the liquid resin is forced through the preform, thus forcing out the air before it. The curing of the liquid plastic resin takes place immediately. Fig. 11–3 shows the fiberglass method of making a bucket-type chair seat. The fiberglass preform is placed in the mold cavity, and liquid polyester resin is poured into the cavity on the preform. The press is closed and the chair is molded into shape.

Premix Molding

The term "premix" implies that components are mixed mechanically before molding. Premix (sometimes referred to as bulk molding compound) is a compound created by mixing chopped glass, resin, and filler, resulting in the flow of glass as well as the resin and filler in molding. This compound is generally used in smaller parts that have fairly heavy wall sections and do not require the better impact and tensile strength offered by preform-type molding. Premix compounds are sometimes used in conjunction with preform molding to obtain ribs and bosses. The outstanding features of premix compounds are high flexural modulus, low cold flow or low creep, and good impact. It is obvious that this type of plastic material and molding process should be used where rigidity or stiffness and load bearing properties are desirable features. Premix molded parts have very low elongation and undergo only small angular deflections before breaking. Therefore, premix molded parts should not be used where they will be subjected to severe deformations. Premix parts should be designed for maximum rigidity and broad distribution of unavoidable deflection.

Sheet Molding Compound

Sheet molding compound (SMC) is a thermosetting polyester resin mixture reinforced with fiber glass strands and formed into a sheet that can be handled easily, cut to shape, and charged into a compression mold.

The resin paste, containing all ingredients except the fiber glass strands, is doctored onto two moving polyethylene films (Fig. 11–1c). The fiberglass roving is chopped into individual strands, usually one-inch lengths, and deposited on the bottom resin film. The other carrier film is coated with resin paste and is placed on top of the sandwich construction. The sandwich is then kneaded under pressure rolls to wet out the fiberglass. The compound in the form of a sheet, is wound under tension and wrapped in a suitable monomer barrier film. The key ingredients of SMC are: fiber glass, filler, thickener, catalyst, release agent, and low shrink or low profile additives.

Filament Winding

In this process continuous fiberglass reinforcement is impregnated by drawing through a resin bath (epoxy or polyester) and winding under tension on to a rotating mandrel (Fig. 11–1d). The filament winding machine traverses the reinforcement over the length of the mandrel in a pre-determined pattern to give maximum strengths in the directions required. After sufficient thickness has been built up, the laminate on the mandrel is cured at room temperature or in an oven, and is then stripped from the mandrel. Soluble mandrels may be used to permit winding of closed-end structures. Pressure pipe and tanks are made by this method. Fig. 11–4 shows a fiberglass epoxy filament wound water well casing. The casing uses a double-keyed coupling that requires no adhesives, or screws for attachment. The slot for the keyed coupling is parallel to the helically wound filaments, which also contributes to the retention of pipe tensile strength by leaving a maximum number of uncut filaments. The key strips are made from acetal resin.

Continuous Pultrusion

In this process continuous reinforcements (generally fibrous glass) are drawn through a resin bath for impregnation, and into a heated, hardened steel die of the cross-sectional shape desired (Fig. 11–1e). The die orients the reinforcement, sets the final shape of the laminate, and controls its resin content. Cure may be completed either within the die or by infrared heat upon exiting the die. Linear speed is determined by a pulling device. The resins used are generally polyester or epoxy. Parts made by the pultrusion process include fishing rods, ladders, electrical pipe-line hardware, dunnage bar equipment, etc.

DESIGN CONSIDERATIONS IN MATCHED-DIE MOLDING OF REINFORCED THERMOSETTING PLASTICS

Wall Thickness

In preform molding, the wall thickness can vary from 0.030 to 0.250 in. (Fig. 11–5a). If appearance is important, it is advisable to have a constant thickness over the entire molded part. Small ribs and bosses are possible, if shallow and slight shrink marks on the outside of the part are not objectionable. Heavy thick walls are subject to blister

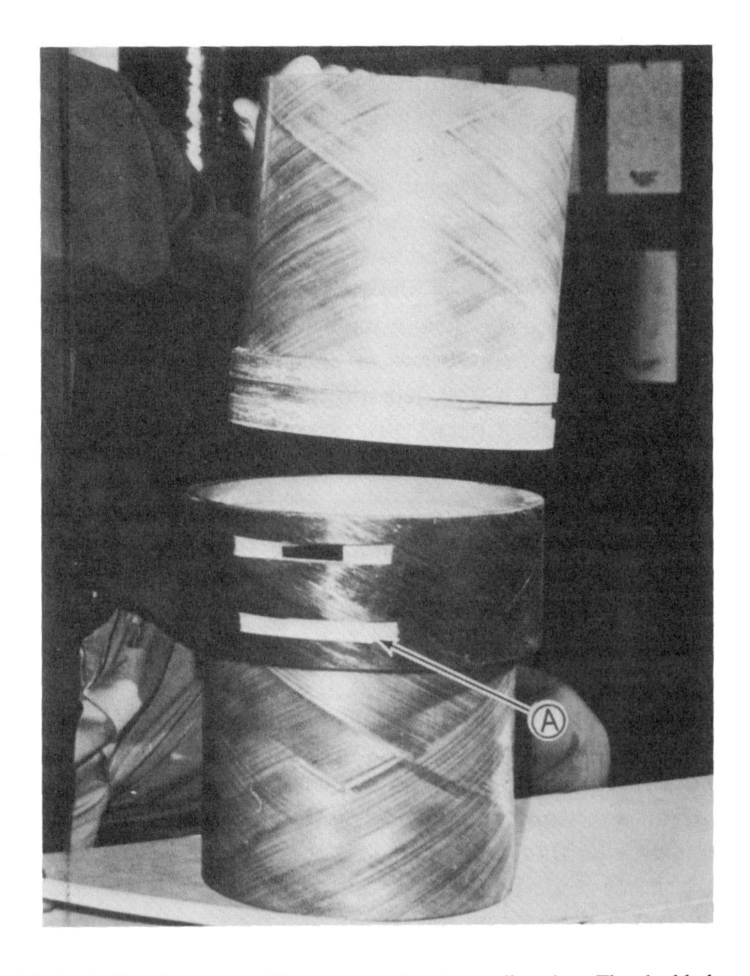

Figure 11–4. A fiberglass epoxy filament wound water well casing. The double-keyed coupling can be quickly joined without adhesives. A flexible key strip (A) slides into the conduit formed by matching grooves in the pipe end and the coupling.

cracking, and warping, because of the non-uniformity of the reinforcing fibers in the preform make-up. Heavy sections can be built up by using multiple layers of mat or two nesting preforms.

In premix molding, the wall thickness can vary from 0.070 to 1.0 in. (Fig. 11–5a). Walls as thin as 0.050 in can be molded if the premix molding material is not required to flow a long distance in the mold.

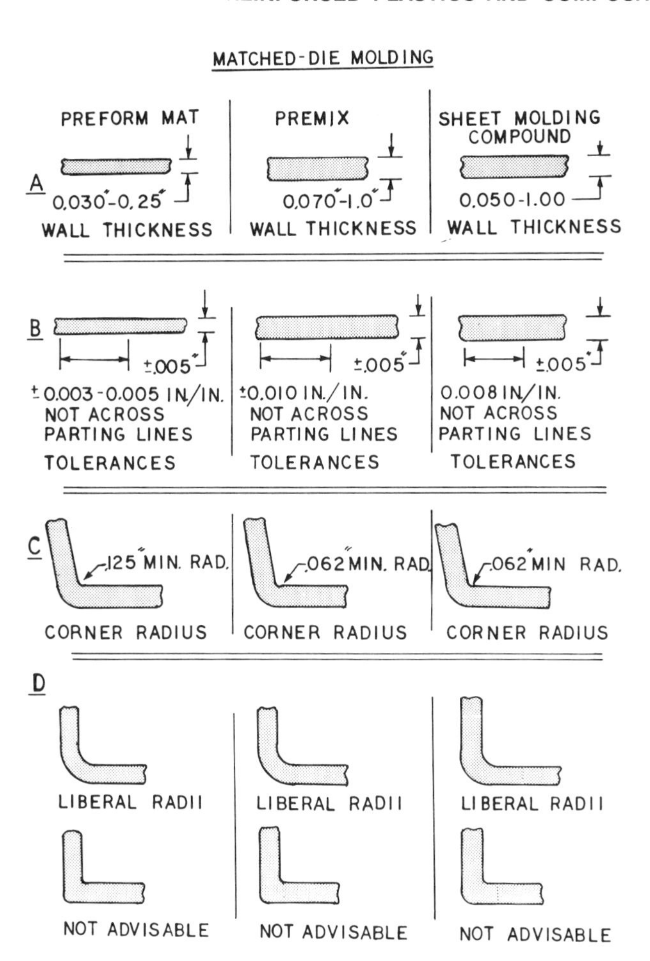

Figure 11–5. Comparative information on matched-die molding of preform mat, premix compound, and sheet molding compound. (A) Wall thickness. (B) Tolerances. (C) Corner radius. (D) Radii.

Thin walls use less material and require shorter curing times, thus allowing faster production rates. Thin walls are less likely to cause problems with blisters, cracking, and warpage. The limiting factor for minimum thickness is dictated by the flow requirements of the fiber and resin and filler mix. The longer the flow, the greater the thickness must be in the molded part.

In sheet molding compound the maximum wall thickness is 1.00 in. and the minimum is 0.050 in.

Tolerances

The tolerances on the wall section in preform molding can be held to plus or minus 0.005 in. per in. Other tolerances can be held to plus or minus 0.003 to 0.005 in. per in., except across parting lines (Fig. 11–5b). Tolerances on contour and thickness may be held easily to 0.010 in. per in., and warp and shrinkage may be predicted and compensated for in the dies. All tolerances should be kept as wide as possible to keep down mold costs. Uneven distribution of glass and resin throughout the part prevent tight tolerances.

In premix molding, the tolerance on the wall thickness can be held to plus or minus 0.005 in. per in. Other tolerances can be held to plus or minus 0.010 in. per in., except across parting lines (Fig. 11–5b). For critical dimensions across the flash line, and allowance of from plus or minus 0.005 to 0.010 in. must be added to the normal tolerance. Very close tolerances on large parts are easier to achieve by designing assemblies that are bonded in jigs and fixtures.

In sheet molding compound the tolerance on the wall thickness can be held to plus or minus 0.005 in. Other tolerances can be held to plus or minus 0.008 in. per in. except across parting lines (Fig. 11–5b).

Corner Radius

In preform molding, a large radius is always required, because it is difficult to deposit glass on a sharp male corner during the making of the fiberglass preform. A minimum radius of 0.125 in. is required (Fig. 11–5c). Sharp female corners become bridged with glass fibers and are subjected to cracking. An outside radius should be designed to keep the wall thickness uniform around the curve. In premix molding the minimum radius should be 0.062 in. and the same is true for sheet molding compound (Fig. 11–5c).

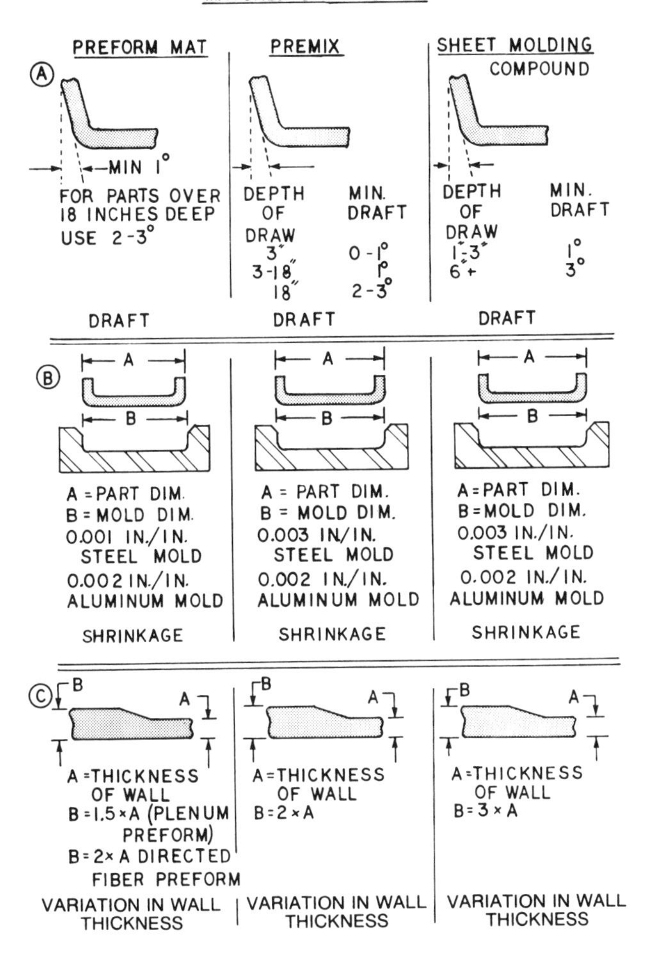

Figure 11–6. Comparative information on matched-die molding of preform mat, premix compound, and sheet molding compound. (A) Draft. (B) Shrinkage. (C) Variation in wall thickness.

Draft

In matched-die molding, the amount of draft is dependent on the depth of draw. Generally, a draft of at least one degree is required (Fig. 11–6a). Greater draft is required on interior walls to prevent shrinkage from binding the part to the mold. Some parts have been molded successfully with a 1/10 degree draft per side, but the die costs rise, and the molding cycle gets longer.

In matched-die molding, a sufficient draft must be allowed for easy removal of the molded part from the mold. The amount of draft required is dependent on the depth of draw. A textured surface mold will require more draft than a mold that is highly polished. The low-shrink or low-profile polyester resins require a little more draft than the conventional ployesters.

Shrinkage

Shrinkage in matched-die molding is the difference between the mold dimension and the part dimension, both measured at room temperature. Fig. 11–6b gives the designer a guide in determing shrinkage in steel and aluminum molds. It should be remembered that glass and filler content affect the shrinkage. Also, uneven shrinkage due to variations in the glass to resin ratio, result in unwelcome warpage.

Shrinkage is dependent on a combination of the following factors: (1) the chemical reaction of the type of thermosetting resin that is used; (2) the temperature range over which the plastic cools after its shape has been permanently established in the mold; and (3) the degree to which the premix molding material has been compressed during molding. To help minimize the effect of shrinkage it is desirable to keep the wall thicknesses as uniform as possible. Uneven shrinkage, due to variations in thickness, may result in unwelcome warpage, crazing, and sink marks on the molded part.

Variation In Wall Thickness

In matched-die molding, abrupt changes in wall thickness and flow paths from thin to thick sections should be avoided. Changes in wall thickness should be tapered by running the taper from the center out to the edges. Fig. 11–6c illustrates the ratio for adjoining sections in preform mat, premix, and sheet molding compound. It should be remembered that the reinforcing fibers in matched-die molding do not move easily in the flow pattern from thick to thin sections.

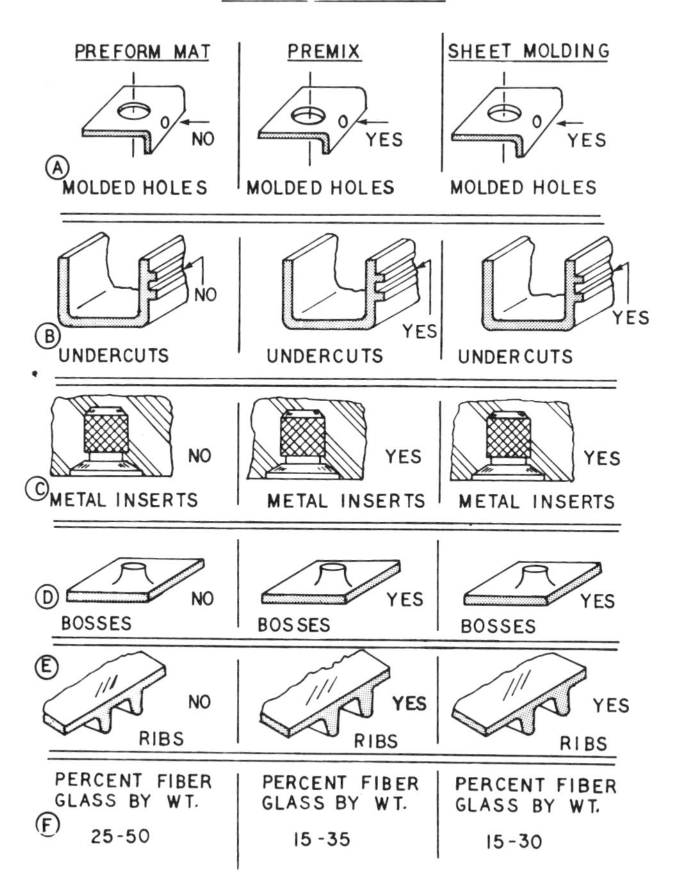

Figure 11–7. Comparative information on matched-die molding of preform mat, premix compound, and sheet molding compound. (A) Molded holes. (B) Undercuts. (C) Metal inserts. (D) Bosses. (E) Ribs. (F) Percent fiberglass by wt.

Undercuts

Internal and external undercuts should be avoided, if possible, in matched-die molding. Split molds can be designed for parts with undercuts, but these molds are expensive. Multiple-piece molds are not recommended, because resin might flow into the joints. Raised lettering or designs should be located on a surface perpendicular to the mold opening. This will eliminate any chance of an undercut in the lettering or design. Internal undercuts are relatively impossible to mold. They require split plungers or removable pieces in the plunger. It is better to machine the internal undercut after molding. There is a very fine distinction between undesirable and impossible undercuts. Fig. 11–7b illustrates the possibility of having undercuts in matched-die molding.

Molded-In Openings

In preform mat molding, molded-in holes or openings on vertical side walls should be eliminated. When necessary, such openings should be formed partially through the part to allow a better flow of the glass reinforcement. After molding, the opening can be completed by machining. If a core pin is used, it should be located far enough from the edge to allow the material to knit after flowing around the pin.

In premix molding and sheet molding, molded-in openings on vertical walls are not recommended. However, where openings in side walls are required, it is advisable to use the kiss-off or mash-off construction in the die (Fig. 11–8). This will eliminate the need for sliding cores. The original cost of a mold with a sliding core is high, as are the maintenance costs.

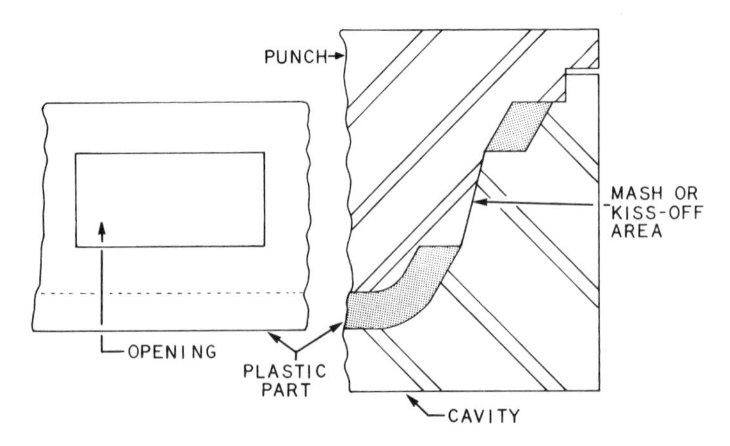

Figure 11–8. A large opening in a side wall that is close to the edge of the part may be made by the kiss-off or mash-off method.

If a large opening in a side wall is located close to the edge of a part, and a kiss-off construction can be used, it may be desirable to use a sliding core. The movement of the sliding core should be operated on a time delay, and it "comes home" after the die is closed to help reduce knitting of the material at the edge of the part. Fig. 11–9 shows openings that were molded-in during the molding process.

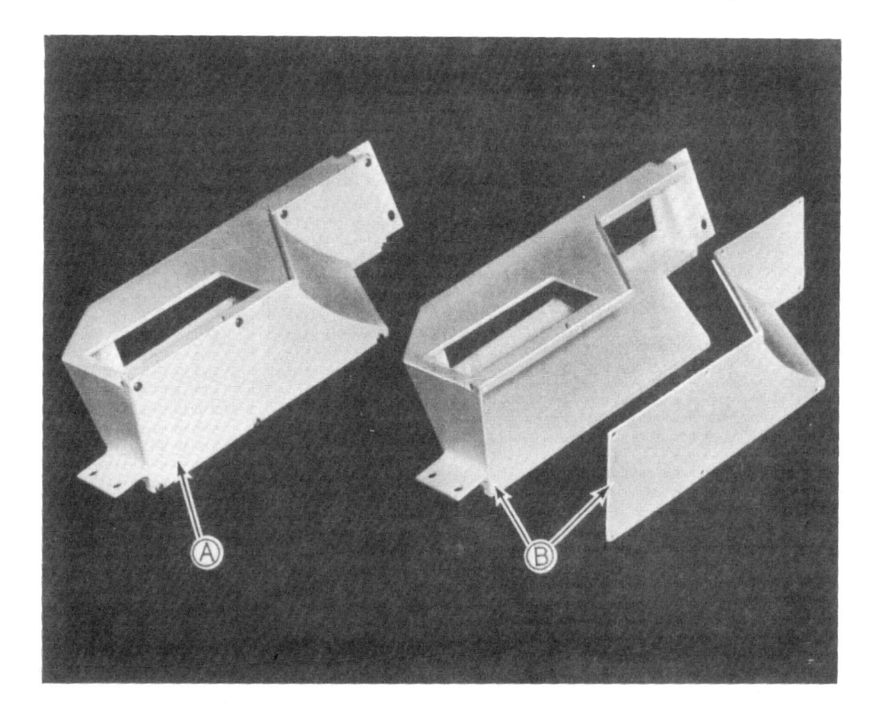

Figure 11–9. This is a picture of an air duct for an automobile. It is made from a premix compound of low profile polyester resin and fiberglass. A matched metal die was used. Note the molded in openings in the part. (A) Assembled duct. (B) Two molded parts that make up the assembly.

RECOMMENDED DEPTHS OF HOLES IN PREMIX MOLDED PARTS

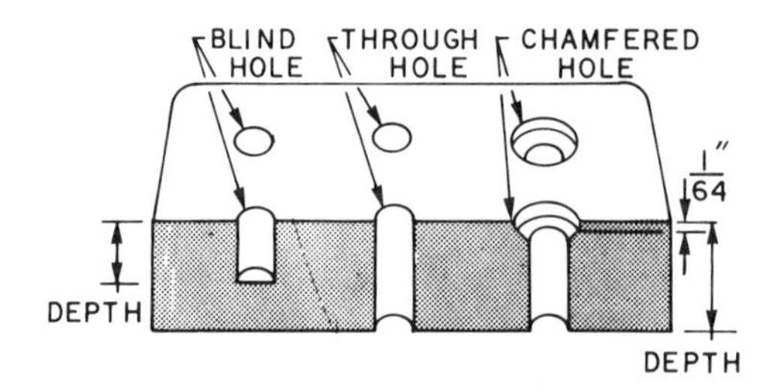

HOLE DIA. INCH	MAX. BLIND HOLE DEPTH INCH	MAX. THRU HOLE DEPTH INCH
0.125	0.500	0.750
0.250	1.250	1.875
0.312	1.500	2.250
0.375	1.875	2.812
0.500	3.000	4.500
0.750	4.500	5.000
1.000	6.000	6.000

Figure 11–10. Recommended hole sizes for through and blind holes in premix molding.

Molded-In Holes

Holes are not recommended to be molded-in if the preform molding process is used. However, in premix molding and sheet molding, the holes can and are molded in during the molding process. Fig. 11–10 shows the recommended depths of holes in premix and sheet molded parts. Fig. 11–11 illustrates the recommended distances between holes and the distances of holes from the side wall in premix and sheet molded parts.

RECOMMENDED DISTANCES BETWEEN HOLES, AND DISTANCES OF HOLES FROM SIDE WALL, IN PREMIX MOLDED PARTS

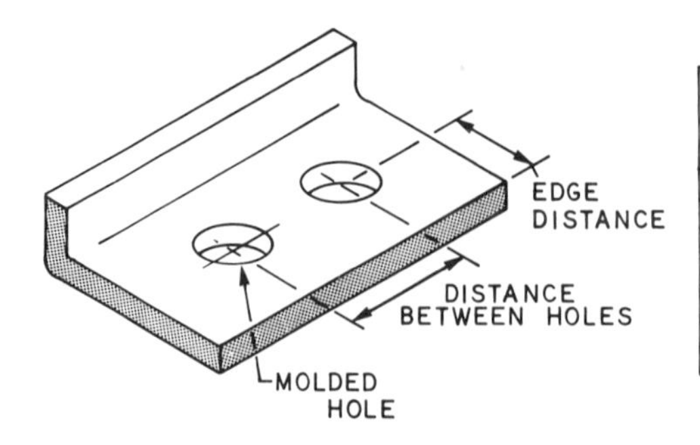

HOLE DIA. (INCH)	MIN. DISTANCE FROM EDGE (INCH)	MIN. DISTANCE FROM EACH OTHER (INCH)
0.125	0.187	0.187
0.250	0.250	0.187
0.312	0.312	0.250
0.375	0.375	0.281
0.500	0.500	0.375
0.750	0.500	0.562
1.000	0.500	0.750

Figure 11–11. Recommended distances between holes and distances of holes from side wall in premix molding.

Wall Variation Design

In all types of molded plastic parts, the designer should strive for a uniform wall section throughout the part. Also avoid large flat areas. Due to uneven shrinkages in reinforced plastic materials, the flat area will tend to warp. Molded in ribs (Fig. 11–12a) will tend to eliminate this condition. Also, a double wall thickness at the edge of a part will increase stiffness and reduce warpage (Fig. 11–12c).

Flanges Used In Reinforced Plastics

An offset flange is used in premix molded parts to increase stiffness (Fig. 11–13 a–a). This method is preferred to beads, because less material is used and the wall thickness is kept constant. Warpage of long flange areas or of rectangular openings can be reduced by thickening these areas or by introducing additional angular blends as close to the edge of the part as possible (Fig. 11–13 b–b).

A molded flange that is used with a mounting bolt should be in compression whenever possible. Fig. 11–14 illustrates a mounting construction that can be used against a flexible gasket.

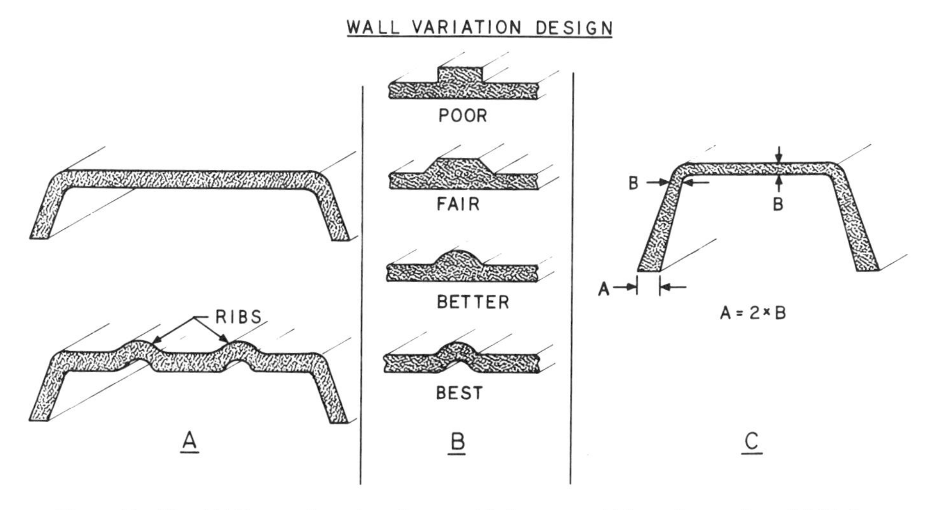

Figure 11–12. (A) In premix and preform molded parts avoid large flat sections. Molded-in ribs will give a more rigid part and reduce warpage. (B) Rib sections with even wall thickness are the best. (C) It is possible to have double wall thickness at the edges of a part. This will increase stiffness and reduce warpage.

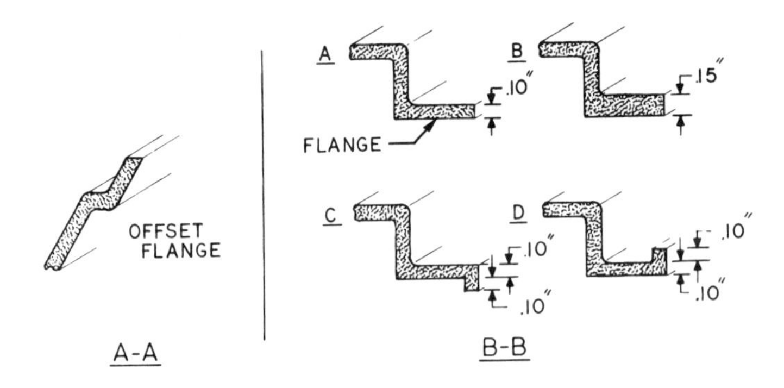

Figure 11–13. Flanges used in premix molding. (A-A) An offset flange is used to increase stiffness. (B-B) This illustrates steps that can be taken to reduce the warpage in long flanges. C and D are the preferred construction.

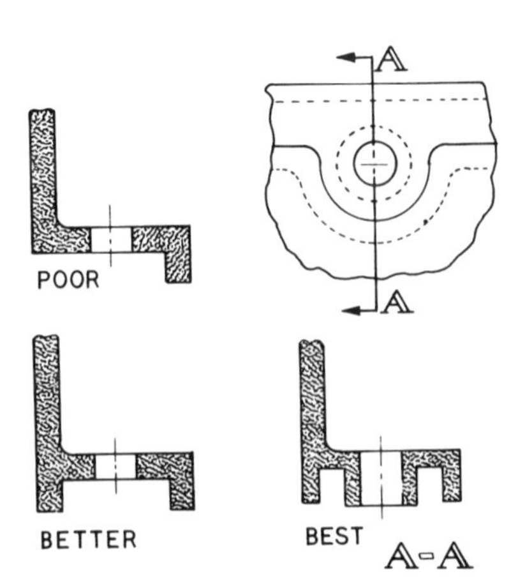

Figure 11–14. This illustrates a flange mounting construction. A flange mounting hole to be mounted on a flexible gasket should be in compression whenever possible

Curved vs. Flat Surfaces

Curved or round parts are less subject to warpage than flat molded parts. Also, narrow lips and flanges help restrict edge warping and increase stiffness. In preform molding, the uniformity of the fiberglass in the preform helps to prevent warping in the molded part. If a molded part is placed on a cooling fixture, as soon as it comes from the mold, the warpage can be controlled and held to a minimum

Fig. 11–15 illustrates a slightly curved surface on a bathroom shower receptor. The receptor is made from matched metal dies by the fiberglass preform method and contains 40% fiberglass. The pigmented polyester resin insures the same color through out the part. The two receptor parts are slightly curved and have an offset flange around the edge that helps in keeping the warpage to a minimum.

Figure 11–15. Two matched metal dies and the molded parts for a bathroom shower receptor. The two identical parts are for the ceiling and floor on the shower. The parts were preform mat molded from polyester fiberglass (A) The molded parts. (B) The cavity. (C) The punch.

Joining Molded Parts

Reinforced thermosetting plastic molded parts may be fastened to each other or other materials by fasteners or by adhesive bonding. Mechanical fasteners of all types have been used with reinforced molded parts. The particular type of fastener depends on the strength required, part thickness, appearance and the need for disassembly. Bolts, rivets, and staples are the most common types of fasteners used (Fig. 11–16). Blind rivets or staples are often convenient in a production operation to hold parts being adhesive bonded.

Surfaces to be bonded are roughened by sanding and then wiped clean with a suitable solvent. The sanding and solvent cleaning also removes any mold release that might be present. The adhesives used are usually polyesters or epoxies with chemical and thermal properties similar to the material being bonded. The best way to determine the method of fastening to be used is to make end use tests.

In designing molded plastic parts that are to fit together with parts of dissimilar materials such as metals, allow for difference in shrinkage and thermal expansion. Fig. 11–16 shows different types of bonded joints. The plain lap joint is good for use with thin moldings. The offset lap joint is good for thick moldings. Adhesive bonding is always recommended over tapped holes when permanent fastening with stresses are encountered.

REINFORCED THERMOPLASTICS

Reinforced thermoplastics, sometimes referred to as (RTP), are thermoplastic compositions in which fibrous reinforcements are imbedded in the thermoplastic resin. Almost any thermoplastic material can have reinforced fibers in it. The most common reinforced fiber is glass, although many other specialty fibers such as boron fibers, carbon fibers, and ceramic fibers have been used. Fiberglass is still the industry's workhorse reinforcement, particularly in large volume end products.

The addition of fibers to any plastic material serves the same purpose as steel rods added to concrete. The fibers or rods increase and improve the tensile and flexural strength. Fillers, on the other hand,

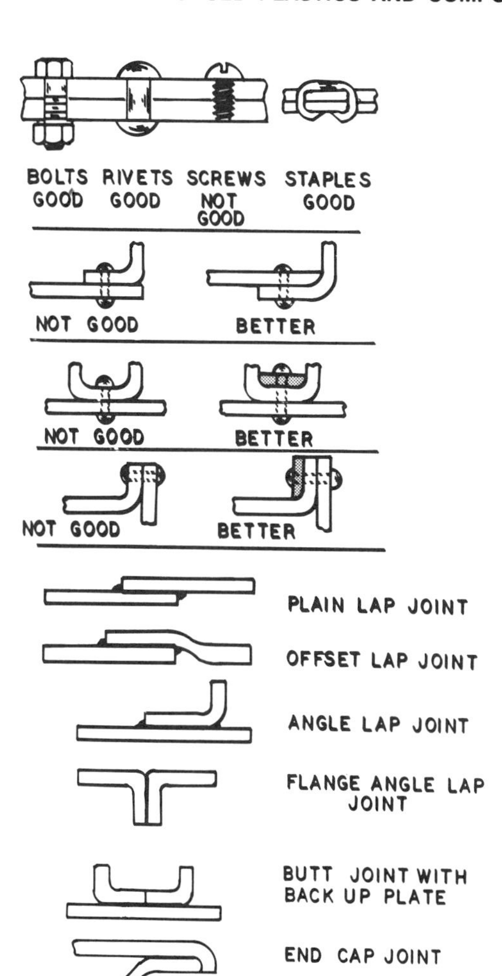

Figure 11–16. Reinforced molded plastic parts can be fastened together by mechanical fasteners or by adhesive bonding.

do not increase these physical properties, but merely act as extenders. The resin actually bonds the reinforcements together, transmitting loads to the fibers so that optimum fiber stresses can be developed before failure. A reinforcing fiber must have a strong adhesive bond to the plastic resin that is used. Coupling agents are generally added to the reinforcing fiber to increase this adhesive bond. Two basic finishes have provided optimum bonds for fiberglass, i.e., chrome finishes and silane finishes.

The tensile strength of a thermoplastic can be at least doubled by the addition of glass reinforcement (Table 11–1). Unlike thermosetting reinforced plastics, fiberglass reinforced thermoplastic compounds can be pelleted and used in conventional molding equipment. Reinforced thermoplastics produce increases in strength and rigidity and marked decrease in the coefficient of thermal expansion. The most significant effect in thermoplastics is the retention of izod impact strength at very low temperatures. Deflection temperatures is improved most markedly in nylon. Other beneficial effects may include increase in hardness and abrasion resistance and decrease in mold shrinkage, creep, and dimensional changes with humidity.

Some of the undesirable effects of fiberglass reinforced thermoplastics include opacity, occasional fiberglass appearance on the surface,

TABLE 11–1. THIS ILLUSTRATES THE INCREASE IN PROPERTIES OF THERMOPLASTIC MATERIALS WHEN FIBER GLASS IS ADDED.

PLASTIC	SPECIFIC GRAVITY	MOLD SHRINKAGE IN/IN.	TENSILE STRENGTH 10^3 PSI	FLEXURAL MODULUS 10^6 PSI	DEFLECTION TEMPERATURE 264 PSI (°F)	THERMAL EXPANSION 10^{-5} IN/IN. (°F)
ASTM →	D792	D995	D638	D790	D648	D696
ABS →	1.05	.006	6.0	.32	195	5.3
30% GLASS →	1.28	.001	14.5	1.10	220	1.6
ACETAL →	1.42	.020	8.8	.40	230	4.5
30% GLASS →	1.63	.003	19.5	1.40	325	2.2
NYLON 6 →	1.14	.016	11.8	.40	167	4.6
30% GLASS →	1.37	.004	23.0	1.20	420	1.7
NYLON 6/6 →	1.14	.018	11.6	.41	170	4.5
30% GLASS →	1.37	.004	26.0	1.30	490	1.8
PPO →	1.06	.005	9.5	.36	265	3.3
30% GLASS →	1.27	.002	21.0	1.30	310	1.4
POLYCARBONATE →	1.20	.006	9.0	.33	265	3.7
30% GLASS →	1.43	.001	18.5	1.20	300	1.3
POLYESTER T. P. →	1.31	.020	8.5	.34	130	5.3
30% GLASS →	1.52	.003	19.5	1.40	430	1.2
POLYETHYLENE H. D.	0.95	.020	2.6	.20	120	6.0
30% GLASS →	1.17	.003	10.0	.90	260	2.7
POLYPROPYLENE →	0.91	.018	4.9	.18	135	4.0
30% GLASS →	1.13	.004	9.8	.80	295	2.0
POLYSTYRENE →	1.07	.004	7.0	.45	180	3.6
30% GLASS →	1.28	.001	13.5	1.30	215	1.9
POLYSULFONE →	1.24	.007	10.0	.40	340	3.1
30% GLASS →	1.45	.003	18.0	1.20	365	1.4
SAN →	1.08	.005	9.8	.50	200	3.4
30% GLASS →	1.31	.001	17.4	1.50	215	1.8

lower gloss, some difficulty in electroplating, and loss of mechanical flexibility. The extrusion of fiberglass reinforced thermoplastic material is not generally used, because it causes a non-random orientation of the glass fibers.

Processing Fiberglass Reinforced Thermoplastics

There are three basic methods of reinforcing thermoplastic materials with fiberglass (Fig. 11–17). The methods are: (1) reinforced pellets; (2) super reinforced pellet concentrates; (3) prechopped fiberglass.

Reinforced pellets. In making fiberglass reinforced thermoplastic pellet materials, continuous glass roving is first sized to increase interfacial adhesion between the glass and resin. After the sizing operation, the fiberglass is then impregnated and coated with thermoplastic resin and finally chopped into pellets. These pellets contain core bundles of parallel glass fibers impregnated and surrounded by a sheath of thermoplastic resin (Fig. 11–18). The pellets contain from 20 to 30% fiberglass by weight. The fiberglass in the pellets can be obtained in 1/8 in. lengths or 3/8 to 1/2 in. lengths. The long glass fibers give higher impact strength. In this type of reinforced pellets, there is virtually no limit to the types of plastics available in reinforced grades.

Prechopped fiberglass. In the third method, the prechopped fiberglass is mechanically metered into the resin flow. The method is more adaptable to large machines that have large screw diameters in the injection machine. The polystyrene bead or powder is the best type of material to use in this process. It is not economically feasible to compound reinforced thermoplastic materials at the press, unless the operation can be set up and run the year around.

Part Design

Fiberglass reinforced thermoplastic resins freeze or set quicker in the mold than unfilled resins, but the glass fibers hold sufficient heat to allow even flow into thin areas. The current theory is that the lower thermal conductivity permits enough retention of heat to maintain additional flow. Higher heat distortion of reinforced materials, permits shorter clamp cycles than are obtainable with unreinforced

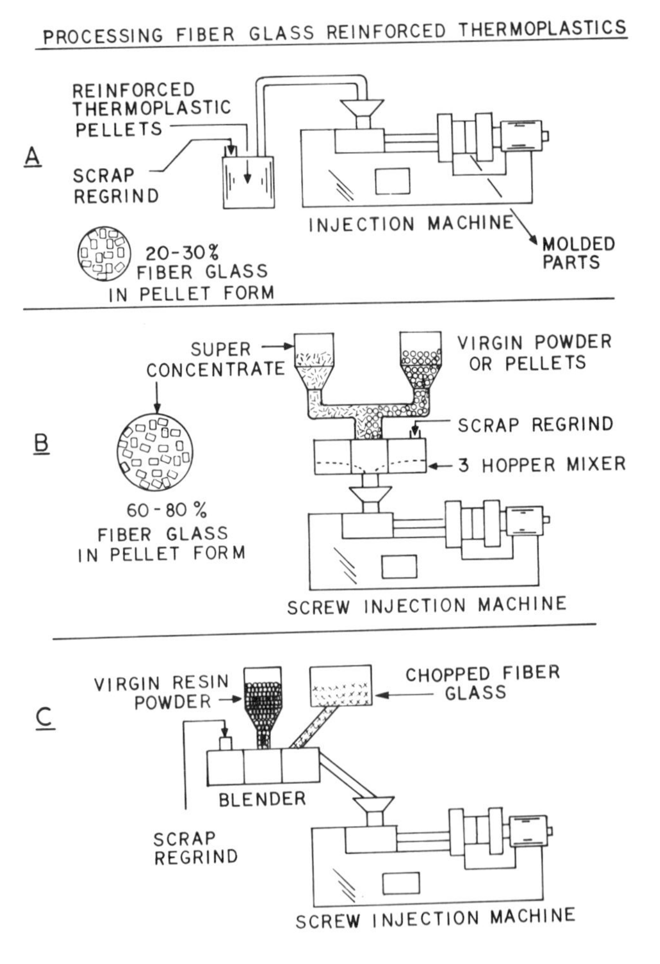

Figure 11–17. This illustrates the three methods of processing fiber glass reinforced thermoplastics: (A) reinforced pellets, (B) super reinforced pellet concentrates. (C) prechopped fiber glass.

materials, because the part may be ejected from the mold at a higher temperature, without fear of distortion. Thermoplastic materials that use fiberglass as reinforcement have low shrinkage and high stiffness, as compared to unreinforced compounds.

Runners for injection molds should be large, full round or trapezoidal, and well polished (Fig. 11–19). This gives high volume to surface

Figure 11–18. This picture illustrates fiberglass reinforced thermoplastic pellet molding material and a molded part. (A) The cylindrical reinforced fiberglass pellet. (B) An injection molded disc made from the reinforced material.

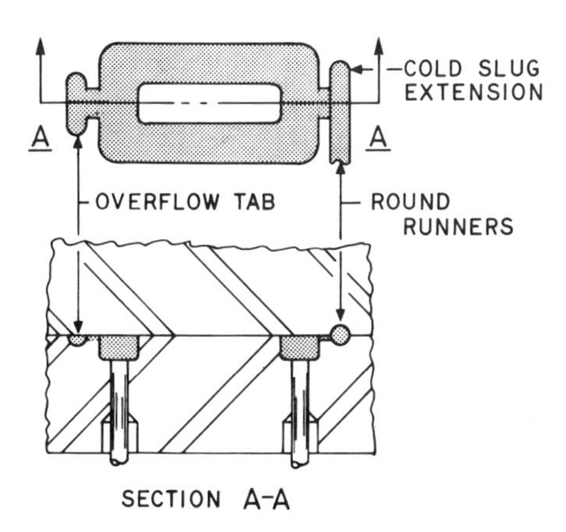

Figure 11–19. Runners for molding reinforced fiber glass thermoplastic materials should be full round. An overflow tab will help eliminate weld lines.

area, and thereby reduces heat losses. It is suggested that gates for fiberglass reinforced thermoplastic materials enter the piece at, or as close as possible to, a heavy section. Positive ejection of molded parts is recommended. Knock-out pins should be located on ribs, bosses, etc. The molded part should be pushed out of the mold, rather than pulled out. Extreme undercuts must be avoided, because of low shrinkage and stiffness, but slight undercuts can be utilized to assist the part in staying on cores (or forces) until the knock-out mechanism is actuated. Fig. 11–20 illustrates an injection-molded side check valve for an automatic washing machine. Note that the valve was molded in two parts and ultrasonically welded together. Thus, a complicated die was eliminated that would have been required to mold or core the four openings. The material used to make this valve was 30% fiberglass reinforced SAN.

LAMINATED PLASTICS

Laminated plastics are one form of "reinforced plastics." The term "reinforced plastics" is used extensively since it includes molded parts, in which the reinforcing is not usually in the laminated form and both thermoplastic and thermosetting resins. Laminated plastics

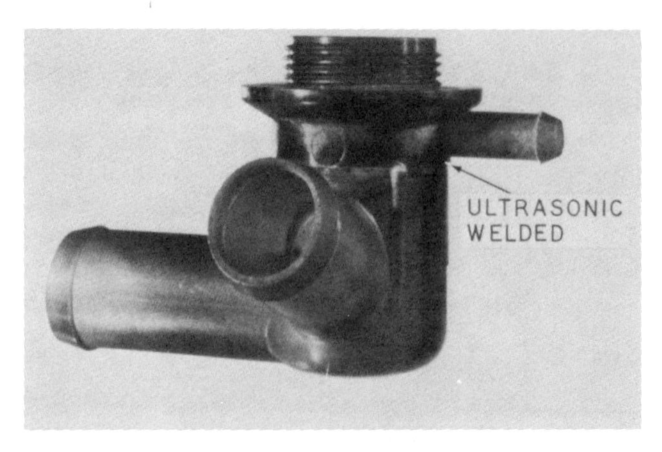

Figure 11–20. This is a side check valve for an automatic washer. It is injection molded in two parts and ultrasonic welded together. The material is fiberglass reinforced styrene-acrylonitrile (SAN).

are plies of sheet material (bases), usually impregnated with a thermosetting resin (binder), and bonded together by means of heat and pressure to form sheets, tubes, rods, or molded shapes.

The principal resins used in laminated thermosetting products are shown in Table 11–2. Phenolics are low cost and have good electrical and physical properties. Another class of thermosetting resins include the melamines. Melamines are more costly but offer flame resistance and have excellent electrical properties. The polyesters are low in cost with average properties. The epoxies are high in chemical resistance and are extremely moisture resistant. Silicones are used primarily for their retention of mechanical and electrical properties, even at very high temperatures, when their higher costs can be justified.

Laminating process

Basically, the laminating process is carried out by impregnating base sheet stock with the liquid thermosetting resin. If a flat laminate is to be made, the correct number of sheets (stacked one upon the other) are simultaneously subjected to heat and pressure between two polished plates in a laminating press. Heat and pressure are applied to the layers, causing the resin to flow and harden the laminate into one solid mass. The equipment for this consists of a platen press that is capable of squeezing the laminate together with sufficient force, and some means of applying heat, usually through the platens of the press. Usually a multiple platen is used so that more than one sheet can be laminated at one time. The stock to be laminated is placed between highly polished plates that are then placed between press platens. Heating is accomplished by passing steam through cores in the platen or by using electricity to heat the press platens. Fig. 11–21 shows a typical high-pressure laminating press. The term "high-pressure laminate" is normally confined to those laminates molded and cured in their final form at pressures no lower than 1000 psi and more commonly in the range of 1200 to 2000 psi. If the pressure is under 1000 psi, the product is called low-pressure laminates. Those laminates made with little or no pressure, such as hand lay-ups, are sometimes called contact pressure laminates. The term laminate is sometimes used to include composites of resins and fibers that are not in distinct layers, such as filament wound structures and spray-ups.

TABLE 11–2. THERMOSETTING RESINS USED IN LAMINATED PRODUCTS.

PROPERTY	PHENOLICS	MELAMINE	POLYESTER	EPOXY	SILICONES
SPECIFIC GRAVITY	1.3	1.48	1.3	1.25	1.3
COST OR PRICE	LOW	MEDIUM	LOW	MEDIUM-HIGH	HIGH
HEAT RESISTANCE	EXCELLENT	EXCELLENT	GOOD	FAIR	EXCELLENT
PHYSICAL PROPERTIES	GOOD	GOOD	GOOD	GOOD	FAIR
ELECTRICAL PROPERTIES	EXCELLENT	EXCELLENT	GOOD	EXCELLENT	EXCELLENT
WATER RESISTANCE	GOOD	FAIR	GOOD	EXCELLENT	GOOD
MACHINING QUALITIES	FAIR TO GOOD	FAIR TO GOOD	FAIR TO GOOD	GOOD	GOOD
MOLDING PRESSURES	LOW TO HIGH	HIGH	LOW	LOW TO MEDIUM	LOW TO HIGH
MOLDING QUALITIES	EXCELLENT	GOOD	EXCELLENT	FAIR	GOOD
ADVANTAGES	GOOD ALL ROUND PROPERTIES	GOOD ALL ROUND PROPERTIES	MANY TYPES AND PROPERTIES	SHRINKAGE NIL	HEAT RESISTANCE

Figure 11–21. A large hydraulic, steam-heated, plate molding press. (A) This indicates where sheets of resin-impregnated cloth or paper are laminated between chrome-plated steel pressing plates.

After the compressed sheets have been cured into a solid state by the heat of the platens, and the platens have been cooled, the press is opened, and the sheets are removed. Cooling the sheets before removing them from the press helps to prevent warpage. After the sheets are trimmed at the edges, they are ready for fabricating.

Thermosetting rods and tubes are treated differently from laminated sheets. Solid laminated rods are made by winding the impregnated filler web on a very thin mandrel, which is withdrawn before molding. The center channel is filled-up when pressure is applied in the metal mold. The mold, as it comes together, closes the center hole and flash develops at the lands of the mold. The flash is removed by centerless grinding. In making a tube, the mandrel is left in the tube, and the tube is molded in the same manner as the rod.

Articles with irregular shapes, such as gears and bearings, are often formed by cutting an uncrued impregnated sheet to a pattern and then stacking and molding. Where small parts are required, as for spacers cams, contact arms and levers, it is economical to saw them or punch them from stock sheets, rods, or tubes.

Classification of Laminates

The high-pressure laminated grades of sheet stock, rods, and tubes are classified by the National Electrical Manufacturers' Association, commonly referred to as NEMA. In as much as laminates have their chief use in the electrical field, it is essential that the electrical properties be classifed along with the physical. Table 11–3 lists some of the

TABLE 11–4. STANDARD TOLERANCES FOR PUNCHED HOLES AND SLOTS.

	BASE MATERIAL	RESIN	SPECIFIC GRAVITY	THICKNESS	HARDNESS	WATER ABSORPTION	CONTINUOUS NO LOAD TEMP. F.	TENSILE STRENGTH P.S.I. LW	CW
X	PAPER	PHENOLIC	1.36	.010"-2"	110	6	225	20,000	16,000
XXP	PAPER	PHENOLIC	1.32	.015-.250	100	1.8	250	11,000	8,500
XXX	PAPER	PHENOLIC	1.32	.015-2.0	110	1.4	250	15,000	12,000
CE	COTTON	PHENOLIC	1.33	.031-2.0	105	2.2	250	9,000	7,000
LE	COTTON	PHENOLIC	1.33	.015-2.0	105	1.95	250	12,000	8,500
AA	ASB. FAB.	PHENOLIC	1.70	.062-2.0	103	3.00	275	12,000	10,000
G-3	CONT. GL.	PHENOLIC	1.65	.010-2.0	100	2.7	290	23,000	20,000
G-5	CONT. GL.	MELAMINE	1.90	.010-3.5	120	2.7	300	37,000	30,000
G-7	CONT. GL.	SILICONE	1.69	.010-2.0	100	.55	400	23,000	18,000
G-9	CONT. GL.	MELAMINE	1.90	.010-2.0	120	.80	325	37,000	30,000
G-10	CONT. GL.	EPOXY	1.75	.010-1.0	110	.25	250	40,000	35,000
N-1	NYLON	PHENOLIC	1.15	.010-1.0	105	.60	250	8,500	8,000
GPO-1	GL. MAT	POLYESTER	1.7	.062-2.0	100	1.0	250	12,000	10,000

more important standard industrial laminates. The following list of laminates explains some of their characteristics and applications.

NEMA GRADE

X	A laminate made with high-strength kraft paper with phenolic resin as the binder. It is used in mechanical parts where electrical properties are of secondary importance. It is used in household appliances, insulating washers, and coil forms.
XP	A paper base laminate made with phenolic resin. Intended for hot punching. Punch cold at .062 in. thick and punch hot at .125 in.
XPC	A paper base laminate made with phenolic resin. Can be punched and sheared cold at .125 in. thick.
XX	A paper-base phenolic laminate. The electrical and mechanical properties make it suitable for usual electrical applications, except where low losses or high humidity are involved. It is used in instrument panels and machined washers, barriers, relays, and switch bases.
XXP	A paper-base phenolic laminate. This laminate is a general-purpose hot punching stock. It is used in terminal boards, insulating washers, and switch parts.
XXX	A paper-base phenolic laminate. It is a low-cost electrical grade for high-voltage and radio-frequency uses. It has good dimensional stability under humid conditions. This type of laminate is used for jack spacers, coil forms, radio and TV parts, and high-voltage switchgear.
XXXP	A paper-base laminate with a plasticized phenolic resin. It is used in hot punching applications. It has high insulation resistance and low dielectric losses at high frequencies. It is used in terminal boards and radio and TV panels.
XXXPC	A paper base laminate made with phenolic resin. Recommended for hot punching. Good electrical properties under severe humidity conditions.

C	A strong cotton fabric laminate bonded together with phenolic resin. Items fabricated from this type laminate include gears, cams, pinions, bearings, and structural parts. It is not to be used for electrical parts.
CE	A medium-weave, canvas-base phenolic laminate. It is used in panel boards, electrode supports, switches, small gears, and small bearings.
L	A laminate made from fine-weave cotton fabric and phenolic resin. It is used for fine machined parts such as pinions, gears, breaker arms, and communication equipment.
LE	A laminate made of cotton-linen base with phenolic resin as the binder. It is used for terminal blocks and strips, radio parts, and ball-bearing retainer rings.
A	A laminate made from asbestos paper and phenolic resin. It has good dimensional stability under humid conditions and has good flame resistant properties. It is used for heat controls for household ovens, electrical ranges, and furnace parts.
AA	A laminate made from asbestos fabric and phenolic resin. It is used in clutches, in machine tools, rotor vanes, and insulation gaskets.
G-3	A laminate made from a continuous-filament type glass cloth and phenolic resin. It is used in armature slot wedges and structural parts requiring good electrical properties.
G-5	A laminate made from a continuous-filament type glass fabric with melamine resin as the binder. It has excellent electrical properties under dry conditions. It is used in switch parts in electrical and communications apparatus.
G-7	A continuous filament glass cloth laminate made with silicone resin. Excellent flame, heat, and arc resistance. Good physical strengths. It is used in motor slot wedges, slot liners, and high-frequency radio and radar insulators.

G-9 A continuous filament glass cloth laminate with melamine resin. Highest mechanical strength and one of the hardest laminates. Excellent electrical properties.

G-10 A laminate made from a continuous-filament-type glass cloth with epoxy resin as the binder. It is used for printed circuits and electronic appliances.

G-11 A continuous filament glass cloth epoxy laminate. It has extremely high mechanical strength at room temperature. Good electrical properties under both dry and humid conditions.

N-1 A nylon cloth base with phenolic as the binder. It has excellent electrical properties under high humidity conditions. Good physical properties.

FR-4,5 A continuous glass cloth epoxy laminate. It has a flame resistance of at least Class 1.

GPO-1,2 A glass mat polyester resin laminate. This is suitable for general purpose work. GPO-2 is good for low flammability and self-extinguishing properties.

DESIGN CONSIDERATION OF LAMINATES

Punching of Laminates

The term "punching" is used to describe the production methods of making laminated parts by blanking, piercing, and shaving, or a combination of these operations. Holes of nearly every geometrical shape have been punched in laminated sheets (Fig. 11–22). Certain basic principles and limitations of design, however, must be followed if the part is to be produced successfully.

Punchable laminates can be cold punched in maximum thicknesses ranging from .031 to .125 in., and hot punched in maximum thicknesses ranging from .093 to .125 in. All laminates do not punch alike, and the only test of whether a laminate is a punchable stock is actually to punch parts from the laminate in question. Punch-grade laminates can be divided into hot and cold punching groups. The heating range of laminates can be from 100 to 280° F. The heating may be accomplished by plates heated by steam or electricity, ovens, hot liq-

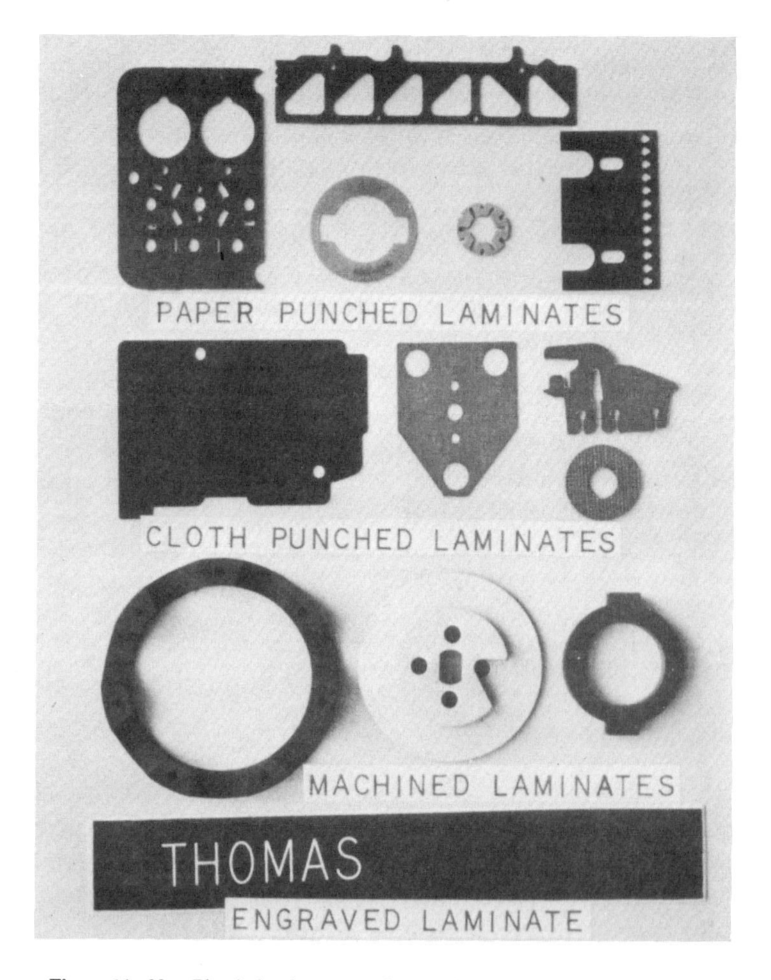

Figure 11–22. Plastic laminates can be punched, machined, and engraved.

uid baths, infrared lamps, or dielectric heating equipment. The laminated material should be heated as rapidly and uniformly as possible and should not be kept hot longer than necessary.

Punching is the forcing of a hole in a sheet, and blanking is the knocking (punching out) of a piece from a sheet. Fig. 11–23a illustrates a pierce, or punch and blank die and a typical part produced

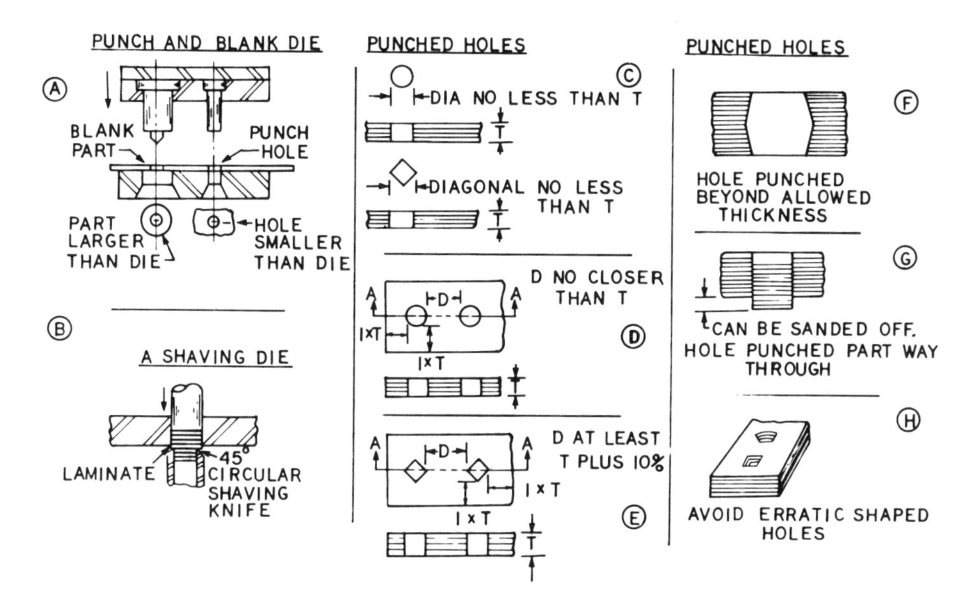

Figure 11–23. (A) The design of a punching and blanking die. (B) A shaving die. Punched holes in laminated sheet stock have limited sizes compared to the thickness of the laminate as shown in C, D, and E. (F) A hole punched beyond the allowable thickness. (G) A hole punched part way through. (H) Erratic shaped holes should be avoided.

by this die. Note that the part first is punched and then is blanked. Because of the resilience (yield) of laminates, punched holes tend to be smaller than punches, and blanks tend to be larger than blanking dies. With hot punching grades, thermal contraction results in holes and blanks smaller than the corresponding punches or dies. The die designer must rely on his knowledge of the characteristics of a certain type of laminate to be used and design the die accordingly.

Parts having thicknesses greater than the maximum allowed for punching may be shaved without damage to the piece or die. The part to be shaved is first rough cut to shape by blanking, by fly cutting, or band sawing. The rough cut part is then placed in a shaving die. Fig. 11–23b illustrates a shaving die. The die is mounted in the die shoe under the punching die and shaves approximately .015 to .020 in. from the edge of the blanked part. The shaving die finishes the rough blanked part to size and produces a very smooth clean surface.

The shaving tool is made by machining the tool to the shape of the part. The shaped tool is then ground on a 45° angle from the edge. This leaves a knife edge at the contour end of the shaving die.

The diameter or diagonal of a hole to be punched should be no less than the thickness of the stock (Fig. 11–23c). Holes in laminate parts should be no closer to each other or to the edge of the part than the thickness of the material (Fig. 11–23d). Square or rectangular holes should have their corners even farther apart than the thickness of the stock, in order to prevent cracking between the holes (Fig. 11–23e). If it is necessary to place holes closer together and closer to the edge than the limits specified, the holes may be drilled.

Tolerances between holes and on the diameters of the holes vary with the thickness of the stock being punched. Table 11–4 gives what is considered to be good shop practice on tolerances for punched holes and slots.

Punching beyond the maximum allowable thickness not only wears the die rapidly, but also results in a poor piece (Fig. 11–23f). The face sheet on both sides of the laminate have been pinched or pulled together at the edge, and this has resulted in squeezed-out laminations in the core. With some designs, it may be desirable to punch a part with an offset wall section or a hole part way through (Fig. 11–23g). When laminated sheet stock is used, the part is made by forcing the punch only part way through the stock. Such a technique can be used with any good grade or punching stock, but it must be remembered that the laminations will be broken, and that this will encourage water absorption and arcing through the laminations, if the part is subjected to electric current. The part should be considered only as a simple insulator and a dust cap. Erratic shaped holes should be

TABLE 11–4. STANDARD TOLERANCES FOR PUNCHED HOLES AND SLOTS.

MAXIMUM PUNCHING TOLERANCES ON SHEET STOCK

MATERIAL THICKNESS IN INCHES	DISTANCE BETWEEN HOLES				SIZE OF SLOT OR DIAMETER OF HOLES	OVERALL DIMENSIONS
	UNDER 2"	2" TO 3"	3" TO 4"	4" TO 5"		
UNDER .062"	±.003	±.004	±.005	±.006	±.0015	±.008
.062" TO .093"	±.005	±.006	±.007	±.008	±.003	±.010
.093" TO .125"	±.006	±.007	±.008	±.009	±.005	±.015

avoided, unless the laminate stock is thin enough to allow the holes to be punched (Fig. 11–23h). Irregular holes in thick laminates require special tools and additional operations.

Dies Used For Punching

Dies used for punching laminates are of the same general types as those used for punching metals, except that the clearances should be less than those normally employed in punching metals. Progressive dies are usually preferred, because they are more economical to make and permit higher production rates. For clean punching of laminated parts it is recommended that close tolerances between the punch and die be held. For standard tolerances, the die hole should be no more than .004 in. larger than the punch, giving .002 in. clearance all around. Deviation from standard tolerances in the die construction can be permitted if several thousandths of an inch tolerance on the size of the hole is allowed.

Milling Laminates

The milling of plastic laminated materials is very similar to that of milling brass. Because of the laminated structure of the plastic laminate, climb or down milling is always used to prevent any tendency toward delamination (Fig. 11–24a). Cutters should have a negative rake of about 10°. The width of a ridge between two milled slots should be greater than the depth of the slots. This will keep the ridge from breaking. Slot ends should be designed to be round instead of square. This makes a stronger part and will permit the use of an end mill, to make the rounded end, instead of hand filing. Corners on machined or punched laminated parts should be rounded to prevent delamination and fractures.

Laminated plastics can be machined on standard wood or metal working equipment. On glass-base laminates, the cutting tools dull rapidly. Diamond or tungsten-carbide tools give a more satisfactory working life.

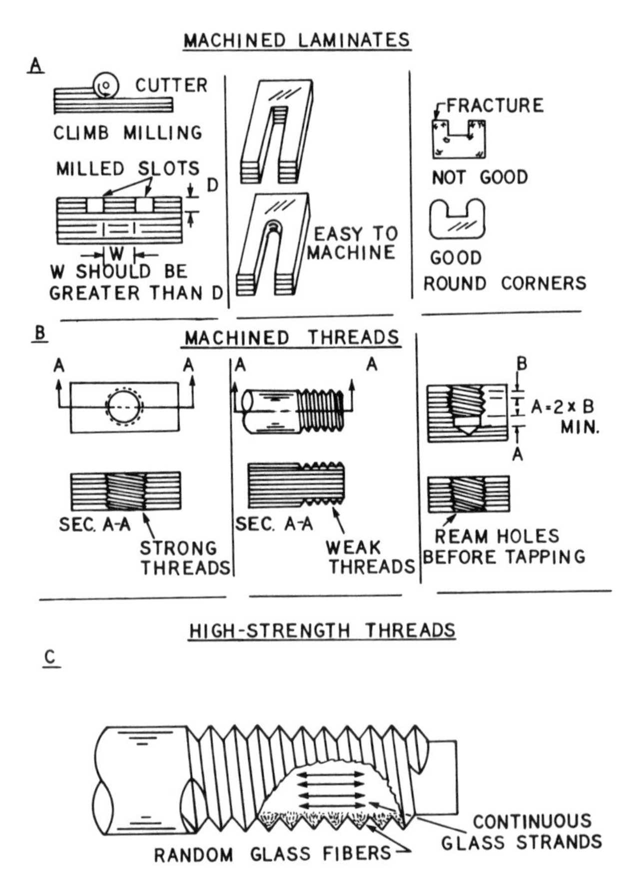

Figure 11–24. (A) Machined laminates. (B) Machined threads. (C) Engineered high-strength machined threads.

Threads Machined In Laminates

Threads are generally always machined and not molded in plastic laminated materials. Machined threads do not start abruptly, nor do they come to an abrupt end, as do molded threads. Threads start and stop with a feather edge. The feather edge in laminated materials is weak. If the axis of a thread is perpendicular to the lamination (Fig. 11–24b), the thread will have maximum strength. Assume that a thread is cut on the outside diameter of a tube or rod. If the mating part exerts strain in a direction parallel to the axis of the tube or rod, the threads of the laminated part will tend to strip or delaminate.

When drilling or tapping parallel to the laminations, always clamp the workpiece between two supports to prevent splitting. This is not necessary when working perpendicularly to the lamination, although a backup plate to prevent chipping makes a cleaner hole. A blind tapped hole should have a clearance at the bottom in order to prevent stripping of the thread or delamination. Though tapped holes should be reamed before tapping, as the edge of the cut thread will be much smoother.

All threads cut into laminated materials should be specified as U.S. Standard with thread tolerance no closer than Class 2 fit.

High-strength Threads

High-strength plastic threads can be made by using continuous strands of glass rovings to form a core and then wrapping multiple spiral layers of glass mat around the core (Fig. 11–24c). The core is first made by the pultrusion method. Polyester or epoxy resin can be used as the plastic binder. After the rod has been made it is then machined to size and threaded. This type of thread construction is used in areas requiring high strength, corrosion resistance, and design versatility.

COMPOSITES

A composite may be classified as an article containing or made up of two or more substances. In the plastic industry, composites apply generally to structures of reinforced elements. There are five general classifications of composite materials: (1) laminar; (2) fiber; (3) flake; (4) separate particles; and (5) filled. This is illustrated in Fig. 11–25.

Laminar Composites

This is the first composite that was made by man. It is made of layers of different materials bonded or fastened together with different plastic resins. In plywoods, the layers are generally of the same type

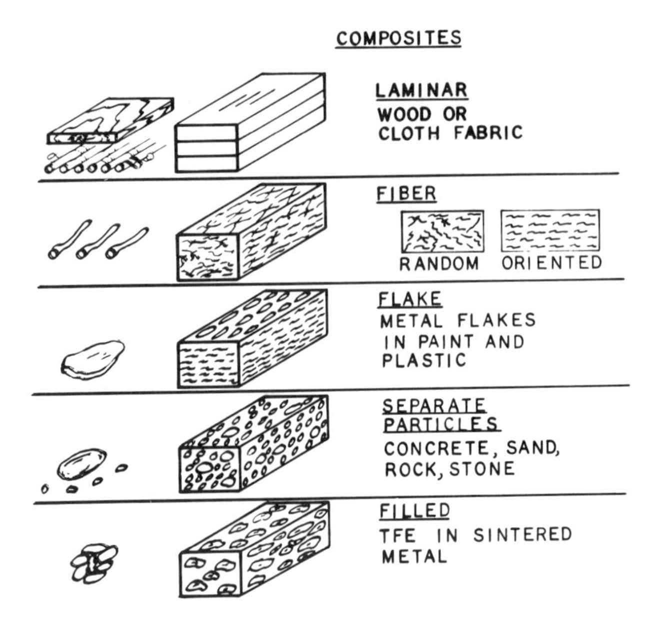

Figure 11–25. The five general classes of composites.

of wood, but the orientation of the layers differs. Plywood is made with the grain of alternating plies at right angles. The wood layers are bonded together with an adhesive (generally plastic) to make a solid, rigid article. Plywood is made from many species of hardwoods and softwoods, ranging chiefly from .250 to .875 in. in thickness, and with plastic adhesives that differ in moisture resistance. Other composite materials made of wood consist of resin-impregnated woods, softboard, hardboard, and particle board.

Laminar composites in the plastic field consist of layers of resin-impregnated fabrics, paper, glass cloth, etc., which possess high strength-to-weight ratios. If highly directional materials like wood, woven fabrics, or bundles or layers of continuous filaments are employed, the resulting laminate is highly directional in its mechanical properties.

Fiber Composites

Practically every type of plastic, rubber, elastomer, and ceramic has been reinforced with fibers. In selecting a reinforced fiber for a composite, the following should be considered: (1) fiber orientation; (2) length, shape, and composition of the fibers; (3) mechanical properties of the matrix; and (4) integrity of the bond between fibers and matrix.

Fibers are arranged in a random, unidirectional, and bidirectional pattern. Continuous filaments are generally used in the filament-winding process

Flake Composite

A very pleasing decorative effect can be obtained by using small metallic flakes in paints and plastics. The flakes produce a brilliant metallic highlight. The highly reflective flake pigments are particles of either aluminum or copper and copper alloys. The aluminum pigments reflect and produce brilliant blue-white highlights. The copper-based pigments produce a bronze or gold color.

Numerous problems have been encountered in using metallic pigments in plastic materials. In injection molding, the orientation of the flake in the plastic material is upset during the molding process at flow and knit lines, producing an unsightly appearance. This condition can be improved by using a large fan-type gate.

Separate Particle Composite

The oldest most widely used particle composite is concrete. The gravel and sand are the particles and cement is the matrix. The cement and water form a paste that hardens by chemical reaction into a strong stone-like mass. No more cement paste is used than is necessary to coat all the aggregate surfaces and fill all the voids. The quality of the paste formed by the cement and water largely determines the character of the concrete.

In the plastic industry, decorative panels for store fronts are made from sand and polished pebbles with polyester resin as the matrix.

The sand is thoroughly mixed with the catalyzed polyester resin and cast into a slab mold. The polished stone pebbles are partially imbedded on the surface of the sand resin mixture. After the polyester resin composite mixture has polymerized and becomes hard, it is removed from the mold. The cast slab is coated with a silicone resin to protect the composite from the outside elements. This type of slab composite is used only for a dccorative item for building fronts.

Filled Composites

A filled composite is an open matrix or sketal structure filled with another material. It might be described as similar to filling the voids in a sponge with a solid material such as plaster. A filled honeycomb is another example

Materials that are porous and can be filled are metal castings, powder metal parts, ceramics, graphite, and foams. Powdered metal parts can be impregnated with a PTFE fluoroplastic to provide a good combination of bearing properties. The PTFE acts as a lubricating medium. Porous castings of aluminum can be sealed with liquid plastic resin such as polyester or epoxy. The liquid resin is forced into the porous metal by air pressure or a vacuum. The liquid resin fills all of the voids and is polymerized to a 100% solid.

Fillers and Reinforcements

Almost all of the known plastics today can be filled, reinforced, or both filled and reinforced with other materials. A filler is a material added to a plastic resin in order to obtain mechanical, chemical, or electrical properties not possessed by the plastic resin itself. Reinforcing fillers or fibers are added to plastic resins to enhance mechanical strengths. Table 11–5 shows various fillers and fibers used in thermosetting plastic resins. Table 11–6 shows various fillers and fibers used in thermoplastic resins.

Carbon and Graphite Fibers

Carbon and graphite are allotropic forms of elemental carbon. The term "carbon" applies to the amorphous form of the element while the term "graphite" refers to the black crystalline form.

TABLE 11-5. VARIOUS FILLERS AND FIBERS USED IN THERMOSETTING RESINS.

THERMOSETS	ALUMINA	CALCIUM CARBONATE	CARBON BLACK	CLAY	COTTON FLOCK	GLASS BUBBLES	GLASS FIBERS	GRAPHITE	MICA	QUARTZ	TALC	WOLLASTONITE SILICATE	WOODFLOUR
ALKYDS		●		●			●		●		●	●	●
DAP		●		●			●				●		
EPOXY	●	●	●	●		●	●			●		●	
PHENOLIC	●	●	●	●	●	●	●	●	●		●	●	●
POLYESTER		●		●		●	●					●	●
MELAMINE					●		●		●		●	●	
UREA				●	●								●
SILICONE	●			●			●			●			
URETHANE	●	●	●	●		●	●			●		●	

TABLE 11-6. VARIOUS FILLERS AND FIBERS USED IN THERMOPLASTIC RESINS.

THERMOPLASTICS	ALUMINA	CALCIUM CARBONATE	CARBON BLACK	CLAY	GLASS BUBBLES	GLASS FIBERS	GRAPHITE	QUARTZ	TALC	MOLYBDENUM DISULFIDE	WOLLASTONITE SILICATE
ABS					●	●					●
ACETAL						●					
ACRYLIC						●					
NYLON				●	●	●	●	●		●	●
POLYCARBONATE						●					
POLYESTER TP						●					
POLYETHYLENE	●	●	●			●		●			●
POLYPROPYLENE		●	●			●		●	●		●
POLYSTYRENE					●	●			●	●	●
POLYSULFONE						●					
POLYURETHANE	●	●		●	●			●			●
PPO MODIFIED						●					
PVC RIGID	●	●	●	●	●	●	●		●		●

Both carbon and graphite exhibit metallic properties in electrical and thermal conductivity. Unlike either ceramic or metal, graphite and carbon increase in strength with increases in temperature.

An organic fiber is the starting point in the manufacturing of high-modulus graphite (HMG). Rayon was used at one time but has been replaced by polyacrylonitrile (PAN), which has a more ordered micro-structure. Carbon fibers derived from petroleum pitch are a much cheaper material but have a lower strength than those made from PAN.

PAN is converted into graphite in a two-phase process. It is first burned in air in a relatively low-temperature furnace. Then it is car-

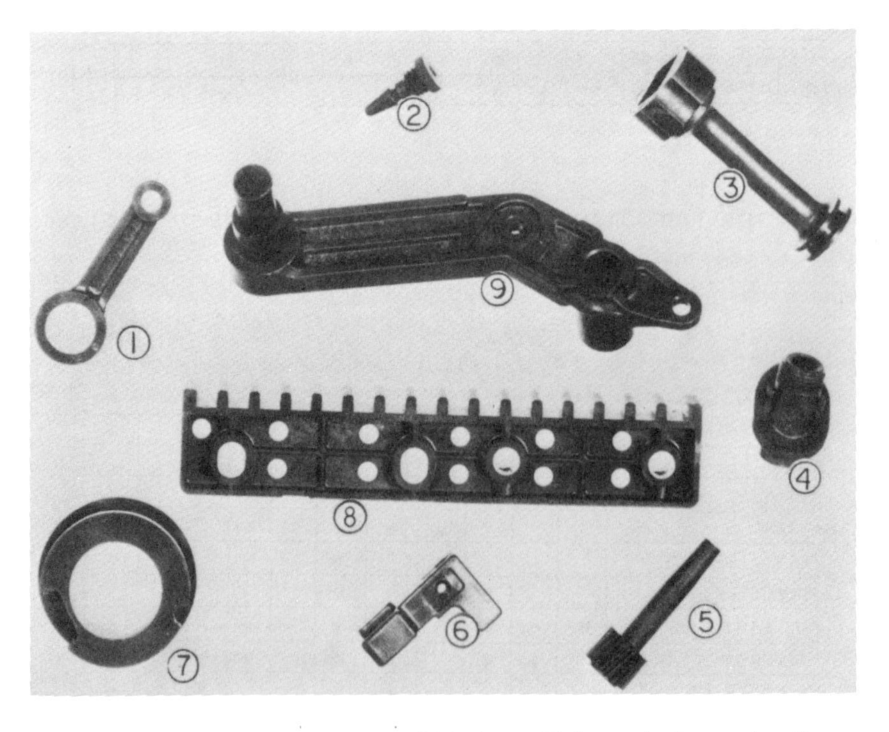

Figure 11–26. This illustrates many small injection molded parts that have carbon fibers as reinforcements. (1) Connecting rod, power tool. (2) Electrode, vending machine. (3) Piston, electric power tool. (4) Wiper pivot, automotive. (5) Gear, automotive speedometer. (6) Middle hook, heddle frame. (7) Governor, gasoline engine. (8) Guide rack, heddle frame. (9) Chain-tension device. *(Courtesy Union Carbide Corp.)*

bonized in an inert atmosphere, and at much higher temperatures (2000–5000° F). The final result is a smooth black fiber, usually seven to eight microns in diameter. Once the fibers are made, they can be woven into cloth, or formed into yarns, mats, tapes, or long bundles of untwisted fibers called tows.

Most uses for carbon fibers have been in high temperature, or high strength structural applications such as reinforcement for parts in missiles and rockets, aircraft, sporting equipment, etc. Fig. 11–26 shows many small injection molded parts that have carbon fibers as a reinforcement.

Microspheres

Hollow glass spheres or microspheres are used as an extender to save weight in thermosetting epoxies and polyester resins. The hollow bubbles can be manufactured from epoxy, glass, phenolic and silica. Glass and silica are used mostly.

The microspheres have an average thickness of 60 microns, are white in color, and pour like a liquid. They come in a wide range of densities, normally .15 to 38 gms./cm^3. The biggest advantage to using microspheres is weight reduction of plastic parts. Also improvements in mechanical and physical properties are noted.

12 Tests and Identification of Plastics

TESTING PLASTIC MATERIALS

The American Society for Testing and Materials (ASTM) has developed standard methods for determining the essential properties of plastic materials. The data published by material manufacturers on the properties of their materials is generally based on ASTM test methods. For more detailed information on the procedures, see the ASTM Standards, Plastics-Methods of Testing.

The test values obtained from plastic materials charts are starting points in selecting a plastic material for a particular application. There is a natural variance in the strength properties, depending on such influences as manufacturing methods, processing conditions, shape and design of the part, and environment conditions. Even if the material is chosen initially on the basis of adequate properties, it is generally necessary to make end-use tests. The end-use or evaluation tests are generally made under more severe conditions than the product is likely to encounter in actual use. It is not possible to establish fixed conditions for determining the expected service life of the plastic components in all products. However, by taking into account the many variables that might exist, standard tests may be modified to obtain satisfactory results. A simplified description of the more important physical, mechanical, and electrical tests used to evaluate properties of plastic materials is explained in the following pages.

Tensile Properties

Tensile properties refer to the behavior of a plastic material when it is subjected to forces that tend to pull it apart. These properties are the most important indication of strength in a material.

ASTM tensile properties are almost always determined on a standard shaped specimen similar to that shown in Fig. 12–1. The most common test sample is defined by ASTM D-638. The over all length is 8.5 in., and the center section is 2.25 in. long with constant cross-sectional dimensions of .50 by .125 in. In the center section, a 2 in. length is marked off. The marks are called "gauge marks". It is this portion of the sample within the gauge marks that is tested.

The test sample is tested in a machine giving straight tensile pull, without any bending or twisting. Two items are recorded or measured when the sample is being pulled apart–tensile stress and tensile strain. The tensile stress is the strength of pull in the area between the gauge marks. It is based on the original cross-sectional area and is expressed in terms of psi. The tensile strain is a measure of how much the test sample has been stretched by the pull. It is calculated by dividing the extended length between the gauge marks by the original

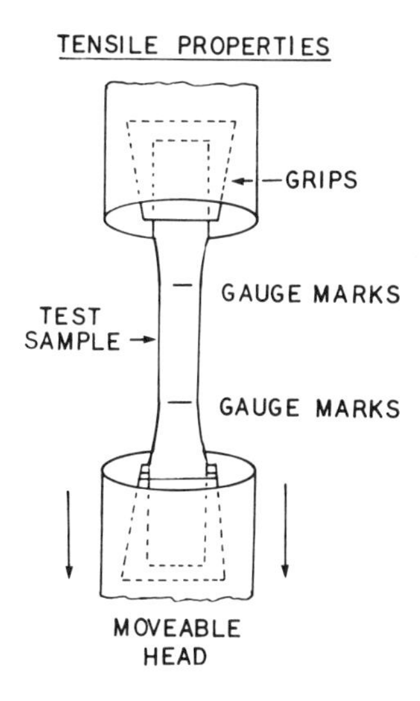

Figure 12–1. Apparatus used in tensile testing.

length. The elastic modulus (modulus of elasticity or tensile modulus) is obtained by dividing the stress by the change in length. This is useful data, because parts should be designed to accommodate stresses to a degree well below this. Tables 12–1 and 12–2 show the relative tensile strength of some thermoplastic and thermosetting materials.

Compressive Properties

A one-in. cube of a rigid plastic material is placed on a firm foundation, and a load is gradually applied to the top of the cube until the

TABLE 12–1. AVERAGE TENSILE STRENGTH OF SOME THERMOPLASTIC MATERIALS AT ROOM TEMPERATURE.

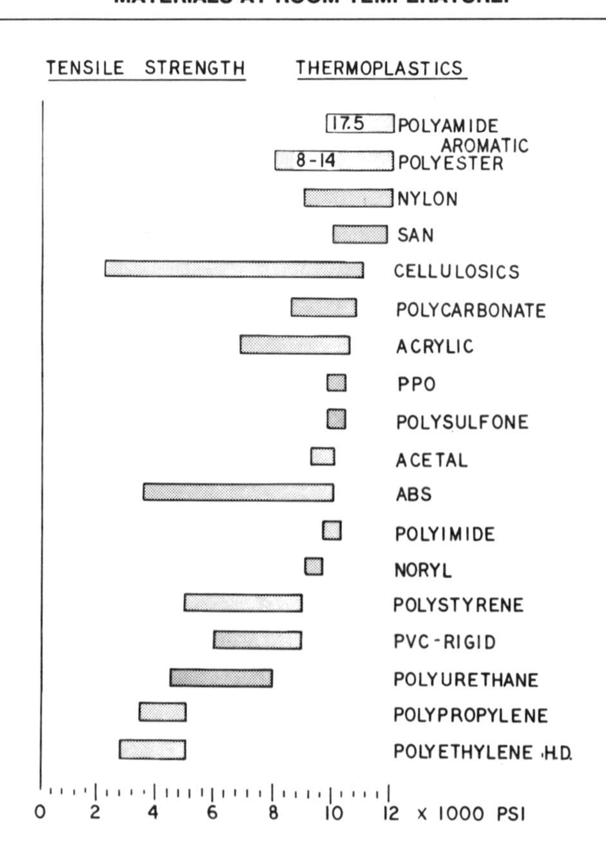

TABLE 12–2. AVERAGE TENSILE STRENGTH OF SOME THERMOSETTING PLASTIC MATERIALS AT ROOM TEMPERATURE.

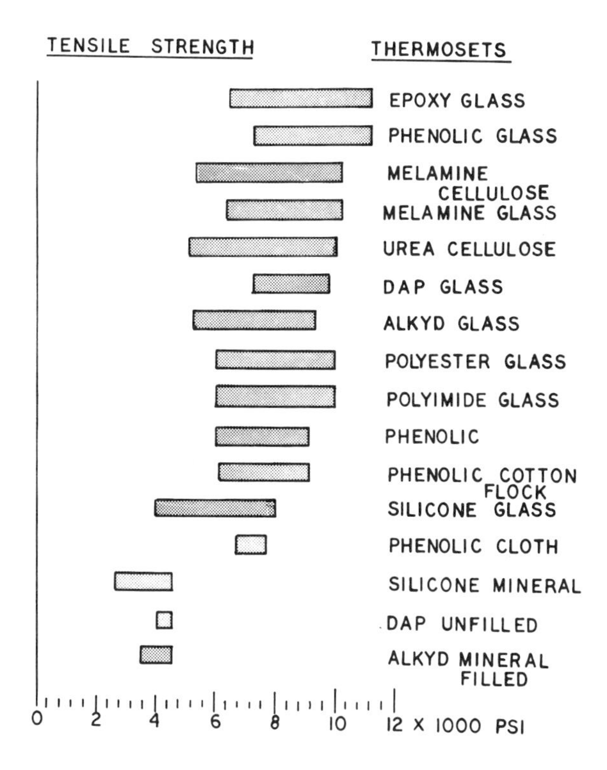

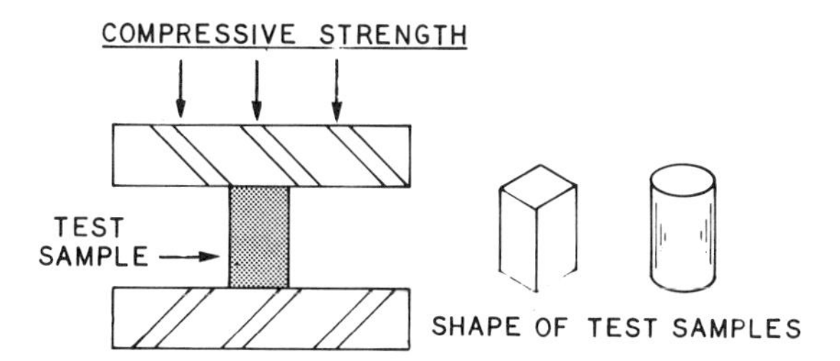

Figure 12–2. This shows the method used to obtain compressive strength data.

point of material collapse; this indicates the compressive strength of the material (Fig. 12–2). The compressive strength is calculated as the pounds per square inch required to rupture or deform the test sample. It may be expressed in psi either at rupture or at a given percentage of yield.

Compressive deformation is the change in length produced in a longitudinal section of the test sample by a compressive load. Tables 12–3 and 12–4 illustrate the compressive strength of some thermoplastic and thermosetting materials. This is an ASTM D–695 test.

Flexural Properties

Flexural strength is the resistance of a material to rupture under a bending stress applied at the center of a rectangular test specimen

TABLE 12–3. AVERAGE COMPRESSIVE STRENGTH OF SOME THERMOSETTING PLASTIC MATERIALS AT ROOM TEMPERATURE.

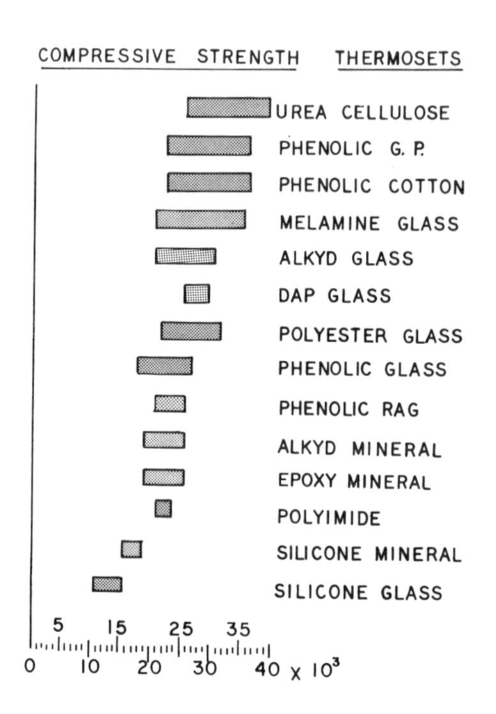

TABLE 12–4. AVERAGE COMPRESSIVE STRENGTH OF SOME THERMOPLASTIC MATERIALS AT ROOM TEMPERATURE.

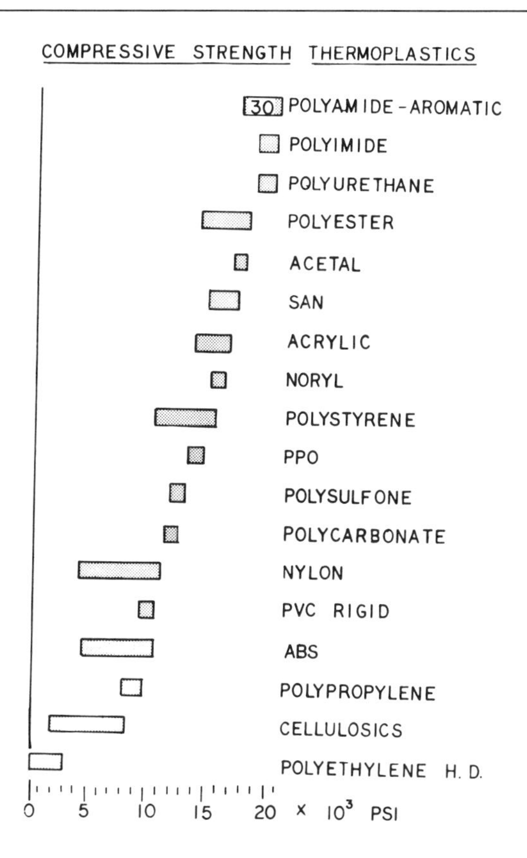

supported at both ends. The flexural strength in pounds per square inch is calculated from the load recorded at the breaking point and the crosssectional area of the bar and the distance between the bar supports. In this test, the upper half of the bar is being compressed while the lower half is under tension (Fig. 12–3).

The test sample generally used is a .50 by .125 in. bar with 2 in. span. This is a ASTM D–790 test. During the test, the force of the load and the amount of the deflection (bending) are measured.

Flexural modulus (modulus, modulus of elasticity or elastic modulus) is the ratio of flexural stress to flexural strain before permanent

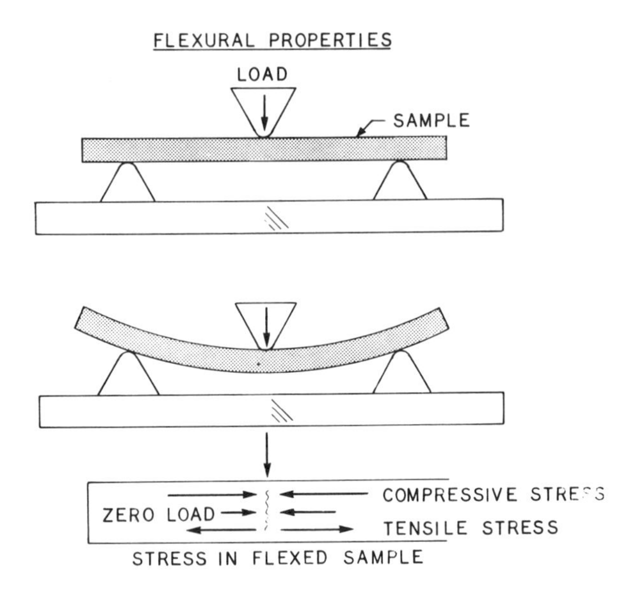

Figure 12–3. The test method used for flexural properties.

deformation has taken place. Modulus has units or pounds per square inch. High values indicate a stiff material and low values a limp material. The word modulus is derived from Latin and means "a small measure of." Flexural modulus means "a measure of flexing or bending." Modulus is a number that expresses a measure of some property of a material. Flexural modulus might better be called "modulus of elasticity in bending." Tables 12–5 and 12–6 show the flexural strength of some thermoplastic and thermosetting materials.

Impact Tests

Impact strength is a measure of the energy required to fracture a material by a sharp blow or fall.

Izod Impact

A widely used impact test is the izod. In this test, a pendulum striker is swung from a fixed height and strikes a test sample in the form of a

TABLE 12–5. AVERAGE FLEXURAL STRENGTH OF SOME THERMOPLASTIC MATERIALS AT ROOM TEMPERATURE.

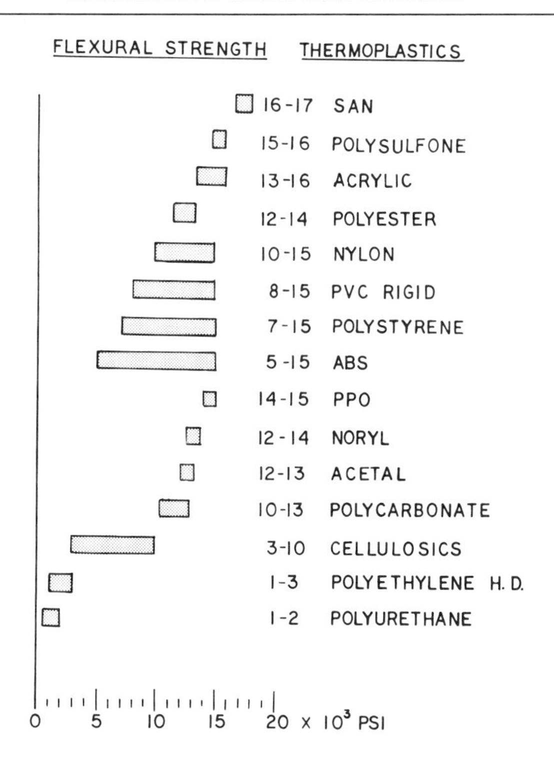

notched bar mounted as a cantiliver beam (Fig. 12–4a). After the breaking of the sample, the arm follows through in the same direction and moves a pointer that gives a reading. The test consists of breaking the test bar and calculating the amount of energy required to break it. The results are reported as foot-pounds per inch of notch. The test bar measures 2.5 x .50 x .50 in. and is notched across the edge. The notch is milled into the sample with a special milling cutter at a specified angle and to a depth of 0.1 in. Tables 12–7 and 12–8 show test results of some thermoplastic and thermosetting materials. This is an ASTM D–256 test.

TABLE 12–6. AVERAGE FLEXURAL STRENGTH OF SOME THERMOSETTING PLASTIC MATERIALS AT ROOM TEMPERATURE.

FLEXURAL STRENGTH ($\times 10^3$ PSI)	THERMOSETS
15-24	MELAMINE GLASS
14-20	POLYESTER GLASS
14-20	POLYIMIDE GLASS
13-19	SILICONE GLASS
17-19	DAP GLASS
10-18	UREA CELLULOSE
11-18	EPOXY MINERAL
9-14	ALKYD GLASS
10-15	MELAMINE CELLULOSE
10-17	PHENOLIC GLASS
10-13	PHENOLIC CLOTH
9-12	PHENOLIC COTTON FLOCK
8-12	PHENOLIC G.P
9-11	ALKYD MINERAL FILLED
7-10	DAP UNFILLED
6-8	SILICONE MINERAL

0 5 10 15 20 25 30 x 10^3 PSI

Charpy Impact

The Charpy type of tester uses a test sample in the form of a beam supported at both ends. The beam is broken by a blow delivered midway between the supports. In this test, the sample can be either plain or notched. The results are reported in terms of foot-pounds per inch (Fig. 12–4b). This is an ASTM D 256 test.

Falling Dart Impact Test

This test consists of a weighted metal spherical dart that is dropped from a height of 30 in. or more (Fig. 12–4c). The weight of the dart can vary from a few ounces to several pounds. The test does not carry

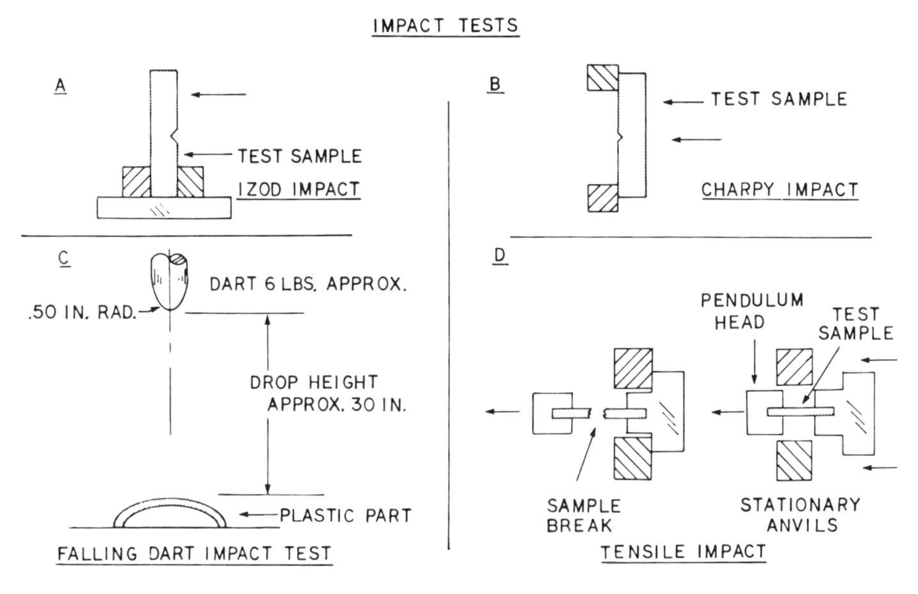

Figure 12–4. Four types of impact tests: (A) Izod, (B) Charpy, (C) Falling Dart, and (D) Impact.

an ASTM number and is used for comparative data only. The energy is expressed in terms of the weight (mass) of the missile falling from a specified height which would result in failure of the part being tested.

Tensile Impact

Tensile impact tests are generally made with a Charpy modified type of impact machine. One end of the test sample is attached to the swinging pendulum. The other end of the test sample is attached to the pendulum head (Fig. 12–4d), in a horizontal position. Upon impact with the stationary anvils, the test sample is broken with a single blow delivered in the direction of its longitudinal axis, and the energy absorbed is reported in foot-pounds. This is an ASTM D 1922 test.

Shear Strength

Shear strength is related to the force required to punch a hole through a sheet of plastic. A punch-type shear fixture is used (Fig. 12–5). The shear strength is calculated as the force per area sheared.

TABLE 12–7. AVERAGE IMPACT STRENGTH OF SOME THERMOPLASTIC MATERIALS AT ROOM TEMPERATURE.

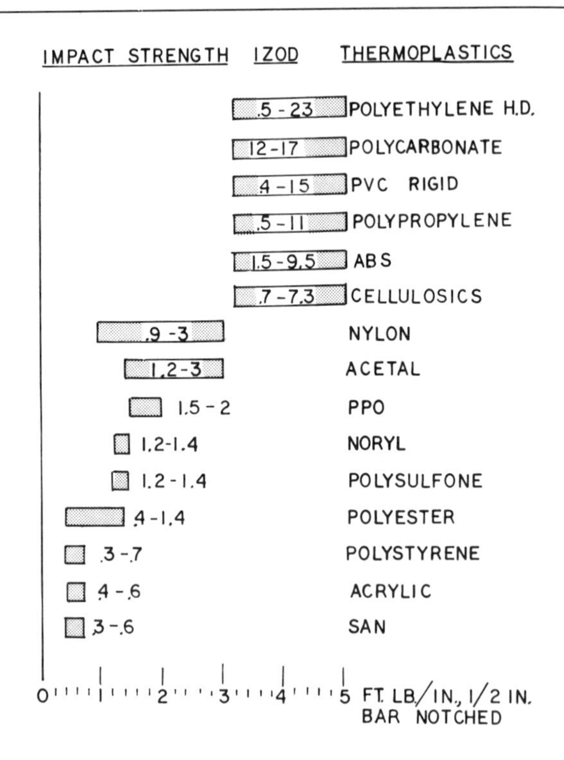

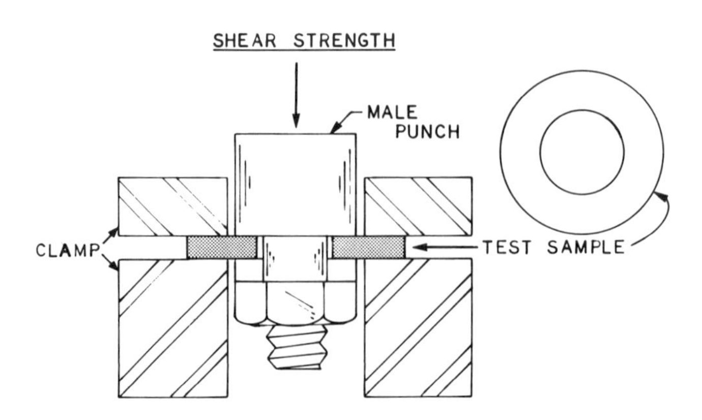

Figure 12–5. This illustrates the test method used to obtain shear strength data.

TABLE 12–8. AVERAGE IMPACT STRENGTH OF SOME THERMOSETTING PLASTIC MATERIALS AT ROOM TEMPERATURE.

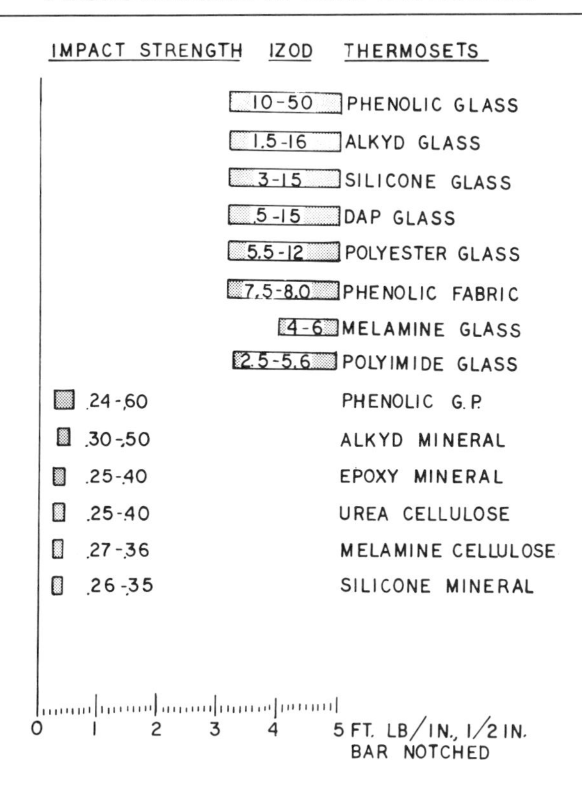

The shear strength in pounds per square inch, is determined by dividing the load by the area of the sheared edge. This area is the product of the thickness of the test sample and the circumference of the punch. This is an ASTM D 732 test.

Stiffness in Flexure

Stiffness in flexure is a measure of the flexibility or rigidity with respect to bending of a plastic. This test provides a means of deriving an index of stiffness of a material by measuring the force and angle of bend of a cantilever beam (Fig. 12–6).

The test is a measure of the relative stiffness of various plastics and,

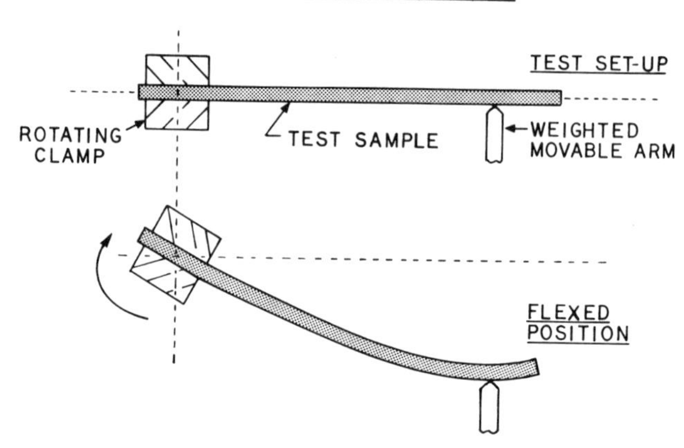

Figure 12–6. The test method used for stiffness in flexure.

if taken with other appropriate property data, is useful in selecting materials. This is an ASTM 747 test.

Creep

Creep is defined as the deformation of a plastic part that takes place over extended periods of time when the part is supporting a load. There is no established method of determining creep. The ASTM test D–674 is actually not a test method but "a recommended practice for creep tests."

Creep tests are carried out by placing test bars in a constant temperature chamber and attaching a load to the sample (Fig. 12–7). Careful measurements of the deflection or extension of the sample are made over a period of time, starting with hourly, daily, and weekly measurements

The term "creep" is sometimes used as cold flow to cover distortion under compression only. More accurately, creep is the first stage in which the deformation rate changes with time. When it becomes constant, cold flow is taking place. The amount of such deformation depends on stress, time, and temperature.

Hardness

The hardness of a plastic material is the resistance of the material to compression, indentation, and scratching. It is purely a relative term.

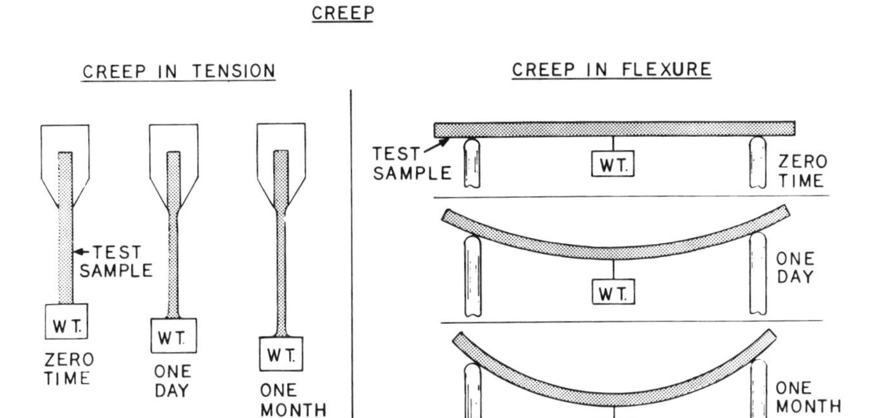

Figure 12–7. This illustrates two methods used to obtain creep data.

This does not mean that the harder the plastic material, the better will be its mar resistance. Hardness herein, is defined as the resistance to indentation. It is the indentation produced by some specific indenter, under some load, or that indentation remaining after the removal of some specific load.

Barcol Hardness

The Barcol hardness tester or impressor is a portable, hand-held instrument that is placed on the surface of the material to be measured. A slight pressure against the instrument drives a spring loaded indenter into the material to be tested. A conical needle with a 0.006 in. diameter, flat at its tip, penetrates the surface. The depth of penetration is automatically converted to a hardness reading, in arbitrary Barcol degrees, on a dial on the back of the instrument. The reading ranges from 0 to 100. The instrument is used in the reinforced plastic industry as a simple means of determining the degree of cure of polyester resins.

Shore Durometer

In this test, hardness is measured with a small hand instrument, called a durometer. The instrument consists of a spring-loaded indenter point protruding through a hole in the base. The base of the tester is held against the specimen, and in turn, the indenter is pushed into the sample by a spring. A scale is read to obtain the resulting durometer hardness. The scale readings range from 0 to 100. If the scale reading is 100, the indenter has penetrated zero ins. If the scale reading is zero the indenter has penetrated 0.100 of an in. The Shore "A" instrument employs a sharp indenter point with a load of 56 to 822 grs. This is the type of durometer used for rubber and some plastic materials. The Shore "D" instrument is used for very hard plastics. It has a blunt point, and a load of 10 lbs. is used.

Rockwell Hardness

This hardness test is a measure of the resistance of a material to indentation by a hardened steel ball under load. The hardness is expressed as a number derived from the net increase in depth of impression as the load on the indenter (a ball in plastic testing) is increased from a fixed amount to a major load, and then returned to the minor load. Fig. 12–8 illustrates the steps taken to obtain a Rockwell hardness reading on a plastic part. (1) This shows the setup of the machine and the part to be tested. (2) The dial is set at zero and the minor load is applied. The surface is indented to a depth "A". (3) The major load is applied and held for 15 secs. The surface is indented to depth "B". (4) After 15 secs. the major load is removed, and partial recovery from the indentation takes place. Then another 15 secs. (a total of 30 secs.), the hardness is read on the dial. The surface is indented to depth "C". The minor load is fixed at 10 kg. The major load varies with the different size balls used (Table 12–9).

For certain types of materials, particularly those having creep and recovery, the time factor involved in the applications of high and low loads have a considerable effect on the results of the measurements. It is not a good idea to compare hardness of various kinds of plastic entirely on the basis of Rockwell hardness. This hardness test can differ-

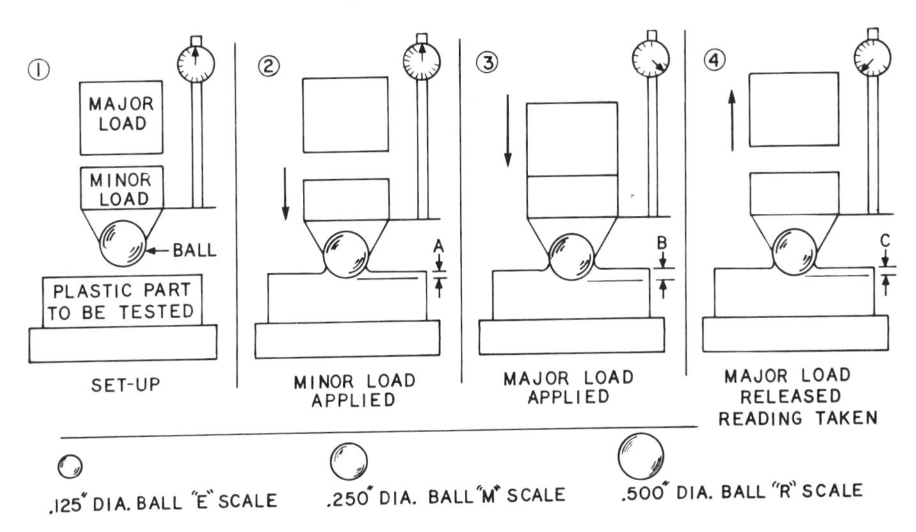

Figure 12–8. This illustrates the steps that are taken to obtain a Rockwell hardness test on a plastic material.

entiate relative hardness of different types of a given plastic, but the elastic recovery is involved, as well as hardness. This is an ASTM D–785 test.

Scratch Hardness

Scratch hardness is the resistance of a material to scratching by another material. The test most often employed for plastics is the

TABLE 12–9. VARIOUS METHODS OF TESTING HARDNESS BY INDENTION.

Name	*Indenter*	*Force*	*Load*	*Reading*
Barcol	Sharp Needle	Hand	Rapidly	At Once
Durometer	Dull Needle	56 to 822 gr. By spring	Rapidly	At Once
Rockwell-E	Ball .125 in.	100 kg.	15 sec.	15 sec.
Rockwell-M	Ball .250 in.	100 kg.	15 sec.	15 sec.
Rockwell-R	Ball .500 in.	60 kg.	15 sec.	15 sec.
Rockwell-A	Ball .500 in.	60 kg.	15 sec.	At once
Rockwell-L	Ball .250 in.	60 kg.	15 sec.	15 sec.
Sward Rocker	Runners	93 gr.	———	———

Bierbaum test. The Bierbaum scratch apparatus is a precision instrument used to measure the width of a scratch produced by a diamond point as it is drawn across the surface of the sample. The diamond point is in the shape of a corner of a cube and is so mounted that the diagonal of the cube is perpendicular to the test surface, and one edge is in line with the direction of the scratch. The standard load is 3 grs. The width of the scratch is measured by a microscope, and the hardness is expressed as the load in kilograms divided by the square of the width of the scratch in millimeters. The Bierbaum number is highest for the hardest materials.

Mohs' scale of hardness was originally devised to classify various minerals according to their resistance to scratching. In Mohs' original scale, talc was chosen as number 1, rocksalt as number 2, calcite as number 3, etc., up to diamond as number 10. It is of little value, however, for differentiating between the scratch resistances of the various plastics, since practically all of them, including both thermosetting and thermoplastic types, are in the range of 2 to 3 Mohs.

It has been found that a series of drawing pencils offer a rough but practical method of classifying the scratch resistance of plastics. The Koh-I-Noor series of 13 pencils, numbered 2B to 9H, approximately cover the range of 2 to 3 Mohs. In use, the pencils are sharpened to a short smooth point and considerable pressure is used in drawing them across the polished surface of the plastic. The pencil carbon will be broken down if the plastic surface is appreciably harder than the carbon. A furrow will be plowed in the surface if the pencil is appreciably harder than the plastic. With some practice, a technique can be developed that enables actual differentiation between a number of the thermoplastics. Table 12–10 shows the hardness of various plastic materials.

Deflection Temperature Under Load (Heat Distortion Point)

The heat distortion temperature of a plastic material indicates the point at which it begins to yield. This is not to be confused with its heat resistance, which is the maximum temperature at which it can be used continuously over a period of days or weeks. This is sometimes called the "safe operating temperature." The heat distortion point is

TABLE 12–10. HARDNESS OF VARIOUS PLASTIC MATERIALS.

HARDNESS OF VARIOUS PLASTICS

PLASTIC MATERIALS	ROCKWELL			DUROMETER	BARCOL
	M	R	E	SHORE D	
ABS HIGH IMPACT		85-109			
ACETAL	94	120			
ACRYLIC	85-105				49
CELLULOSICS		30-125			
DAP GLASS FILLED			80-87		52-65
EPOXY G. P.	75-108				
EPOXY GLASS FILLED	100				
PTFE FLUOROPLASTIC				50-65	
PFEP FLUOROPLASTIC		25		50-60	
PCTFE FLUOROPLASTIC		75-95		76	
VF_2 FLUOROPLASTICS				80	
SURLYN IONOMER				60	
MELAMINE CELLULOSE	116-124				
PPO MODIFIED	78	119			
NYLON 66		108-120			
NYLON 6		120			
PHENOLIC G. P.	45-60				
POLYCARBONATE	70	116			
POLYESTER RIGID	65-115				30-50
POLYETHYLENE H. D.				60-70	
POLYETHYLENE M. D.				50-60	
POLYETHYLENE L. D.				41-46	
POLYPROPYLENE		90-110		75-85	
POLYSTYRENE G. P.	68-80				
VINYL RIGID		117		65-85	
VINYL PLASTICIZED				40-100 SHORE A	
UREA CELLULOSE	110				
SILICONE	84				
POLYSULFONE	69	120			
POLYESTER TP	80				

merely an arbitrary value of deformation under a given set of test conditions. In ASTM D–648, it is defined as a total deflection of 0.010 in. in a rectangular bar supported at both ends under a load of 264 or 66 psi. The temperature of the oil bath is increased 2° C per minute. Fig. 12–9 illustrates the test method and Table 12–11 shows some of the results of thermoplastic and thermosetting materials.

Heat Resistance

The effect of heat on the properties of plastic materials is of great importance. The maximum safe continuous operating temperature of a

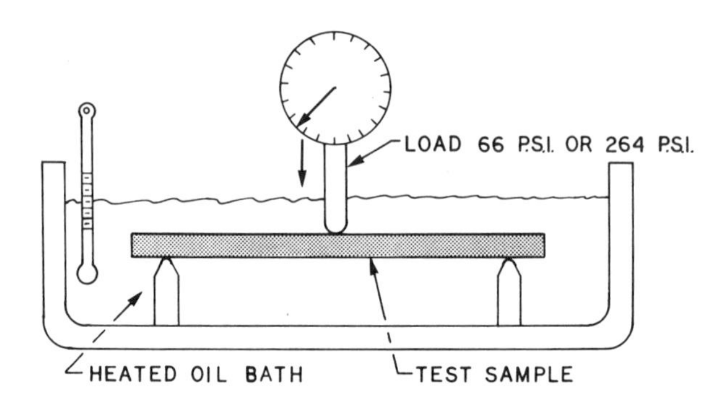

Figure 12–9. This is a schematic showing the method used to obtain the heat distortion point of plastic materials.

plastic part is that temperature that the material will withstand over a wide variety of service conditions, without blistering, shrinking, distorting, or failing to function as required.

As a general rule, plastics are low in heat resistance. The fluoroplastic PTFE, will withstand a temperature of around 550° F, but will not stand up under very much stress at that point. Some reinforced thermosetting plastic materials (the silicones, phenolics, and epoxies) will withstand temperatures between 300 and 400° F. Most thermosetting materials can be safely used at temperatures around 250° F. Usually all thermoplastic materials are ruled out where high heat resistance is needed. Polyphenylene oxide, polysolfone, and polycarbonate are the best for temperatures between 250 to 300° F. The great majority of thermoplastic materials are not serviceable above 200° F. Table 12–12 illustrates the range of recommended continuous heat resistance temperatures for most plastics.

Vicat Softening Point of Thermoplastics

The Vicat softening point is the temperature at which a flat-ended needle of one-square-millimeter circular or square cross-section will penetrate a thermoplastic test sample to a depth of one mm. A speci-

TABLE 12–11. THE APPROXIMATE RANGE OF DEFLECTION TEMPERATURE UNDER LOAD OF SOME PLASTIC MATERIALS.

DEFLECTION TEMPERATURE OF PLASTICS

MATERIAL	264 PSI.	66 PSI.
SILICONE GLASS FILLED	500-900°F	
ALKYD GLASS FILLED	330-540	
ALKYD MINERAL FILLED	320-540	
MELAMINE CELLULOSE FILLED	350	
MELAMINE GLASS FILLED	400	
DAP UNFILLED	310	
DAP GLASS FILLED	330	
DAP ACRYLIC FIBER FILLED	225-266	
EPOXY MINERAL FILLED	250	
EPOXY GLASS FILLED	250	
PHENOLIC G.P.	260	
PHENOLIC GLASS FILLED	300	
POLYESTER GLASS FILLED	150-450	
THERMOPLASTICS		
POLYSULFONE	345	
POLYCARBONATE	265	
POLYCARBONATE GLASS FILLED	295	
PPO MODIFIED	265	
PPO MODIFIED GLASS FILLED	280	
ABS	220	
ACETAL	212	
ACETAL GLASS FILLED	315	
POLYSTYRENE G.P.	170	
POLYSTYRENE GLASS FILLED	180	
POLYSTYRENE HIGH HEAT	190	
CELLULOSICS	100-190	
POLYETHYLENE H. D.		140-190
POLYPROPYLENE	125	200-230
POLYPROPYLENE GLASS FILLED	230	305-310
NYLON 6	150	
NYLON 6 GLASS FILLED	320	
NYLON 6/6	150	
NYLON 6/6 GLASS FILLED	320	
PTFE FLUOROPLASTIC		258
PCTFE FLUOROPLASTIC		250
PVF_2 FLUOROPLASTIC	195	300

fied load on the needle and a uniform rate of temperature rise is necessary (Fig. 12–10).

To run the Vicat test, the sample is placed in the temperature controlled oil bath. The oil bath should be about 50° C below the anticipated softening point of the plastic. The temperature of the oil bath is raised 50° C per hour. A load of 1000 gr. is placed on the needle

TABLE 12–12. THE APPROXIMATE RECOMMENDED CONTINUOUS HEAT RESISTANT TEMPERATURE FOR SOME PLASTIC MATERIALS.

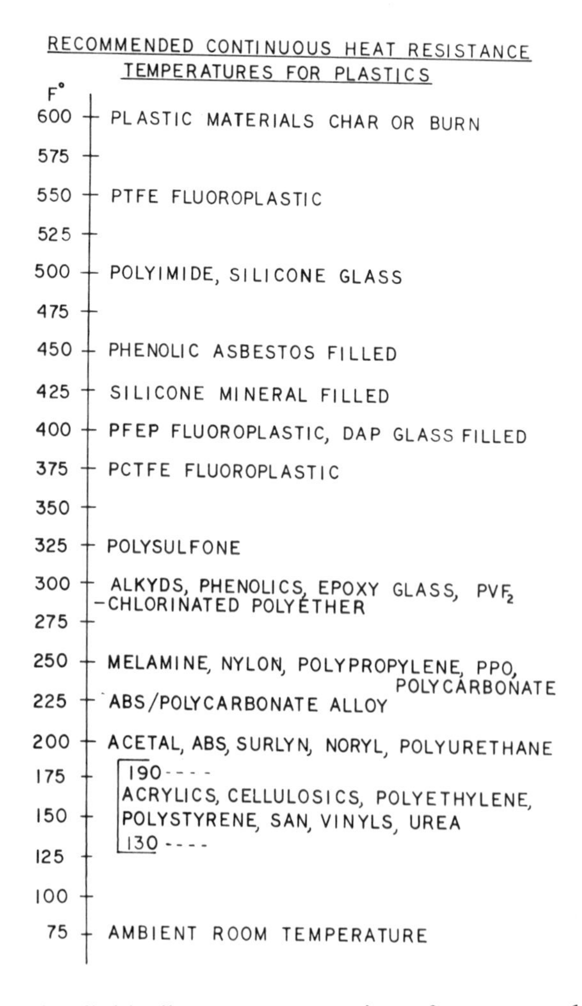

shaft. When the dial indicates a penetration of one mm., the temperature is recorded as the Vicat softening temperature for that material. This is used mostly on highly crystalline type thermoplastic materials. It is not accurate with all noncrystalline or amorphous thermoplastic materials. Some of these materials will tend to creep before the softening temperature is reached. This is a ASTM D–1525 test.

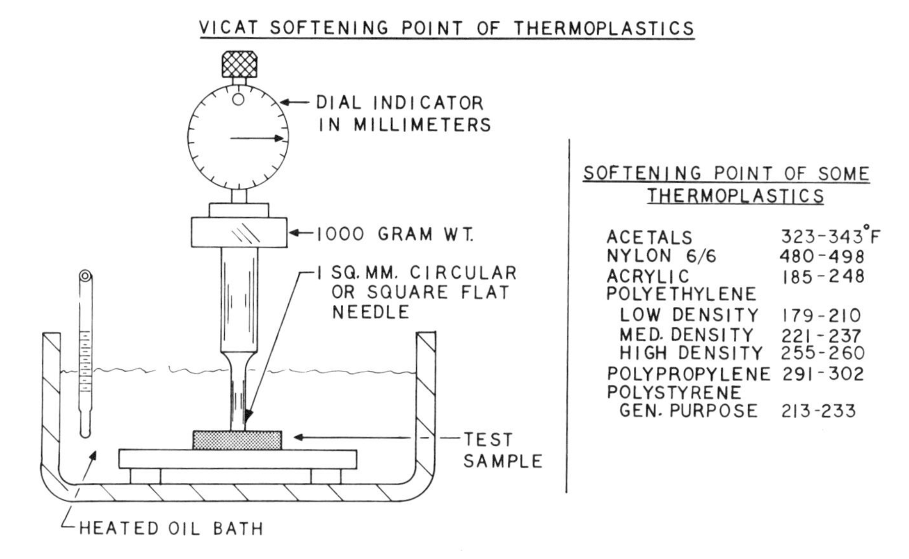

Figure 12–10. This illustrates the Vicat method of obtaining the softening point of thermoplastics.

Melting-Point Test Methods

Crystalline thermoplastic materials have a very sharp melting point, while non-crystalline or amorphous thermoplastic materials melt over a wider range.

There are three test methods used to determine the melting point of thermoplastic materials. The tests are: (1) a heated metal bar; (2) the Fisher-Johns melting point with a heated fluid, ASTM D–789, and (3) the Kofler hot stage with optical birefrigence, ASTM D–2117.

Heated Metal Bar

The heated metal bar methods of determining the melting point of a thermoplastic material is not considered to be the best, but it is an inexpensive method of testing (Fig. 12–11a). A metal bar is heated at one end, and the heat travels along the entire length in a declining manner. The sample plastic pellets are placed on the metal bar, starting at the cool end and continuing toward the hot end. As soon as a pellet melts, the temperature of the bar at that point is taken by the

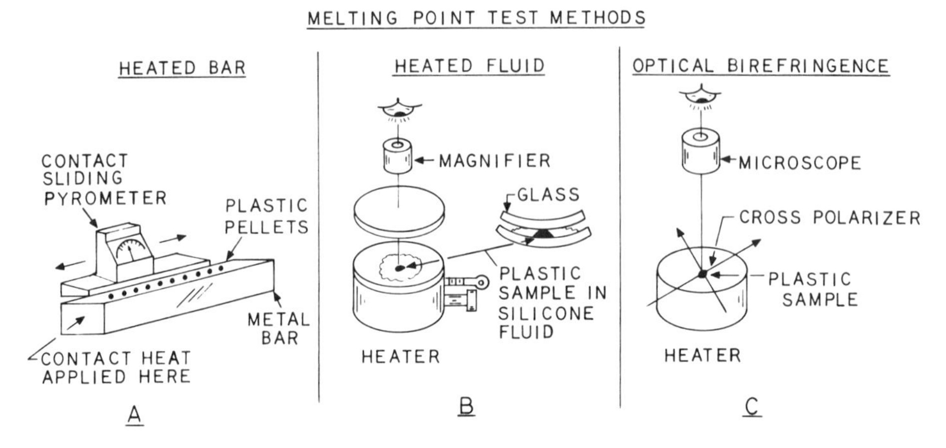

Figure 12–11. Three methods of obtaining the melting points of thermoplastic materials.

sliding pyrometer. This temperature is considered to be the melting point.

Fisher-Johns with a Heated Fluid

In this melting-point test (Fig. 12–11b), a sample pellet and a few drops of silicone oil are placed together between cover glasses. The cover glasses are placed on an aluminum stage (heater) that is heated by a current passing through a variable controlled resistance. The oil forms a miniscus on the glass. The temperature is raised until the sample pellet melts or softens enough to deform and the glass moves downward slowly. The temperature at which the miniscus moves is the melting point of the plastic. This test is described in ASTM D 789.

Kofler Hot Stage with Optical Birefrigence

This test method of determining the melting point of thermoplastic semicrystalline materials is considered to be the best of the three (Fig. 12–11c).

A sample plastic pellet is placed on a heated stage between crossed polarizers and viewed through a microscope. As the temperature rises, the crystalline plastic material melts, and the double refraction from the crystalline pellet disappears. The stressed areas in the plastic pellet yields, and the typical rainbow color viewed through the microscope disappears. It is at this moment that the melting point is taken. The test is described in ASTM D 2117.

ELECTRICAL TESTS

Arc Resistance

Arc resistance is the ability of a plastic material to resist the action of a high-voltage electrical arc. Arc resistance is usually stated in terms of time required to render the surface conductive because of carbonization by the arc flame. ASTM Test D 495 describes continuous and intermittent arcs as applied to the surface with pointed electrodes to measure the arc resistance (Fig. 12–12a). Consideration of this electrical property is important in plastic application where there is likely to be momentary flashover. A typical example may be a switch or a circuit breaker.

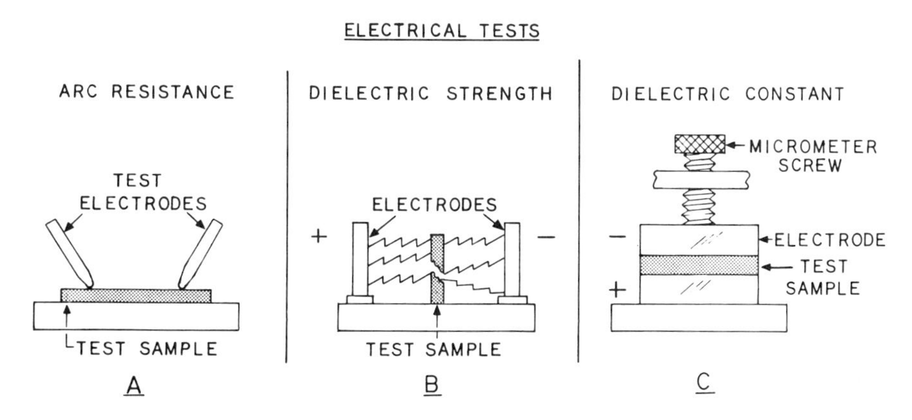

Figure 12–12. Three electrical tests: (A) Arc resistance, (B) Dielectric strength, and (C) Dielectric constant.

Dielectric Strength

If metal electrodes are placed on either side of a piece of plastic and an increasing rate of voltage is applied to the electrodes, a stage is reached where the electrical force will pass through the plastic. The voltage at which this occurs is called the "dielectric strength" and is expressed in volts per mil (one mil is one thousandth of an inch). This is illustrated in Fig. 12–12b and is an ASTM D 149 test. Sixty cycles per sec. current is assumed, since this is the standard for alternating current in this country. Higher frequencies give different results.

In practice, three methods of expressing the dielectric strength of an unknown are used. In the instantaneous method, the voltage is increased at a slow uniform rate until breakdown occurs. In the step-by-step method, the voltage is increased in steps until breakdown occurs. In the endurance method, the plastic is subjected to a high voltage for a long time. Dielectric strength represents the electrical insulating value of a plastic.

Dielectric Constant

When free air is used as the dielectric in a condenser, it is found to possess a dielectric constant of one. When air is replaced by a plastic material, values above one are obtained, and this value (in relation to air) is expressed as the dielectric constant. The dielectric constant of most plastics are very high. For example, if a plastic material had a dielectric constant of four, this would mean that the dielectric constant would be four times as high as air.

Air is used as the standard for dielectric constant, since it has the best dielectric constant property of any known material. The values for dielectric constant may vary with variation in electrical frequency. The test is illustrated in Fig. 12–12c and is an ASTM D 150 test.

Dissipation Factor

The expression "dissipation factor" (power factor), when applied to electrical insulating materials under test, is a measure of that small amount of electrical energy that is absorbed in the insulator and dissipated in the form of heat. The test is conducted at the one-megacycle frequency, except where otherwise stated.

The power factor of a dielectric is a measure of the energy loss in an alternating electric field, as determined by ASTM D 150. Since it is expressed as a percentage, it is more convenient for engineers than dissipation faction values.

Water Absorption

Water absorption is the total amount of water a plastic material will take up on immersion. It is usually reported as a percentage of the original weight. This is an ASTM D 570 test. The test sample for molded plastics is in the form of a disk two in. in diameter and .125 in. in thickness. The test sample for a sheet is three in. long, one in. wide, and the thickness of the material (Fig. 12–13a). The test sample is immersed in water for 24 hrs. A correction is made for any soluble material extracted by the water. The water absorption of plastic materials is important, because of the effect that water has on physical and electrical properties. The plastic material may warp, swell, crack, and become degraded in toughness, rigidity, and impact strength.

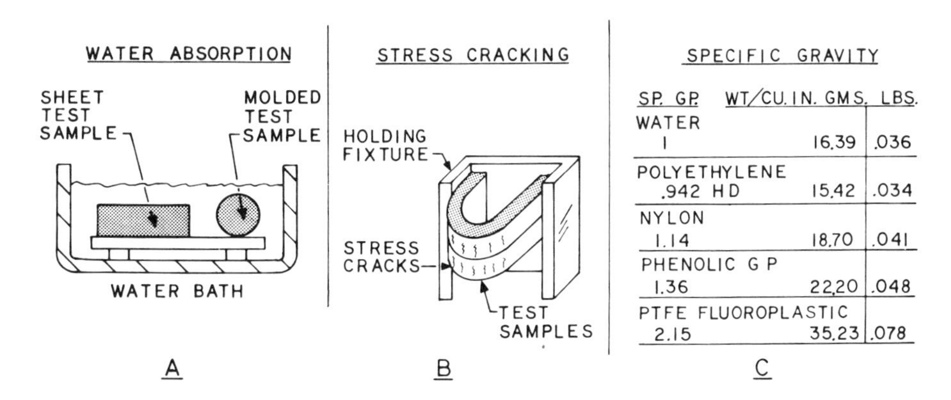

Figure 12–13. This illustrates the methods used to obtain data on water absorption, stress cracking, and specific gravity.

Stress Cracking

This is the highest stress a material can sustain without failure. The stresses that cause cracking may be present internally or externally or may be combinations of these stresses. The appearance of a network of fine cracks on a surface of a plastic material is called crazing.

Fig.12–13b shows a simple stress cracking test method. Small strips of the plastic material to be tested are bent longitudinally and inserted in the channel holding fixture. Generally, four types of results are obtained: (1) environmental stress cracking; (2) thermal stress cracking; (3) static fatigue failure; and (4) mechanical stress cracking. Environmental stress cracking is due to the chemical attack of the plastic material. Thermal stress cracking is due to elevated temperatures. Static fatigue stresses might be considered due to aging. Mechanical stress cracking is due to alternate stressing or straining of the part. This is an ASTM D 1693 test.

Fig. 12–14 shows three injection-molded polystyrene handles. The (A) handle illustrates thermal stress cracking. The (B) handle has not been exposed to any stresses. The (C) handle has been exposed to chemical stress cracking.

Specific Gravity

This is a measure of the comparative weight of a material. It is the ratio of the weight of a given volume of a substance to that of an equal volume of water at the same temperature. Generally, water is taken as a standard, so that the ratio of the weight of a material to that of an equal volume of water as a specified temperature is its specific gravity. A cubic centimeter of water at plus 4° C weights one gr. This natural standard of volume and weight is the basis with which the weight of any other material at equal volume and temperature is compared. Fig. 12–13c compares the specific gravity of water with that of a few common plastic materials and shows the different weights in grams and pounds.

The specific gravity of plastic materials can be employed as a control on compounding and fabricating conditions, as well as the identification property for the plastic material. This is an ASTM test number D 792.

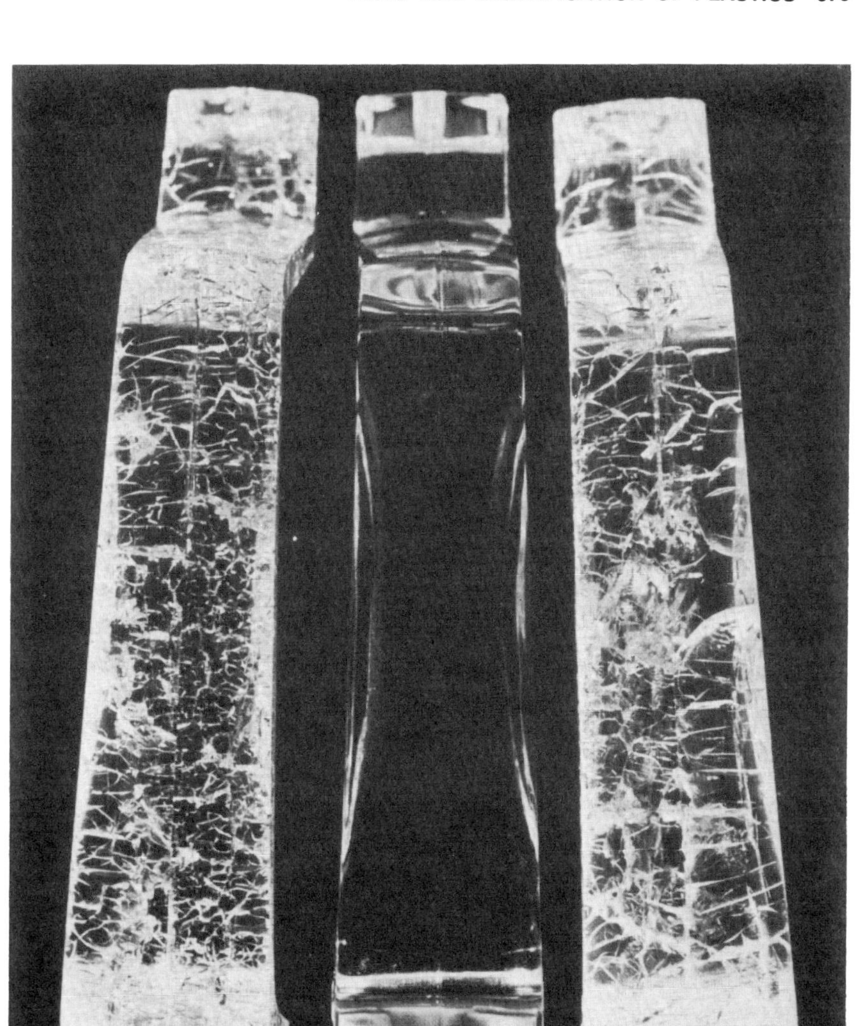

Figure 12–14. Three injection molded polystyrene handles. Handle (A) shows thermal stress cracking. Handle (B) has not been exposed to any stresses. Handle (C) has been exposed to chemical stress cracking.

Resin Moisture Test

This is a process of very quickly determining whether moisture-sensitive thermoplastic pellets are dry and ready for processing.

An electric hot plate and two glass slides are used (Fig. 12–15). The plastic pellets are placed between slides and pressure is applied until the melted pellets flatten out to about .500 in. in diameter. If bubbles appear moisture is present. If no bubbles appear the plastic pellets are dry and ready for molding. One or two small bubbles may be only trapped air.

Toughness

Toughness is important in plastic design engineering, but the term has not been clearly defined. It is bound up with many physical tests. It may be measured by the energy required to rupture a specimen. The design engineer should select a plastic material for a particular application on an over-all physical test basis. Fig. 12–16 illustrates the fundamental makeup of plastic materials from rigid to flexible. It should be noted that a material with high tensile strength will be rigid, and a material with a high percentage of elongation will be flexible. The designer should strive for an overall balance of material properties. Perhaps a tough plastic material will be better than either a rigid or flexible material.

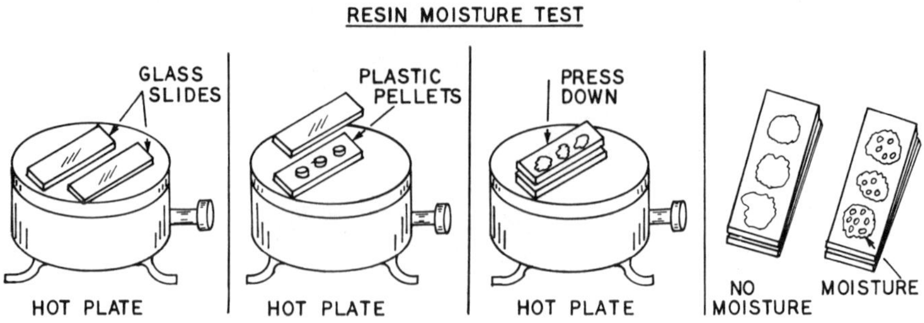

Figure 12–15. This illustrates a process of determining the presence of moisture in thermoplastic pellets.

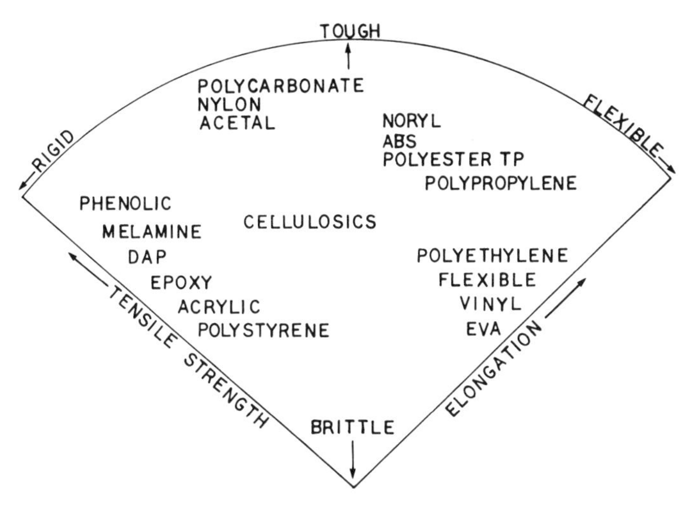

Figure 12–16. A comparison of many plastic materials is shown in relation to their physical make-up.

Polariscope

Most clear plastic materials exhibit stress-optical properties. This means that they show an internal pattern when subjected to internally or externally applied stress and viewed in polarized light.

A very simple and inexpensive polariscope can be made to study the stress concentrations in transparent fabricated or molded plastic parts (Fig. 12–17a). The polarizing filters can be purchased from an engineering material supply company. Two 8.5 x 11 in. sheets in the film form are adequate to make the polariscope. A 200-watt light bulb can be used for the unpolarized light source.

In a simple polariscope, the unpolarized light passes through the first polarized filter and becomes plane polarized. This means that the light vibrates in one plane only. The light next passes through the stressed plastic molded part, which causes the light to become doubly polarized. This means that the light is split into two components, one of which is generally out of phase. If there is a difference, it is caused by the stresses in the plastic part. The two components impinge on the second polarizing filter. This polarizing filter is the same as the first, except that it is turned 90°. The observer will see a fringe pattern where the stress concentrations are located in the plastic part.

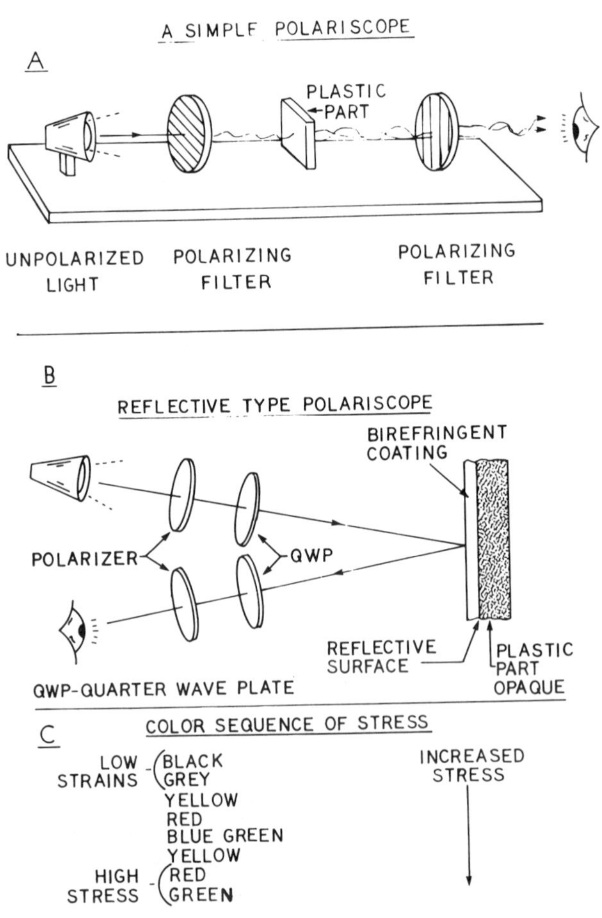

Figure 12–17. This illustrates the general arrangement of a simple polariscope and a reflective type polariscope. The color sequence of stress is shown.

Photoelastic analysis is now being used to physically test thermoplastics as well as thermosets, by using a polariscope. Mechanical failures of a plastic part can generally be determined in design, processing, and assembly, by using this type of stress analysis. A reflective coating is applied first to the surface of the plastic part (Fig. 12–17b). Next the transparent coating is applied over the reflective coating. Since the transparent coating is bonded securely to the surface of the plastic part, the coating will experience the same strains as does the surface of the part. When typical service stresses are applied to an actual part, the optical properties of the transparent coating change in direct proportion to the strains developed in the part. The results can be analyzed under the colorful patterns developed by a polariscope.

The photoelastic color sequence shows stress distribution in the part. In order of increasing stress, the sequence is black, gray, yellow, red, blue-green, yellow, red and green (Fig. 12–17c). The black and gray areas show low strains while a continued repetition of red and green color bands indicates extremely high concentrations of stress. As the stress concentration increases, the number of black bands in a fringe order also increases. The wider the rainbow color the greater the stress.

COEFFICIENT OF THERMAL EXPANSION

All materials change in dimension as the temperature changes. This property represents the increase in size, length, area, or volume of plastic per unit of temperature rise. The thermal expansion values for most plastic materials are relatively high. They expand and contract with changes in temperatures about five times as much as mild steel.

In ASTM Test D 696, the change in length of a bar or rod is measured in a dialatometer over the desired temperature range. The change in volume (coefficient of cubical expansion) is determined in ASTM Test D 864 in a dialatometer using mercury. Table 12–13 shows the approximate range of the coefficient of linear thermal expansion of some plastics. Note that with the addition of fiberglass, the thermal expansion in all cases is reduced. The average plastic material will expand from 0.01 to 0.2 mils per inch per degree rise Centigrade.

Thermal Conductivity

Thermal conductivity is the ability of a given substance to conduct heat. Not all substances are good conductors of heat. Plastic materials are excellent thermal insulators. They have low thermal conductivity. Metals like silver and copper transmit over 2000 times as much

TABLE 12–13. APPROXIMATE RANGE OF THE COEFFICIENT OF THERMAL EXPANSION OF SOME PLASTICS. NOTE THAT WITH THE ADDITION OF FIBER GLASS THE THERMAL EXPANSION IN ALL CASES IS REDUCED.

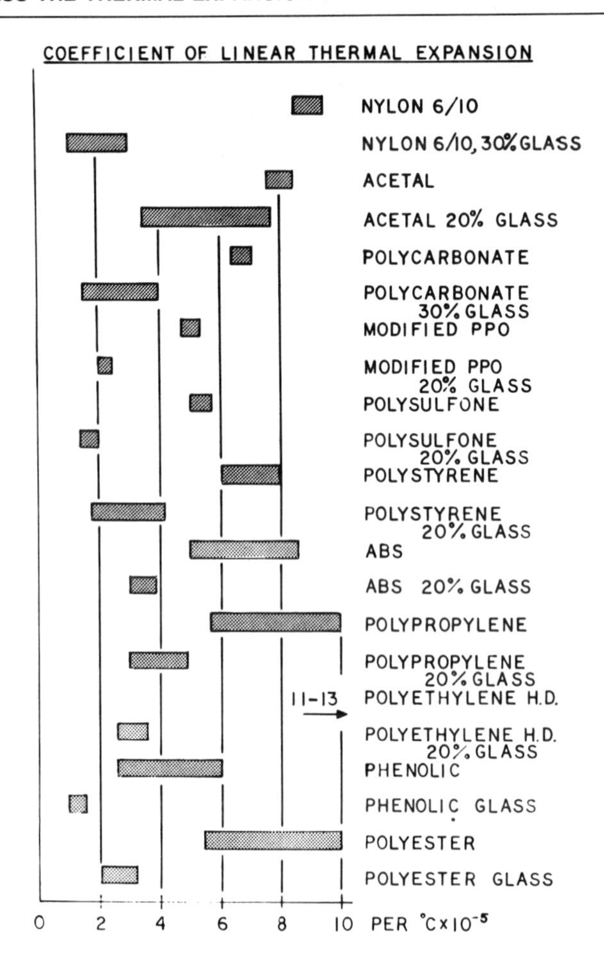

heat as most plastics, asbestos four times as much, and wood and paper about twice as much. Since plastic materials are excellent thermal insulators, unlike metals, they are pleasant to the touch when cold. Also, a plastic handle on a hot pan transmits relatively little heat from the pan to the hand. The test for thermal conductivity of materials is ASTM C 177.

OPTICAL PROPERTIES

Plastic materials that have high-light transmission and low-haze properties are used in optical applications. The advantages that plastic materials have over glass in optical uses are low density, high impact strength, and high light transmission. Their low density makes it possible to produce binoculars that are 30% less in weight than equivalent binoculars with glass prisms. The acrylic resins are the most widely used in optical applications, although polystyrene and allyl resins (CR–39) are used.

LIGHTING WITH ACRYLICS

The optical properties of methyl methacrylate (acrylics) are the main reasons for its wide acceptance. It is transparent with excellent clarity. Acrylic resins are more transparent than the best glass and transmit more light. The index of refraction of acrylic resin is 1.48. Refractive index is the ratio of the velocity of light in a vacuum (or air) to the velocity of light in the acrylic material. Glass has an index of refraction of 1.52 to 1.89 and air or vacuum is 1.00.

Acrylic plastic parts can be edge lighted, back lighted, flood lighted, or illuminated by "piping" the light through the part. Edge lighting is fundamentally the same effect as light piping, but is used in a different way. Light entering one side of a sheet stays in an acrylic sheet until it reaches the opposite edge, unless the polish of the surface along the way is broken. Any point where the polish is disturbed, light will escape. If a design is placed on the acrylic sheet either painted, engraved, sanded, or carved, the light will be emitted from the design and it will glow.

The excellent internal reflection and transmission of light, makes acrylic material good for pipe lighting. Its ability to "pipe" light around curves, plus its edge-lighting characteristics, can be used to advantages where lighting effects are important. Light pipes are not suitable for transmission of image, but only for quantities of light. Refer to Fig. 12–18, which illustrates some of the following principles of lighting with acrylic resins.

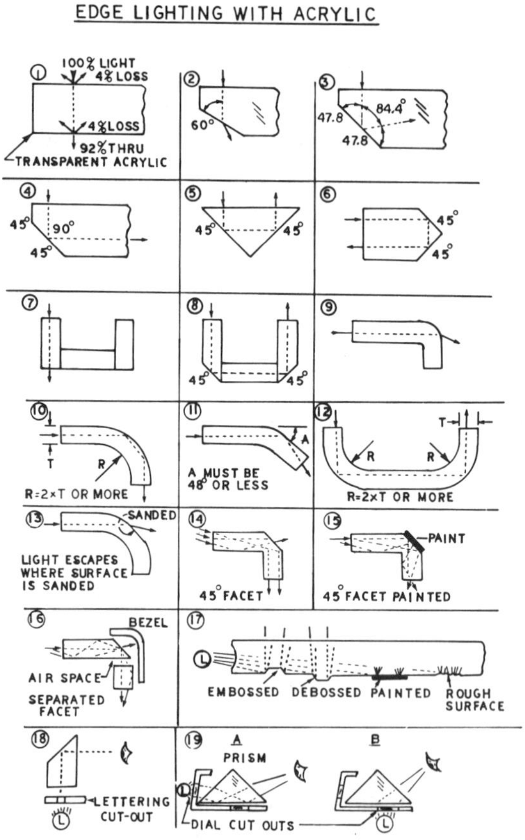

EDGE LIGHTING WITH ACRYLIC

Figure 12–18. Edge lighting with acrylics. Light can be piped through methyl methacrylate in many different ways.

1. If a light ray strikes an acrylic surface at right angles, approximately 92% of the light will emerge from the opposite surface; 4% will be lost through scattering at each interface.
2. If a ray of light strikes an acrylic surface as shown in the drawing, it will pass on through.
3. This is the first opaque angle in acrylics. Any ray of light within acrylic plastic that encounters an air interface at the critical angle (an angle of 42.2° of the normal or greater) will not be transmitted. It will be totally reflected back into the material at an equal and opposite angle.

4. This is most often used for transmitting light.
5. This illustrates a facetted prism. It is sometimes called the "diamond" principle.
6. A horizontal two-faced right-angle facet.
7. In this example, three pieces of acrylic are placed together in the shape of a box. Light enters the top edge and escapes at the base.
8. In this example, three pieces of acrylic are placed together in the shape of a box. The three pieces of plastic will pipe light, if the two outside members have 45° facets.
9. In order to pipe light around a corner, a definite curved section must be used. A sharp right angle will not pipe light. The radius must be as shown in example 10.
10. In order to pipe light around a radius, the minimum inside radius must be at least twice the thickness of the part.
11. Acrylic material can pipe light around angles and curves with great efficiency, but the limiting angle is 48°, as shown in the example.
12. Light may be piped in and out of an acrylic material, if the radii are two or more times the thickness of the piece.
13. A roughened or sanded surface will not reflect light like a polished surface. Frosting, sand blasting, or scratching the polished reflected surfaces of an acrylic part will cause the rays to scatter as they strike the locally abraded area.
14. Light can be piped around a sharp corner by using a 45° facet. The facet should be highly polished.
15. Light can be piped around a sharp corner by using a 45° facet. The facet should be painted white to minimize the loss of light.
16. Internal light can be piped around a sharp corner by using a separated facet. A bezel is used to restrict any stray light.
17. Internal light in a plastic part is interrupted, the light will be diffused in a different direction. This principle is used in signs and can be accomplished by having raised or depressed areas, painted surfaces, or roughened surfaces. Embossing, debossing, etching, and faceting are other common techniques used to control internal lighting.

18. A viewer can see an object through an acrylic prism when the light rays strike the beveled surface on the prism at 45° and are reflected internally into the line of sight.
19. Two views from one prism. Two views of different objects can be seen from one prism, as illustrated in the example. In the "A" view, the viewer will see the dial cutout object because of total internal reflection. In the "B" view, the viewer will see the dial cutout object through the front and bottom surfaces of the prism. The viewer is looking from different positions in both cases, but through a single prism.

IDENTIFICATION OF PLASTICS

It is difficult, even with experience, to identify more than a few plastics by visual inspection or a simple mechanical test. A systematic chemical analysis of plastic materials is perhaps the only method of true identification. One of the best test methods is the infrared spectrophotometer. However, some very simple tests have been worked out for quick tentative checks. The best results are obtained when the unknown plastic is a simple plastic type. If the unknown is a copolymer or a mixture of several polymers the results may be ambiguous. Fillers, plasticizers, and other additives all have some effect upon the test results.

In order to identify an unknown plastic material, it is always a good idea to use a control material of a plastic whose identities are known. The control sample should be run along with the unknown and comparison made of the behavior of this known control with that of the unknown plastic sample.

The first step is to determine whether the unknown is thermosetting or thermoplastic. The best method to make this distinction is to heat a small piece of the plastic on a hot plate or in a test tube. If the sample darkens and decomposes without softening, it must be a thermosetting plastic. If the plastic sample melts, it probably is a thermoplastic. Another method is to press a hot glass stirring rod or soldering iron firmly against the sample. The hot glass rod will indent the plastic surface if it is a thermoplastic resin, but will not soften the thermosetting resin.

Ordinary techniques and simple, inexpensive, readily available tools can be used to conduct the following tests:

1. Burning
2. Melting point
3. Specific gravity
4. Solubility
5. Copper wire test

Almost all plastics will burn when exposed to a flame. The burning characteristics to be noted are: odor, color of the flame, and melt behavior. A Bunsen burner or small propane torch can be used. Hold the unknown test sample just to the edge of the nonluminous flame until it ignites. Note the character of the flame, and cautiously smell any odor produced after the flame from the burning plastic has been extinguished.

The melting point can be determined by using a Fisher-Johns melting apparatus (see Fig. 12–11b). A control sample is helpful, because all plastic materials do not have the same melt characteristics. Crystalline plastic materials have a sharp melting point, while amorphous materials have a gradual melting point. Fluoroplastic resins are difficult to observe, because the melt does not become fluid. Thermosetting materials do not melt or have melting points.

The specific gravity of an unknown plastic can be determined by using an analytical balance and a beaker of water. The operating temperature should be around 23° C.

The solubility test will usually serve to determine what is anticipated from earlier tests. Each thermoplastic material will be attacked or dissolved in a definite chemical solvent. Place the plastic sample into a test tube and pour in enough solvent to cover the sample. It may require a few minutes to an hour to dissolve some plastic materials.

Plastic materials that contain chlorine may be identified by using a flame and a copper wire. Touch the heated wire to the polymer in such a way as to retain some of the polymer on the wire, then return the wire to the flame. A green flame indicates the presence of chlorine. Chlorinated polyether, vinyl chloride, vinyl chloride-vinyl acetate, vinylidene chloride polymers, and fluorochlorocarbon polymers give positive results in this test.

IDENTIFICATION TEST RESULTS OF THERMOPLASTIC MATERIALS

Acetals

This rigid material will burn with a pale blue flame and give off an odor of formaldehyde. The dripping may also burn. The crystalline melting point is around 347° F. The specific gravity is 1.42.

Acrylic (Methyl Methacrylate)

This plastic polymer will burn slowly with a blue flame with a yellow tip. It has a fruit-like odor. The melting point is 374° F. The specific gravity is 1.18. It is soluble in acetone, benzene, and ethyl acetate.

Acrylonitrile Butadiene Styrene (ABS)

This rigid plastic material is opaque and will burn with a yellow flame, black smoke, and drip. It has an acrid odor with an additional smell of burning rubber. It has a specific gravity of 1.04. It is soluble in acetone, ethylene dichloride, and trichloromethane. When SAN is burned, it gives off a characteristic odor of styrene with an additional smell which is bitter.

Cellulose Acetate Butyrate

This cellulosic material burns with a blue flame and gives off an odor of rancid butter. It has wide range of melting points, depending upon its composition. The melting point is around 356° F, and it has a specific gravity of 1.24. It is soluble in acetone and trichloromethane.

Fluorocarbons or Fluoroplastics

This family of plastic materials does not burn in a flame, but it does deform. The fluoroplastics are extremely inert, and the best method of identification is by fluorine detection. PTFE has a melting point of

621° F and is the only fluoroplastic that will not flow above its melting point. The other fluoroplastics will flow above their melting points. PTFE has a specific gravity of 2.14 to 2.20 and will volatilize at about 750° F. This material has a milky, waxy appearance and a "slippery" feel. PFEP has a melting point of 554° F and a specific gravity of 2.12 to 2.17. PCTFE has a melting point of 410° F and a specific gravity of 2.1 to 2.2. PVF_2 has a crystalline melting point of 340° and a specific gravity of 1.75 to 1.78. When exposed to a flame, it degrades essentially to hydrogen fluoride gas and a carbon residue, without dripping. It is attacked by fuming sulfuric acid at room temperatures. It will swell in acetone.

Nylon

Nylons or polyamides burn with a blue flame with a yellow tip. They give off an odor of burned wool, burned hair, or burning vegetation. The melting point can range from 293 to 489° F, and the specific gravity can range from 1.04 to 1.17. Nylons are unaffected by common solvents and are attacked by strong acids only. It is soluble in phenol and in formic acid.

Polycarbonate

This thermoplastic material burns with difficulty, has a yellow flame, and is self extinguishing. It has a weak odor of phenol. It has a melting point around 430° F and a specific gravity of 1.2. It is soluble in methylene chloride and ethylene dichloride.

Polyimide

This plastic material is classified as a thermoplastic, but it has all the characteristics of a thermoset. It has a specific gravity of 1.43. It is non-burning and is attacked by strong alkalies, nitrogen tetra oxide, and hydrazine. It gives off a faint odor of benzene when removed from a flame.

Polyvinyl Chloride

This plastic material burns with a yellow flame with green at the edges. It gives off a bright green flame with the copper wire test. It gives off a pungent irritating odor of hydrochloric acid. This polymer has a melting point near 302° F and a specific gravity of 1.16 to 1.72. It is soluble in acetone and cyclohexanone. Polyvinyl chloride has an ignition temperature of 735° F.

Polyethylene

This polyolefin plastic material burns at a fairly rapid rate and melts and drips like wax. The flame is blue with a yellow tip. It has an odor of burning paraffin wax. Low density polyethylene has a melting point of 221° F, and high density polyethylene has a melting point of 248° F. The density of polyethylene is less than one, and it will float in water. It is soluble in hot benzene or hot toluene, and insoluble in the other commonly used solvents. It is easy to identify by its waxy feel and natural milky-white appearance. Polyethylene has an ignition temperature of 645° F.

Polypropylene

This polyolefin plastic material burns slowly with a yellow tipped blue flame. The odor is similar to diesel fumes. It has a melting point around 334° F and a specific gravity of 0.906. This material will float in water. It is soluble in hot toluene. It is more rigid than polyethylene.

Polystyrene

Polystyrene burns with an orange-yellow flame that produces dense smoke with clumps of carbon in the air. It smells like illuminating gas or marigolds. Polystyrene has a melting point of 374° F and a density of 1.09. It is soluble in acetone, benzene, ether, and trichloromethane. Polystyrene will give a metallic ring when dropped or tapped with a hard object. Polystyrene has an ignition temperature of 680° F.

Polyphenylene Oxide (PPO)

This material has a melting point around 504° F and a specific gravity of 1.06. It is self-extinguishing and non-dripping. PPO has an odor of illuminating gas when burned. It is soluble in toluene and dichloroethylene. Modified PPO is opaque and gives off a slight odor of rubber when it is burned. Modified PPO has a melting point around 500° F.

Polysulfone

This thermoplastic material has a melting point around 500° F and a specific gravity of 1.24. It burns with a white flame and gives off an odor of sulfur. It gives a metallic ring when dropped or struck with a hard object.

IDENTIFICATION TEST RESULTS OF THERMOSETTING PLASTIC MATERIALS

Diallyl Phthalate (DAP)

The DAP filled molding compounds have a specific gravity of 1.34 to 1.78. DAP is self-extinguishing to non-burning. (See test of DAP, Alkyd, Phenolic, or Epoxy).

Epoxy

This thermosetting material burns with a yellow flame and in most cases will give off a sharp irritating amine odor. It has a specific gravity of 1.11 to 2.10. It is affected slightly by chlorinated hydrocarbons and ketones. (See test for DAP, Alkyd, Phenolic, or Epoxy).

Phenol Formaldehyde

This material burns with difficulty and is often self-extinguishing, depending on the filler. The material cracks when burned and gives off an odor of phenol and decomposed filler. Phenolic materials are generally black brown, or a mottle of black and brown. Sometimes a dark green color is used. (See test for DAP, Alkyd, Phenolic, or Epoxy).

Urea Formaldehyde

Urea formaldehyde gives a strong odor of formaldehyde and also a musty odor when it is burned. Urea molding materials have a specific gravity of 1.47 to 1.52.

Melamine Formaldehyde

This plastic gives off a fishlike odor when burned. It is self-extinguishing and cracks when burning.

Polyester (Alkyds)

Reinforces polyester molded plastics (unsaturated polyesters) burn with a yellow flame with black soot and have a sour cinnamon odor. Saturated polyesters are widely used in films and fibers. They burn with a yellow flame with black soot and have a mild sweet odor. Polyesters modified with fatty acids are called alkyds. The molded alkyds are difficult to identify, because the burned odor is generally that of the filler that is used. The alkyds are molded in many colors and will show a white mark when scratched with a sharp instrument. (See test for DAP, Alkyd, Phenolic, or Epoxy).

Silicones

When a silicone is burned, it will show a white ash. The flexible silicones have a specific gravity of 1.05 to 1.23. The filled silicones have a specific gravity of 1.18 to 2.82. A silicone material will show the SiO_2 bond in an infrared spectophotometer.

TEST FOR DAP, ALKYD, PHENOLIC, OR EPOXY

The test for an unknown thermosetting material that might be DAP, alkyd, phenolic, or epoxy is determined by the following procedure. The molded unknown sample to be tested should be pulverized into small pieces. Place some of the unknown sample on a hot plate that has a temperature between 500 to 600° F. Over the hot plate, hold a piece of aluminum foil .500 in. from the sample. Attach one end of the aluminum foil to a cooling medium (between two pieces of metal). The other end of the foil is free. A white thin cloudy film will form on the bright surface of the aluminum foil if it is DAP of alkyd. If nothing forms on the aluminum, it is an epoxy or phenolic.

If a white cloudy film has formed on the aluminum foil, part of the sample should be refluxed in a 20 to 30% solution on sodium hydroxide for one to two hrs. If the sample degrades, it is an alkyd. If it does not degrade it is DAP.

If no white cloudy film has formed on the aluminum foil, the unknown is a phenolic or an epoxy. Reflux part of the sample in a 20 to 30% solution of sodium hydroxide for one to two hrs. If the unknown sample degrades, it is a phenolic. If it does not degrade, it is an epoxy.

Fiberglass and Asbestos Fillers

The fiberglass and asbestos fillers in plastic molded articles can be detected by burning or ashing the molded part in a flame. The molded part should be placed on a fine wire mesh screen over a large gas burner. The set up should be placed in a hood, as large amounts of unburned carbon or soot will be given off. After the plastic has been burned, the fiberglass or asbestos filler pattern can be studied by looking at the completely burned sample.

Glossary of Technical Words

Accelerated Aging. Aging by artificial means to obtain an indication on how a material will behave under normal conditions over a prolonged period.

Acetal Resins. A crystalline type thermoplastic material made from formaldehyde. Trade names: Delrin and Celcon

Acrylics. The name given to plastics that are produced by the polymerization of acrylic acid derivatives, usually including methyl methacrylate. An amorphous type thermoplastic material.

Acrylonitrile. A crystalline type thermoplastic. This material is used mostly in the making of a synthetic fiber.

Acrylonitrile, Butadiene, Styrene. (Abbreviated ABS.) A thermoplastic classifed as an elastomer-modified styrene.

Aging. The degradation of a plastic material with time. The environmental conditions must be defined.

Alkyd Resins. The name given to synthetic resins processed from polyhydric alcohols and polybasic acid or anhydrides. Alkyds are thermosetting resins.

Alloy. A word used to denote blends of polymers or copolymers with other polymers.

Amorphous. Plastic materials that have no definite order of crystallinity.

Antistatic Agent. A chemical material added to plastic compounds in order to minimize or eliminate static electricity.

Automatic Mold. A mold or die in injection or compression molding that repeatedly goes through the entire cycle without human assistance.

Back Pressure. In molding, it is the resistance of the material to flow when the mold is closing. In extrusion, it is the resistance of the plastic material to forward flow.

Back Taper. Sometimes called back draft or reverse draft. It is used in a mold to prevent the molded part from drawing freely.

Blind hole. A hole that is not drilled or molded entirely through.

Blister. An undesirable enclosed raised spot on the surface of a plastic piece, resembling in shape a blister on the human skin.

Boss. A small projection above the general surface of a part to form a seating or reinforcement for another part.

Bulk Factor. The ratio of the volume of loose molding powder to that of the volume of the molded article.

Burned. A carbonized condition on the surface of a plastic piece. The plastic will show discoloration and distortion.

Butadiene. A synthetic rubber used in butadiene-styrene, butadiene-acrylonitrile, and acrylonitrile-butadiene-styrene.

Buttress Thread. A type of thread used for transmitting power in only one direction. It has the efficiency of the square thread and the strength of the V-thread.

Casting Resin. A process by which liquid plastic resins are poured into a mold and cured by a catalyst or by a catalyst and heat.

Cementing. A process of joining two like plastic materials to themselves or to unlike materials by means of solvents.

Chlorinated Polyether. A crystalline type thermoplastic. It has extremely good resistance to heat and chemicals.

Chrome Plating. An electrolytic process that deposits a hard film of chrome on the working surfaces of metal molds.

Coefficient of Thermal Expansion. The increase in length or volume of a plastic material for a unit change in temperature.

Cold Flow. A plastic is said to exhibit cold flow when it does not return to its original dimensions after being subjected to stress. The term "creep" is sometimes used.

Color-fast. The ability to resist change in color.

Cooling Channels. Channels or ports in a metal mold through which a cooling medium can be circulated to control temperature on the mold surface.

Cooling Fixture. A block of steel, wood, or composite material that is made similar to the shape of the molded piece so that the hot molded article, just taken from the mold, can be placed on it and allowed to cool, without distorting. Also known as a shrink fixture.

Copolymer. A polymer whose molecules are made up of more than one type monomer.

Core. The portion of a mold that shapes the interior of a hollow molded part.

Core Pin. A pin for forming a hole or opening in a plastic molded piece.

Crazing. Fine cracks that extend in an irregular pattern perpendicular to the surface.

Cross-Linking. The chemical combination of molecules to form thermosetting resins.

Crystallinity. An ordered arrangement and regular recurrence of molecular structures in a polymer.

Cure. That portion of the molding cycle during which the plastic material in the mold becomes sufficiently rigid or hard to permit ejection.

DAP. Abbreviation for diallyl phthalate.

DIAP. Abbreviation for diallyl isophthalate.

Deflashing. The removal of excess plastic material that is forced from a mold cavity during the closing operation. It may also occur between worn mold sections.

Design Stress. A long-term stress, including creep factors and safety factors, that is used in designing structural fabrication.

Die. A metal form in making or punching plastic products. It is used interchangeably with mold.

Dimensional Stability. The ability of a plastic part to retain precise dimensions in which it was molded.

Double-Shot Molding. A method of producing two-color pieces in thermoplastic materials by successive injection-molding operations. The part molded first becomes an insert for the second molding.

Draft. A taper or slope in a mold required to facilitate removal of the molded piece from the mold. The opposite of this is called back draft.

Ejector Pin. A pin (in a mold) that directly pushes the molded part out of the cavity or off the punch. Sometimes called knockout pin. A number of these pins may be used in one mold.

Elastic Deformation. A deformation in which a substance goes back to its original dimensions on release of the deforming stress.

Elastomer. A rubber-like material that stretches at room temperature (under load) to at least twice its length and snaps back to the original length.

Elongation. The increase in length of a material under test, expressed as a percentage difference between the original length and the length at the moment of the break.

Encapsulation. Surrounding of a unit with a uniform coating of resin.

Epoxy Resins. Epoxy resins are thermosetting materials. When converted by curing agents, the thermosetting resins become hard infusible systems.

Ethylene-Vinyl Acetate. A plastic copolymer made from the two monomers of ethylene and vinyl acetate. This copolymer is similar to polyethylene, but it has considerable increased flexibility.

EVA. Ethylene-vinyl acetate.

Extrudate. A word used in extrusion to note the product or material delivered by the extruder. This may be in the form of a film, a rod, a pipe, etc.

FEP. Abbreviation for fluorinated ethylene propylene.

Fillet. A rounded inside corner of a plastic piece. The rounded outside corner is called a bevel.

Flame Treating. A process of treating a thermoplastic surface so that it can be painted, inked, etc. The thermoplastic surface is bathed in an open flame to promote oxidation of the surface.

Flash. The excess plastic material that is forced from a mold cavity during the molding operation. Flash may also occur between worn mold sections.

Flash Line. Marks formed where the excess plastic material flows out of the mold.

Flexible Molds. Molds made of rubber, elastomers or flexible thermoplastics, used for casting thermosetting plastics or non-plastic materials such as plaster. The flexible molds can be stretched to permit removal of cured pieces with undercuts.

Flock. Very short fibers of cotton, wood, glass, etc., that are used as an inexpensive filler.

Flow Lines. Lines that are visible on the surface of a plastic piece made by the meeting of two flow fronts during molding. Sometimes called weld lines.

Fluidized Bed Coating. Tiny particles of thermoplastic resin are suspended in a gas stream (generally air) and behave like a liquid. A heated article is immersed in this fluidized bed of powder. The thermoplastic resin particles melt and fuse to the heated surface, forming a smooth coating.

FEP. Abbreviation for fluorinated ethylene propylene.

Fluorocarbons. Thermoplastic resins made of monomers containing fluorine and carbon. Sometimes called fluoroplastics.

Force. The male part of a mold which enters the female or die section. The "force" (sometimes called the punch) exerts pressure on the material and causes it to flow when the mold closes.

Gate. An orifice or opening through which the plastic melt material enters the cavity. This is in injection and transfer molding.

Graining. This refers to wood graining on plastics. This can be done by hand, roller coating, hot stamping, or printing.

Guide Pins. Generally, round steel pins that maintain proper alignment of the plunger and cavity as the mold closes.

Gusset. An angular piece of material used to support or strengthen two adjoining walls.

Hand Molds. Molds that are removed from the press by the operator, who opens the mold and extracts the part by hand.

HMWPE. Abbreviation for high molecular weight polyethylene.

Hob. A master steel block or punch used to sink a desired mold shape into mild machine steel.

Hydroscopic. Tending to absorb moisture.

Ionomer. A thermoplastic that has polyethylene as its major component. It contains covalent and ionic bonds. The metallic cations are sodium, potassium, and magnesium.

Isocyanate Resins. Generally known as urethane chemistry.

Laminar Flow. The movement of one layer of fluid plastic material past another, with no transfer of matter from one to the other. This type of flow in molding is necessary to duplicate the mold surface.

Land. That part of a mold where the two halves come together flush. The bearing surface along the top of the flights of a screw in an extruder. The surface of an extrusion die that is parallel to the direction of melt flow.

Low-pressure Laminate. Laminated and molded materials cured in the pressure range of 25 to 400 psi.

Mandrel. A metal bar used as a core around which fabrics are wound. In the plastic industry, it is used chiefly in making laminated tubes. In extrusion, it is the central member of a tubing die.

Matched Die Molding. A process of molding reinforced plastic parts. The punch and cavity have a telescoping fit or a matched metal shearing area for the cut-off. Sometimes called matched metal molding.

Melt Fracture. An elastic strain set up in a molten polymer as the polymer flows through the die. It shows up in irregularities on the surface of the plastic.

Methyl Methacrylate. An amorphous type thermoplastic resin. A common name is acrylic resin.

Methylpetene. A thermoplastic resin with a specific gravity of 0.83, the lowest of any plastic.

Micron. A unit of length equal to 10,000 A.

Modulus of Elasticity. The measure of the ratio of the stress to the strain in a material, within the elastic limits of the material.

Molding Pressure. The force necessary to cause the plastic material to flow into a mold.

Mold Shrinkage. Refers not to the shrinkage of the mold, but to the immediate shrinkage of the plastic part after it has been removed from the mold. The mold and molded part are measured at room temperature.

Non-polar. Incapable of having a significant dielectric loss. Polystyrene and polyethylene are non-polar.

Notch Sensitive. A plastic material is said to be notch sensitive if it will break when it has been scratched, notched, or cracked. Glass is considered to be highly notch sensitive.

Nylon. A generic name for polyamides. A crystalline-type thermoplastic.

Organic. Refers to the chemistry of carbon compounds.

Orientation. A process of stretching a heated plastic article to realign the molecular configuration.

Parison. The hollow plastic tube from which a part is blow molded.

Parallel to the Draw. The axis of the cored position (hole) or insert is parallel to the up and down movement of the mold as it opens and closes.

PCTFE. Abbreviation for polychlorotrifluoroethylene.

Phenolic Resin. A family group of thermosetting resins made by the reaction between phenol and aldehydes.

Phenolics. A generic term applied to the entire group of phenolformaldehyde plastic resins.

Plastic. Any synthetic organic compound that can be permanently shaped under the influence of heat and pressure. This does not include rubbers or elastomers.

Plasticizers. An organic material added to a plastic in order to increase its flexibility or workability.

Polyamides. A group of crystalline thermoplastics of which nylon is typical.

Polyimide. Classified as a thermoplastic, it can not be processed by conventional molding methods. The polymer has rings of four carbon atoms tightly bound together. It has excellent resistance to heat.

Polyphenylene Oxide. (PPO) an amorphous type thermoplastic. This material is noted for its useful temperature range from -275° to 375° F.

PTFE. Polytetrafluoroethylene. A crystalline-type thermoplastic. The oldest of the fluorocarbons or fluoroplastics. Common trade name of "Teflon."

PVC. Polyvinyl Chloride.

Polyvinylidene Chloride. A crystalline type thermoplastic material composed of polymers of vinylidene chloride. The polymer is generally made into a copolymer with vinyl chloride. Also known as "Saran."

PVF_2. Polyvinylidene fluoride. A member of the fluorocarbon family. A crystalline type thermoplastic material that can be processed on conventional equipment.

Powder Molding. A general term used to indicate several types of molding. The types may be rotational molding, slush molding, centrifugal molding, fluidized bed coating, and sinter molding.

Prototype Mold. A temporary or experimental mold construction made in order to obtain information on part design, tool design, and market reactions.

PVF. Abbreviation for polyvinyl fluoride.

PVF_2. Abbreviation for polyvinylidene fluoride.

Residual Stress. The stresses remaining in a plastic part as a result of thermal or mechanical treatment.

RTV. Abbreviation for room temperature vulcanizing.

SAN. An abbreviation for styrene-acrylonitrile copolymers.

Shrinkage Allowance. The additional dimensions that must be added to a mold to compensate for shrinkage of the plastic molding material on cooling.

Shrink Fixture. A device on which a molded part is placed immediately after it comes from the mold. The shrink fixture aids in holding closer dimensions on the molded part.

Sink Mark. A depression on the surface of a molded part caused by internal shrinkage.

Terpolymer. Three different monomers polymerized to form one plastic resin.

TFE. Abbreviation for tetrafluorethylene.

Vinyl Resins. A large group of thermoplastic polymeric resins that contain a vinyl linkage. There are over ten different types of vinyl resins.

Weld Mark (Flow Line). A mark formed by the incomplete fusion of two or more streams of plastic flowing together.

Index